Advanced Mathematics: Mathematical Methods, Systems and Applications

Advanced Mathematics: Mathematical Methods, Systems and Applications

Editor: Zayne Young

NY RESEARCH PRESS
New York

Published by NY Research Press
118-35 Queens Blvd., Suite 400,
Forest Hills, NY 11375, USA
www.nyresearchpress.com

Advanced Mathematics: Mathematical Methods, Systems and Applications
Edited by Zayne Young

International Standard Book Number: 978-1-64725-439-1 (Hardback)

Cataloging-in-Publication Data

Advanced mathematics : mathematical methods, systems and applications / edited by Zayne Young.
p. cm.
Includes bibliographical references and index.
ISBN 978-1-64725-439-1
1. Mathematics. 2. Mathematics--Methodology. 3. Applied mathematics. I. Young, Zayne.
QA36 .A38 2023
510--dc23

Contents

Preface

Every book is a source of knowledge and this one is no exception. The idea that led to the conceptualization of this book was the fact that the world is advancing rapidly; which makes it crucial to document the progress in every field. I am aware that a lot of data is already available, yet, there is a lot more to learn. Hence, I accepted the responsibility of editing this book and contributing my knowledge to the community.

The field of study which focuses on numbers, formulas and associated structures, shapes and the spaces which contained them, and quantities and their changes is known as mathematics. Advanced mathematics is involved in the study of problems which cannot be solved using basic mathematical methods. They require a detailed understanding of the fundamentals of mathematics and advanced mathematical models. The advanced mathematical methods are involved in the study of several topics in linear algebra and multivariate calculus. Some applications of these models are in areas such as statistics, operations research, computer science, econometrics, and mathematical economics. This book outlines the mathematical methods, systems and applications used in advanced mathematics in detail. It consists of the contributions made by scientists and leading experts in advanced mathematics. The book is an essential guide for both academicians and those who wish to pursue this discipline further.

While editing this book, I had multiple visions for it. Then I finally narrowed down to make every chapter a sole standing text explaining a particular topic, so that they can be used independently. However, the umbrella subject sinews them into a common theme. This makes the book a unique platform of knowledge.

I would like to give the major credit of this book to the experts from every corner of the world, who took the time to share their expertise with us. Also, I owe the completion of this book to the never-ending support of my family, who supported me throughout the project.

Editor

1

On the Matrix Mittag–Leffler Function: Theoretical Properties and Numerical Computation

Marina Popolizio [1,2]

[1] Department of Electrical and Information Engineering, Polytechnic University of Bari, Via E. Orabona n.4, 70125 Bari, Italy; marina.popolizio@poliba.it

[2] INdAM Research Group GNCS, Istituto Nazionale di Alta Matematica "Francesco Severi", Piazzale Aldo Moro 5, 00185 Rome, Italy

Abstract: Many situations, as for example within the context of Fractional Calculus theory, require computing the Mittag–Leffler (ML) function with matrix arguments. In this paper, we collect theoretical properties of the matrix ML function. Moreover, we describe the available numerical methods aimed at this purpose by stressing advantages and weaknesses.

Keywords: Mittag–Leffler function; matrix function; Schur decomposition; Laplace transform; fractional calculus

---

1. Introduction

The Mittag–Leffler (ML) function has earned the title of "Queen function of fractional calculus" [1–3] for the fundamental role it plays within this subject. Indeed, the solution of many integral or differential equations of noninteger order can be expressed in terms of this function.

For this reason, the accurate evaluation of the ML function has deserved great attention, not least because of the serious difficulties this computation raises. We cite, among the most fruitful contributions, the papers [4–8].

We have recently observed an increasing interest in computing the ML function for matrix arguments (e.g., see [9–15]): this need occurs, for example, when dealing with multiterm Fractional Differential Equations (FDEs), or with systems of FDEs, or to decide the observability and controllability of fractional linear systems.

In this paper, we want to collect the main results concerning the matrix ML function: we will start from the theoretical properties to move to the practical aspects related to its numerical approximation. Our inspiring work is the milestone paper by Moler and van Loan [16], dating back to 1978, dealing with the several ways to compute the matrix exponential. The authors offered indeed a review of the available methods which were declaimed, already in the paper's title, as "dubious" in the sense that none of them can be considered the top-ranked. Due to the great interest of the topic, twenty-five years later, the same authors published a revised version of this paper [17] to discuss important contributions given in the meantime. In this paper, we would like to use the same simple approach to highlight the difficulties related to the numerical approximation of the matrix ML function.

It is worth stressing that the exponential function is a special ML function; however, it has very nice properties that are not valid for any other instance of ML functions. The semi-group property is one of these and the impossibility to apply it enables, for example, the use of local approximations (which, in the case of the exponential, can be generalized to any argument by exploiting the cited property). Moreover, several methods for the matrix exponential computation were deduced from the fact that this function can be regarded as the solution of simple ordinary differential equations. An analog of this strategy for the ML function presents more difficulties since it can be regarded as a solution of the more involved FDEs.

It becomes clear then that the difficult goal of settling the best numerical method for the exponential function becomes even more tough when treating the matrix ML function. However, in this case, we can affirm that a top-ranked method exists; it was recently proposed [18] and is based on the combination of the Schur–Parlett method [19] and the Optimal Parabolic Contour (OPC) method [4] for the scalar ML function and its derivative. Roughly speaking, this method starts from a Schur decomposition, with reordering and blocking, of the matrix argument and then applies the Parlett's recurrence to compute the function in the triangular factor. Since this step involves the computation of the ML scalar function and its derivatives, the OPC method [4] is, with some suitable modification, fruitfully applied, as we will accurately describe in the following.

The paper is organized as follows: in Section 2, we recall the definition of the ML function and some basic facts about it. In Section 3, we collect the main theoretical properties of the ML function when evaluated in matrix arguments. We then move to the description of the numerical methods for the matrix ML function in Section 4 and to the computation of its action on given vectors in Section 5. Finally, some concluding remarks are collected in Section 6.

2. The Matrix ML Function

The ML function is defined for complex parameters α and β, with $\Re(\alpha) > 0$, by means of the series

$$E_{\alpha,\beta}(z) = \sum_{j=0}^{\infty} \frac{z^j}{\Gamma(\alpha j + \beta)}, \quad z \in \mathbb{C}, \tag{1}$$

with the Euler's gamma function $\Gamma(z) = \int_0^\infty t^{z-1} e^{-t} dt$.

It was introduced for $\beta = 1$ by the Swedish mathematician Magnus Gustaf Mittag–Leffler at the beginning of the twentieth century [20,21] and then generalized to any complex β by Wiman [22]. Throughout the paper, we will consider real parameters α and β since they are the most common.

The exponential is trivially a ML function for $\alpha = \beta = 1$.

Even the numerical computation of the scalar ML function is not a trivial task, and several studies have been devoted to it [4–6,23]. They all agree that the best approach to numerically evaluate $E_{\alpha,\beta}(z)$ is based on a series representation for small values of $|z|$, asymptotic expansions for large arguments, and special integral representations for intermediate values of z. Finally, Garrappa [4] proposed an effective code, based on some ideas previously developed in [5], which allows for reaching any desired accuracy on the whole complex plane. It is implemented in Matlab (2019, MathWorks Inc., Natick, MA, USA) and we will use this routine for the numerical tests in the following .

The simplest way to compute the matrix ML function is for diagonal arguments. Indeed, if A is a diagonal matrix with eigenvalues $\lambda_1, \ldots, \lambda_n$, then $E_{\alpha,\beta}(A)$ is also a diagonal matrix, namely $E_{\alpha,\beta}(A) = \text{diag}(E_{\alpha,\beta}(\lambda_1), \ldots, E_{\alpha,\beta}(\lambda_n))$, and only the ML function for scalar arguments comes into play.

There are many equivalent ways to extend the definition of the ML function to matrix arguments, as for more general functions [24]. Here, we recall some of them:

Definition 1. *Let $A \in \mathbb{C}^{n \times n}$, α and β complex values with $\Re(\alpha) > 0$. Then, the following equivalent definitions hold for the matrix ML function:*

- *Taylor series*

$$E_{\alpha,\beta}(A) = \sum_{j=0}^{\infty} \frac{A^j}{\Gamma(\alpha j + \beta)}. \tag{2}$$

- *Jordan canonical form*
  Let $\lambda_1, \ldots, \lambda_p$ be the distinct eigenvalues of A; then, A can be expressed in the Jordan canonical form

$$A = ZJZ^{-1} = Zdiag(J_1, \ldots, J_p)Z^{-1}$$

with

$$J_k = \begin{bmatrix} \lambda_k & 1 & & 0 \\ & \lambda_k & \ddots & \\ & & \ddots & 1 \\ 0 & & & \lambda_k \end{bmatrix} \in \mathbb{C}^{m_k \times m_k} \tag{3}$$

and

$$m_1 + \ldots + m_p = n.$$

Then,

$$E_{\alpha,\beta}(A) = ZE_{\alpha,\beta}(J)Z^{-1} = Zdiag(E_{\alpha,\beta}(J_1), \ldots, E_{\alpha,\beta}(J_p))Z^{-1}$$

with

$$E_{\alpha,\beta}(J_k) = \begin{bmatrix} E_{\alpha,\beta}(\lambda_k) & E'_{\alpha,\beta}(\lambda_k) & \cdots & \frac{E^{(m_k-1)}_{\alpha,\beta}(\lambda_k)}{(m_k-1)!} \\ & E_{\alpha,\beta}(\lambda_k) & \ddots & \vdots \\ & & \ddots & E'_{\alpha,\beta}(\lambda_k) \\ 0 & & & E_{\alpha,\beta}(\lambda_k) \end{bmatrix}.$$

- *Cauchy integral*

$$E_{\alpha,\beta}(A) := \frac{1}{2\pi i} \oint_\Gamma E_{\alpha,\beta}(z)(zI - A)^{-1} dz, \tag{4}$$

where Γ is a simple closed rectifiable curve that strictly encloses the spectrum of A.

- *Hermite interpolation*
If A has the eigenvalues $\lambda_1, \lambda_2, \ldots, \lambda_p$ with multiplicities $m_1, m_2, \ldots, m_p$, then

$$E_{\alpha,\beta}(A) := r(A),$$

where r is the unique Hermite interpolating polynomial of degree less than $\sum_{i=1}^{p} m_i$ that satisfies the interpolation conditions

$$r^{(j)}(\lambda_i) = E_{\alpha,\beta}(\lambda_i), \ j = 0, \ldots, m_i - 1, \ i = 1, \ldots, p.$$

3. Theoretical Properties of the Matrix ML Function

We collect here the main theoretical properties of the matrix ML function [24–26]: the first 11 hold for general matrix functions while the remaining are specific for the ML function.

Proposition 1. *Let $A, B \in \mathbb{C}^{n \times n}$, $\alpha, \beta \in \mathbb{R}$ with $\alpha > 0$. Let I and 0 denote the identity and the zero matrix, respectively, of dimension n. Then,*

1. $AE_{\alpha,\beta}(A) = E_{\alpha,\beta}(A)A$;
2. $E_{\alpha,\beta}(A^T) = (E_{\alpha,\beta}(A))^T$;
3. $E_{\alpha,\beta}(XAX^{-1}) = XE_{\alpha,\beta}(A)X^{-1}$ *for any nonsingular matrix* $X \in \mathbb{C}^{n \times n}$;
4. *the eigenvalues of $E_{\alpha,\beta}(A)$ are $E_{\alpha,\beta}(\lambda_i)$ where λ_i are the eigenvalues of A;*
5. *if B commutes with A, then B commutes with $E_{\alpha,\beta}(A)$;*
6. *if $A = (A_{ij})$ is block triangular, then $F = E_{\alpha,\beta}(A)$ is block triangular with the same block structure of A and $F_{ii} = E_{\alpha,\beta}(A_{ii})$;*
7. *if $A = diag(A_{11}, \ldots, A_{mm})$ is block diagonal, then*

$$E_{\alpha,\beta}(A) = diag(E_{\alpha,\beta}(A_{11}), \ldots, E_{\alpha,\beta}(A_{mm}));$$

8. $E_{\alpha,\beta}(A \otimes I) = E_{\alpha,\beta}(A) \otimes I$, *where* $\otimes$ *is the Kronecker product;*
9. $E_{\alpha,\beta}(I \otimes A) = I \otimes E_{\alpha,\beta}(A)$;
10. *there is a polynomial* $p(t)$ *of degree at most* $n-1$ *such that* $E_{\alpha,\beta}(A) = p(A)$;
11. $E_{\alpha,\beta}(AB)A = AE_{\alpha,\beta}(BA)$;
12. $E_{\alpha,\beta}(0) = \frac{1}{\Gamma(\beta)}I$;
13. $mE_{m\alpha,\beta}(A^m) = \sum_{k=0}^{m-1} E_{\alpha,\beta}(e^{2\pi ki/m}A)$ *for any natural number* $m \geq 1$;
14. $mA^r E_{m\alpha,\beta+r\alpha}(A^m) = \sum_{k=0}^{m-1} e^{2\pi ki(m-r)/m} E_{\alpha,\beta}(e^{2\pi ki/m}A)$ *for any natural numbers* m *and* r *with* $m \geq 1$ *and* $m > r$;
15. $A^m E_{\alpha,\beta+m\alpha}(A) = E_{\alpha,\beta}(A) - \sum_{k=0}^{m-1} \frac{A^k}{\Gamma(\alpha k + \beta)}$ *for* $\beta \geq 0$.

If A *has no eigenvalues on the negative real axis, then*

16. $E_{\alpha,\beta}(A) = \frac{1}{m} \sum_{k=0}^{m-1} E_{\alpha/m,\beta}(e^{2\pi ki/m} A^{1/m})$;
17. $E_{2\alpha,\beta}(A) = \frac{1}{2}[E_{\alpha,\beta}(A^{1/2}) + E_{\alpha,\beta}(-A^{1/2})]$.

4. Numerical Evaluation of the Matrix ML Function

We give now an overview of different methods for the numerical evaluation of the matrix ML function, with emphasis on the strengths and weaknesses of each of them.

4.1. Series Expansion

As for the exponential, the Taylor series expansion (2) may be regarded as the most direct way to compute the matrix ML function. Indeed, in this definition, only matrix products appear thus to make the approach ideally very simple to implement. In practice, once a fixed number K of terms is chosen, one can use the approximation

$$E_{\alpha,\beta}(A) \approx \sum_{j=0}^{K} \frac{A^j}{\Gamma(\alpha j + \beta)}. \tag{5}$$

However, by following exactly the example presented in [16] for the exponential, we show the weakness of this approach. Indeed, we consider the matrix argument

$$A = \begin{bmatrix} -49 & 24 \\ -64 & 31 \end{bmatrix}, \tag{6}$$

whose eigenvectors and eigenvalues are explicitly known,

$$V = \begin{bmatrix} 1 & 3 \\ 2 & 4 \end{bmatrix}, D = \begin{bmatrix} -1 & 0 \\ 0 & -17 \end{bmatrix}.$$

Then, the exact solution can be directly computed as $VE_{\alpha,\beta}(D)V^{-1}$ and $E_{\alpha,\beta}(D)$ is the diagonal matrix of diagonal entries $E_{\alpha,\beta}(-1)$ and $E_{\alpha,\beta}(-17)$. In Figure 1, we relate the relative error, in norm, between the exact solution and the approximation (5) as K varies, for three different values of α and $\beta = 1$.

Figure 1 clearly shows that the numerical approximation (5) can give unreliable results. In this specific example, the impressive growth of the error is due to numerical cancellation; indeed, the summation terms in Equation (5) get larger as j enlarges and they change sign by passing from the jth power to the next one. This means that we sum up terms with very large modulus and opposite sign, and this is an undisputed source of catastrophic errors.

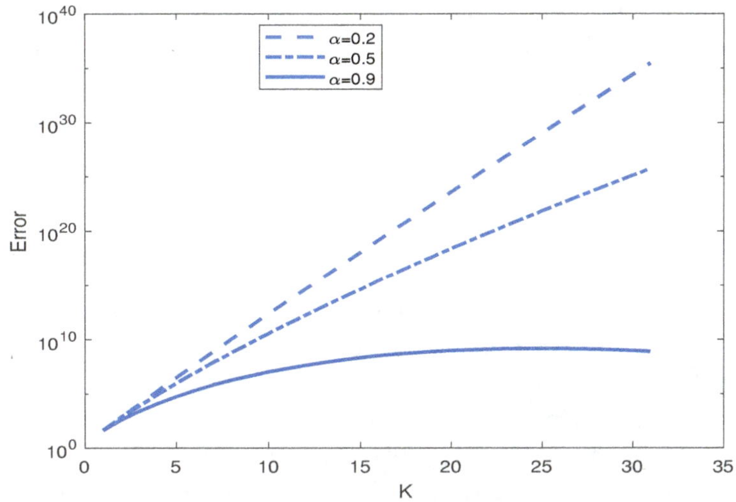

Figure 1. Relative error Vs number of terms K in the series (5) for three values of α, $\beta = 1$ and the matrix A as in Equation (6).

4.2. Polynomial Methods

Methods based on the minimal polynomial or the eigenpolynomial of a matrix have been proposed to numerically evaluate the matrix ML function. This kind of approach is in general poor and we show the weak points (which are exactly the same we encounter in applying it for the matrix exponential [16]).

The first thing to stress is that they require the eigenvalues' knowledge. This is usually not a priori available and numerical methods for their computation are usually very expensive. Thus, their application is limited to the case in which eigenvalues are at one's disposal.

Although in general the minimal polynomial and the eigenpolynomial are very difficult to compute, the latter is simpler to calculate and we focus on this approach.

Let $c(z)$ denote the characteristic polynomial of A with

$$c(z) = \det(zI - A) = z^n - \sum_{k=0}^{n-1} c_k z^k.$$

Then, by means of the Cayley–Hamilton theorem, it is easy to prove that any power of A can be expressed as a linear combination of $I, A, \ldots, A^{n-1}$. Thus, also $E_{\alpha,\beta}(tA)$ is a polynomial in A with analytic coefficients in t; indeed, formula (2) for the matrix ML function reads

$$\begin{aligned} E_{\alpha,\beta}(tA) &= \sum_{j=0}^{\infty} \frac{t^j A^j}{\Gamma(\alpha j + \beta)} = \sum_{j=0}^{\infty} \frac{t^j}{\Gamma(\alpha j + \beta)} \left(\sum_{k=0}^{n-1} p_{jk} A^k \right) \\ &= \sum_{k=0}^{n-1} \left(\sum_{j=0}^{\infty} p_{jk} \frac{t^j}{\Gamma(\alpha j + \beta)} \right) A^k = \sum_{k=0}^{n-1} \tilde{p}_k(t) A^k. \end{aligned}$$

The expression of coefficients p_{jk} is simply obtained once the coefficients c_j are known. However, the weak point is related to their numerical computation since it is very prone to round off error (as shown in [16] already for the 1-by-1 case). For this reason, methods of this kind are strongly discouraged.

4.3. The Schur–Parlett Method Combined with the OPC Method

The third property of Proposition 1 suggests looking for a suitable similarity transformation which moves the attention to the matrix function evaluated in a different argument, hopefully simpler to deal with. In particular, among the best conditioned similarity transformations, one can resort to the Schur one. Indeed, it factors a matrix A as

$$A = QTQ^*$$

with T upper triangular and Q unitary. Then,

$$E_{\alpha,\beta}(A) = QE_{\alpha,\beta}(T)Q^*. \tag{7}$$

The Schur decomposition is among the best factorization one can consider since its computation is perfectly stable, unlike other decompositions, as the Jordan one that we will describe in the following. For this reason, it is commonly employed for computing matrix functions [27,28]. The actual evaluation of Equation (7) requires the computation of the ML function for a triangular matrix factor. This topic has been properly addressed for general functions by Parlett in 1976 [29], resulting in a cheap method. Unfortunately, Parlett's recurrence can give inaccurate results when T has close eigenvalues. In 2003, Higham and Davies [19] proposed an improved version of this method: once the Schur decomposition is computed, the matrix T is reordered and blocked according to its eigenvalues resulting in a matrix, say $\tilde{T}$. Specifically, each diagonal block of $\tilde{T}$ has "close" eigenvalues and distinct diagonal blocks have "sufficiently distinct" eigenvalues. In this way, Parlett's recurrence works well even in the presence of closed eigenvalues of T. Just a final reordering is required at the end of the process to recover $E_{\alpha,\beta}(T)$ from $E_{\alpha,\beta}(\tilde{T})$.

The evaluation of $E_{\alpha,\beta}(\tilde{T})$ starts from the evaluation of the ML function of its diagonal blocks, which are still triangular matrices whose eigenvalues are "close". Let T_{ii} be one of these diagonal blocks and σ denotes the mean of these eigenvalues. Then, $T_{ii} = \sigma I + M$ and

$$E_{\alpha,\beta}(T) = \sum_{k=0}^{\infty} \frac{E^{(k)}_{\alpha,\beta}(\sigma)}{k!} M^k, \tag{8}$$

with $E^{(k)}_{\alpha,\beta}$ denoting the k-th order derivative of $E_{\alpha,\beta}$. The powers of M are expected to decay quickly since the eigenvalues of T_{ii} are close. This means that only a few terms of (8), usually less than the dimension of the block T_{ii}, suffice to get a good accuracy.

Evidently, the computation of (8) involves the computation of the derivatives of the scalar ML function, up to an order depending on the eigenvalues' properties. This issue has been completely addressed in [18], and we refer to it for the details.

In particular, the analysis of the derivatives of the ML function has been facilitated by resorting to the three parameters' ML function (also known as the Prabhakar function)

$$E^{\gamma}_{\alpha,\beta}(z) = \frac{1}{\Gamma(\gamma)} \sum_{k=0}^{\infty} \frac{\Gamma(\gamma+k)z^k}{k!\Gamma(\alpha k+\beta)}, \quad \alpha,\beta,\gamma \in \mathbb{C}, \quad \Re(\alpha) > 0,$$

since

$$E^{(m)}_{\alpha,\beta}(t) = m! E^{m+1}_{\alpha,\beta+\alpha m}(t). \tag{9}$$

The Prabhakar function is an important function occurring in the description of many physical models [30–35].

In practice, as for the scalar ML function, one could compute the Prabhakar function, or equivalently the ML derivatives, by the Taylor series for small arguments, the asymptotic expansion for large arguments and an integral representation in the remaining cases. In [18], however, to obtain the same accuracy for all arguments, the inverse Laplace transform has been used in all cases to obtain the simple expression

$$E^{\gamma}_{\alpha,\beta}(z) = \frac{1}{2\pi \mathrm{i}} \int_{\mathcal{C}} \mathrm{e}^{s} \frac{s^{\alpha\gamma-\beta}}{(s^{\alpha}-z)^{\gamma}} \mathrm{d}s,$$

with $\mathcal{C}$ any suitable contour in the complex plane encompassing at the left the singularities of the integrand. This last issue is quite delicate: indeed, from a theoretical point of view, the contours chosen to define the inverse Laplace transform are all equivalent while they can lead to extremely different results when the numerical evaluation of these integrals comes into play. Then, an accurate analysis is needed to choose the "optimal" contour which guarantees the desired accuracy, minimizes the computational complexity, and results in a simple implementation. The method proposed in [18], grounded on well established analysis [4,5,15], actually fulfills these requirements since the obtained accuracy is in any case close to the machine precision and the computational complexity is very reasonable. The Matlab code `ml_matrix.m` implements this method and will be used in the following for numerical tests.

4.4. Jordan Canonical Form

The expression of the matrix ML function in terms of its Jordan canonical form, as stated in Equation (3), could be a direct way to numerically evaluate it. However, the true obstacle in using it is the high sensitivity of the Jordan canonical form to numerical errors (in general, "there exists no numerically stable way to compute Jordan canonical forms" [36]).

To give an example, we consider the Matlab code by Matychyn [37], which implements this approach. We restrict the attention to the exponential case (that is, $\alpha = \beta = 1$) to have as reference solution the result of the well-established `expm` code by Matlab.

We consider the Chebyshev spectral differentiation matrix of dimension 10. Oddly, even for this "simple" function, the relative error is quite high, namely proportional to 10^{-3}.

The error source is almost certainly the well-known ill-conditioning of the eigenvector matrix. Indeed, the code `ml_matrix` gives a relative error proportional to 10^{-7} since it does not involve the Jordan canonical form.

Now, we consider as example the test matrix in [36]

$$A = \begin{bmatrix} \varepsilon & 0 \\ 1 & 0 \end{bmatrix}$$

as matrix argument; as before, we just consider the simplest case $\alpha = \beta = 1$ as a significant example.

For small values of ε, say $\varepsilon < 10^{-16}$, the code [37] stops running, since, when computing the Jordan canonical form, Matlab recognizes that the matrix is singular. On the other hand, the code `ml_matrix` works very well even for tiny values, say ε equal to the Matlab machine precision.

Analogously, let

$$A = \begin{bmatrix} 0 & 1 & & \\ & \ddots & \ddots & \\ & & \ddots & 1 \\ \varepsilon & & & 0 \end{bmatrix}$$

be a $n \times n$ matrix. For $n = 16$ and $\varepsilon = 10^{-16}$, the code [37] just reaches an accuracy proportional to 10^{-2}, while the code `ml_matrix` reaches 10^{-17}.

From these examples, we can appreciate the high accuracy reached by the code `ml_matrix` described in Section 4.3. Moreover, its computational cost is lower than the code based on the Jordan canonical form and, far more important, it does not suffer from the eigenvalues' conditioning.

4.5. Rational Approximations

Among the nineteen methods to approximate the matrix exponential, Moler and van Loan [16] consider the "exact" evaluation of a rational approximation of the exponential function evaluated in the desired matrix argument. This is indeed a very common approach when dealing with more general functions having good rational approximations (see, e.g., [38–40]).

Indeed, let p_μ and q_ν be polynomials of degree μ and ν, respectively, such that, for scalar arguments z,

$$E_{\alpha,\beta}(z) \approx \frac{p_\mu(z)}{q_\nu(z)}.$$

To evaluate the approximation above in the matrix case, we use a partial fraction expansion of the right-hand side above, leading to

$$E_{\alpha,\beta}(z) \approx \tilde{p_\ell}(z) + \sum_{i=1}^{\nu} \omega_i \frac{1}{z-\sigma_i}, \tag{10}$$

in this way, the computation of $\tilde{p_\ell}(A)$ is trivial while the sum requires the computation of ν matrix inversions, namely $(A-\sigma_i I)^{-1}$, for $i = 1, \ldots, \nu$.

Once the rational approximation is fixed, the sum (10) can be computed by actually inverting the matrices $(A-\sigma_i I)^{-1}$ if A is a small well-conditioned matrix or, if it is large, incomplete factorizations of A can be cheaply applied [40].

For the ML function, the problem is the detection of a suitable rational approximation to use. The Padé and the Chebyshev rational approximation are commonly preferred for the exponential; this choice is mainly due to their good approximation properties, to the fact that they are explicitly known, and the error analysis is well established.

A key feature of the Padé approximation is that it can be used if $\|A\|$ is not too large. This does not represent a restriction for the exponential function since it is endowed with the fundamental property

$$\exp(A) = (\exp(A/m))^m,$$

it allows for computing the exponential of an arbitrarily small argument A/m to then extend it to the original argument A. In general, m is chosen as a power of two in order to require only the repeated squaring of the matrix argument.

The property above is only valid for the exponential function, meaning that, for the ML function, there is no direct way to extend the local approximation to the global case.

Some years ago, a global Padé approximation to the ML function was proposed [8] working on the whole real semiaxis $(-\infty, 0]$. In the matrix case, this restricts the applicability to matrix arguments with real positive eigenvalues. Moreover, the computation of the coefficients is arduous; for this reason, small degrees are considered in [8], which lead to quite important errors.

We now describe the Carathéodory–Fejér approximation of the ML function as an effective tool when a rational approximation is needed.

Carathéodory–Fejér Approximation of the ML Function

As concerns the ML function, Trefethen was the first to address the problem of finding rational approximations when $\alpha = 1$ and $\beta \in \mathbb{N}$ [41]. Later on, the most general case has been deeply analyzed [11] grounding on the Carathéodory–Fejér (CF) theory; this theory is important since it allows for constructing a near best rational approximation of the ML function. In practice, once we fix a given degree ν, the residues $\omega_0, \ldots, \omega_\nu$ and the poles $\sigma_1, \ldots, \sigma_\nu$ are found that define the CF approximation of degree ν of the ML function. Thus,

$$E_{\alpha,\beta}(A) \approx \omega_0 I + \sum_{j=1}^{\nu} \omega_j (A - \sigma_j I)^{-1}. \tag{11}$$

When dealing with real arguments, the sum can be arranged as to almost halve the number of terms to compute. Moreover, since a small degree ν usually suffices to give a good approximation, only a few matrix inverses are actually required. Obviously, this approach is meaningful only for matrix arguments whose inversion can be computed in a stable and reliable way.

5. Numerical Computation of $E_{\alpha,\beta}(A)b$ for a Given Vector b

In many situations, the interest is in the computation of $E_{\alpha,\beta}(A)b$ for a given vector b. Any method described so far can be applied to compute $E_{\alpha,\beta}(A)$ and then multiply it by the vector b. However, when the dimension of the matrix argument A is very large, ad hoc strategies have to be preferred.

The rational approximation (11) is, for example, a good solution; indeed, it reads

$$E_{\alpha,\beta}(A)b \approx \omega_0 b + \sum_{j=1}^{\nu} \omega_j (A - \sigma_j I)^{-1} b$$

and, rather than matrix inversions, the right-hand side requires only solving linear systems. This approach is effective even for small matrix arguments A, in which case direct methods can be applied for the linear systems involved. When the matrix argument is very large, several alternatives are at one's disposal: iterative methods can be, for example, applied (see [11]) and, when preconditioning is needed, the same preconditioner can be computed just once and then applied to all shifted systems. Incomplete factorizations of A can be used for example as preconditioners for the systems involved (we refer to [40] for a deep description of the approach).

Krylov subspace methods are an effective tool for the numerical approximation of vectors like $E_{\alpha,\beta}(A)b$; their first application was related to the exponential and then they have been successfully employed for general functions [38,42,43]. In particular, for the ML function, we refer to [11,12]. The idea is to approximate $E_{\alpha,\beta}(A)b$ in Krylov subspaces defined as

$$\mathcal{K}_m(A,b) \equiv span\{b, Ab, \ldots, A^{m-1}b\}, \ m \in \mathbb{N}.$$

The matrix A is projected in these spaces as $H_m = V_m^T A V_m$, where $V_m \in \mathbb{C}^{n \times m}$ is an orthonormal basis of $\mathcal{K}_m(A,b)$ built by applying the Gram–Schmidt procedure with $b/\|b\|$ as starting vector and $H_m \in \mathbb{C}^{m \times m}$ is an unreduced Hessenberg matrix whose entries are the coefficients of the orthonormalization process.

Then,

$$E_{\alpha,\beta}(A)b \approx \|b\| V_m E_{\alpha,\beta}(H_m) e_1, \tag{12}$$

where e_1 denotes the first column of the identity matrix of dimension $m \times m$.

The potency of these techniques is that usually a small dimension m is enough to get a sufficiently accurate approximation; thus, some classical method usually works to compute $E_{\alpha,\beta}(H_m)$.

The convergence of Krylov subspace methods can be quite slow when the spectrum of A is large; this phenomenon was primarily studied for the exponential [44] and successively for the ML function [12]. In these cases, the Rational Arnoldi method can be successfully used, with superb results already for the "one-pole case" [12,45–47]. The idea of this method, known as Shift and Invert in the context of eigenvalue problems, is to fix a parameter γ and to approximate $E_{\alpha,\beta}(A)b$ in the Krylov subspaces generated by $Z = (I + \gamma A)^{-1}$, rather than A and b.

The computational complexity is larger than for standard Krylov subspace methods since the construction of the Krylov subspaces requires computing vectors of the form $(I + \gamma A)^{-1} y$, that is, solving several linear systems with the same shifted coefficient matrix. However, for suitable shift parameters, the convergence becomes much faster, to thus compensate the additional cost.

We refer to [12] for a comprehensive description of this method applied to the computation of the matrix ML function, together with the numerical tests to show the effectiveness of the approach. Moreover, for completeness, we want to stress that the actual computation of the matrix ML function in [12] was accomplished by combining the Schur–Parlett recurrence and the Matlab code `mlf.m` by Podlubny and Kacenak [48]. However, this approach cannot handle the derivatives of the ML scalar function; therefore, to treat more general situations, as, for example, matrix arguments with repeated eigenvalues, the approach described in Section 4.3 has to be considered within the implementation.

6. Conclusions

This paper offers an overview of the matrix ML function: the most important theoretical properties are collected to serve as a review and to help in the treatment of this function. Moreover, the existing methods for its numerical computation are presented, by following the plot used by Moler and Van Loan [16] to describe the methods for the numerical computation of the matrix exponential.

From this analysis, we may conclude that the approach based on the combination of the Schur–Parlett method and the OPC method is the most efficient: it is indeed cheap, accurate, and easy to implement.

Abbreviations

The following abbreviations are used in this manuscript:

CF	Carathéodory–Fejér
FDE	Fractional Differential Equation
ML	Mittag–Leffler
OPC	Optimal Parabolic Contour

References

1. Gorenflo, R.; Kilbas, A.A.; Mainardi, F.; Rogosin, S. *Mittag–Leffler Functions. Theory and Applications*; Springer Monographs in Mathematics; Springer: Berlin, Germany, 2014; pp. xii, 420p.
2. Gorenflo, R.; Mainardi, F. Fractional calculus: Integral and differential equations of fractional order. In *Fractals and Fractional Calculus in Continuum Mechanics (Udine, 1996)*; Springer: Vienna, Austria, 1997; Volume 378, pp. 223–276.
3. Mainardi, F.; Mura, A.; Pagnini, G. The *M*-Wright function in time-fractional diffusion processes: A tutorial survey. *Int. J. Differ. Equ.* **2010**, *2010*, 104505. [CrossRef]
4. Garrappa, R. Numerical evaluation of two and three parameter Mittag-Leffler functions. *SIAM J. Numer. Anal.* **2015**, *53*, 1350–1369. [CrossRef]
5. Garrappa, R.; Popolizio, M. Evaluation of generalized Mittag–Leffler functions on the real line. *Adv. Comput. Math.* **2013**, *39*, 205–225. [CrossRef]
6. Gorenflo, R.; Loutchko, J.; Luchko, Y. Computation of the Mittag-Leffler function $E_{\alpha,\beta}(z)$ and its derivative. *Fract. Calc. Appl. Anal.* **2002**, *5*, 491–518.
7. Valério, D.; Tenreiro Machado, J. On the numerical computation of the Mittag–Leffler function. *Commun. Nonlinear Sci. Numer. Simul.* **2014**, *19*, 3419–3424. [CrossRef]
8. Zeng, C.; Chen, Y. Global Padé approximations of the generalized Mittag–Leffler function and its inverse. *Fract. Calc. Appl. Anal.* **2015**, *18*, 1492–1506. [CrossRef]
9. Garrappa, R.; Moret, I.; Popolizio, M. Solving the time-fractional Schrödinger equation by Krylov projection methods. *J. Comput. Phys.* **2015**, *293*, 115–134. [CrossRef]
10. Garrappa, R.; Moret, I.; Popolizio, M. On the time-fractional Schrödinger equation: Theoretical analysis and numerical solution by matrix Mittag-Leffler functions. *Comput. Math. Appl.* **2017**, *74*, 977–992. [CrossRef]

11. Garrappa, R.; Popolizio, M. On the use of matrix functions for fractional partial differential equations. *Math. Comput. Simul.* **2011**, *81*, 1045–1056. [CrossRef]
12. Moret, I.; Novati, P. On the Convergence of Krylov Subspace Methods for Matrix Mittag–Leffler Functions. *SIAM J. Numer. Anal.* **2011**, *49*, 2144–2164. [CrossRef]
13. Popolizio, M. Numerical solution of multiterm fractional differential equations using the matrix Mittag–Leffler functions. *Mathematics* **2018**, *1*, 7. [CrossRef]
14. Rodrigo, M.R. On fractional matrix exponentials and their explicit calculation. *J. Differ. Equ.* **2016**, *261*, 4223–4243. [CrossRef]
15. Weideman, J.A.C.; Trefethen, L.N. Parabolic and hyperbolic contours for computing the Bromwich integral. *Math. Comp.* **2007**, *76*, 1341–1356. [CrossRef]
16. Moler, C.; Van Loan, C. Nineteen dubious ways to compute the exponential of a matrix. *SIAM Rev.* **1978**, *20*, 801–836. [CrossRef]
17. Moler, C.; Van Loan, C. Nineteen dubious ways to compute the exponential of a matrix, twenty-five years later. *SIAM Rev.* **2003**, *45*, 3–49. [CrossRef]
18. Garrappa, R.; Popolizio, M. Computing the matrix Mittag–Leffler function with applications to fractional calculus. *J. Sci. Comput.* **2018**, *77*, 129–153. [CrossRef]
19. Davies, P.I.; Higham, N.J. A Schur-Parlett algorithm for computing matrix functions. *SIAM J. Matrix Anal. Appl.* **2003**, *25*, 464–485. [CrossRef]
20. Mittag-Leffler, M.G. Sopra la funzione $E_\alpha(x)$. *Rend. Accad. Lincei* **1904**, *13*, 3–5.
21. Mittag-Leffler, M.G. Sur la représentation analytique d'une branche uniforme d'une fonction monogène-cinquième note. *Acta Math.* **1905**, *29*, 101–181. [CrossRef]
22. Wiman, A. Ueber den Fundamentalsatz in der Teorie der Funktionen $E_\alpha(x)$. *Acta Math.* **1905**, *29*, 191–201. [CrossRef]
23. Garrappa, R. Numerical Solution of Fractional Differential Equations: A Survey and a Software Tutorial. *Mathematics* **2018**, *6*, 16. [CrossRef]
24. Higham, N.J. *Functions of Matrices*; Society for Industrial and Applied Mathematics (SIAM): Philadelphia, PA, USA, 2008; pp. xx, 425p.
25. Horn, R.; Johnson, C. *Topics in Matrix Analysis*; Cambridge University Press: Cambridge, UK, 1994.
26. Sadeghi, A.; Cardoso, J.R. Some notes on properties of the matrix Mittag–Leffler function. *Appl. Math. Comput.* **2018**, *338*, 733–738. [CrossRef]
27. Del Buono, N.; Lopez, L.; Politi, T. Computation of functions of Hamiltonian and skew-symmetric matrices. *Math. Comp. Simul.* **2008**, *79*, 1284–1297. [CrossRef]
28. Politi, T.; Popolizio, M. On stochasticity preserving methods for the computation of the matrix pth root. *Math. Comput. Simul.* **2015**, *110*, 53–68. [CrossRef]
29. Parlett, B.N. A Recurrence Among the Elements of Functions of Triangular Matrices. *Linear Algebra Appl.* **1976**, *14*, 117–121. [CrossRef]
30. Mainardi, F.; Garrappa, R. On complete monotonicity of the Prabhakar function and non-Debye relaxation in dielectrics. *J. Comput. Phys.* **2015**, *293*, 70–80. [CrossRef]
31. Garrappa, R. Grünwald–Letnikov operators for fractional relaxation in Havriliak–Negami models. *Commun. Nonlinear Sci. Numer. Simul.* **2016**, *38*, 178–191. [CrossRef]
32. Garrappa, R.; Mainardi, F.; Maione G. Models of dielectric relaxation based on completely monotone functions. *Fract. Calc. Appl. Anal.* **2016**, *19*, 1105–1160. [CrossRef]
33. Garra, R.; Garrappa, R. The Prabhakar or three parameter Mittag-Leffler function: Theory and application. *Commun. Nonlinear Sci. Numer. Simul.* **2018**, *56*, 314–329. [CrossRef]
34. Giusti, A.; Colombaro, I. Prabhakar-like fractional viscoelasticity. *Commun. Nonlinear Sci. Numer. Simul.* **2018**, *56*, 138–143. [CrossRef]
35. Colombaro, I.; Garra, R.; Garrappa, R.; Giusti, A.; Mainardi, F.; Polito, F.; Popolizio, M. A practical guide to Prabhakar fractional calculus. **2019**, preprint.
36. Horn, R.; Johnson, C. *Matrix Analysis*; Cambridge University Press: Cambridge, UK, 1990.
37. Matychyn, I. Matrix Mittag–Leffler function. In *MATLAB Central, File Exchange, File ID: 62790*; MathWorks, Inc.: Natick, MA, USA, 2017.
38. Lopez, L.; Simoncini, V. Analysis of projection methods for rational function approximation to the matrix exponential. *SIAM J. Numer. Anal.* **2006**, *44*, 613–635. [CrossRef]

39. Trefethen, L.N.; Weideman, J.A.C.; Schmelzer, T. Talbot quadratures and rational approximations. *BIT* **2006**, *46*, 653–670. [CrossRef]
40. Bertaccini, D.; Popolizio, M.; Durastante, F. Efficient approximation of functions of some large matrices by partial fraction expansions. *Int. J. Comput. Math.* **2019**, *96*, 1799–1817. [CrossRef]
41. Schmelzer, T.; Trefethen, L.N. Evaluating matrix functions for exponential integrators via Carathéodory-Fejér approximation and contour integrals. *Electron. Trans. Numer. Anal.* **2007**, *29*, 1–18.
42. Saad, Y. Analysis of some Krylov subspace approximations to the matrix exponential operator. *SIAM J. Numer. Anal.* **1992**, *29*, 209–228. [CrossRef]
43. Frommer, A.; Simoncini, V. Matrix Functions. In *Mathematics in Industry*; Springer: Berlin, Germany, 2008; Volume 13, pp. 275–303.
44. Hochbruck, M.; Lubich, C. On Krylov subspace approximations to the matrix exponential operator. *SIAM J. Numer. Anal.* **1997**, *34*, 1911–1925. [CrossRef]
45. Moret, I.; Novati, P. RD-Rational Approximations of the Matrix Exponential. *BIT Numer. Math.* **2004**, *44*, 595–615. [CrossRef]
46. van den Eshof, J.; Hochbruck, M. Preconditioning Lanczos approximations to the matrix exponential. *SIAM J. Sci. Comput.* **2006**, *27*, 1438–1457. [CrossRef]
47. Popolizio, M.; Simoncini, V. Acceleration techniques for approximating the matrix exponential. *SIAM J. Matrix Anal. Appl.* **2008**, *30*, 657–683. [CrossRef]
48. Podlubny, I.; Kacenak, M. The Matlab mlf code. In *MATLAB Central File Exchange, 2001–2009. File ID: 8738.2001*; MathWorks, Inc.: Natick, MA, USA, 2017.

2

Dynamic Keynesian Model of Economic Growth with Memory and Lag

Vasily E. Tarasov [1,*] **and Valentina V. Tarasova** [2,3]

1 Skobeltsyn Institute of Nuclear Physics, Lomonosov Moscow State University, 119991 Moscow, Russia
2 Faculty of Economics, Lomonosov Moscow State University, 119991 Moscow, Russia; v.v.tarasova@mail.ru
3 Yandex, Ulitsa Lva Tolstogo 16, 119021 Moscow, Russia
* Correspondence: tarasov@theory.sinp.msu.ru.

Abstract: A mathematical model of economic growth with fading memory and continuous distribution of delay time is suggested. This model can be considered as a generalization of the standard Keynesian macroeconomic model. To take into account the memory and gamma-distributed lag we use the Abel-type integral and integro-differential operators with the confluent hypergeometric Kummer function in the kernel. These operators allow us to propose an economic accelerator, in which the memory and lag are taken into account. The fractional differential equation, which describes the dynamics of national income in this generalized model, is suggested. The solution of this fractional differential equation is obtained in the form of series of the confluent hypergeometric Kummer functions. The asymptotic behavior of national income, which is described by this solution, is considered.

Keywords: fractional differential equations; fractional derivative; Abel-type integral; time delay; distributed lag; gamma distribution; macroeconomics; Keynesian model

MSC: 91B02; 26A33; 91B55

JEL Classification: E00; C02

---

1. Introduction

Advanced mathematical methods of fractional calculus [1–5] are a powerful tool for describing the fading memory and spatial non-locality. Fractional derivatives and integrals of non-integer order have different applications in natural and social sciences [6,7].

In this article, we suggest a generalization of one of the most famous models of economic growth, which is associated with the founder of modern macroeconomic theory, John M. Keynes [8–10]. In the suggested generalization, we take into account two types of phenomena: (I) long memory with power-law fading and (II) continuously distributed lag with gamma distribution of delay time.

The continuously distributed lag has been considered in economics starting with the works of Michal A. Kalecki [11] and Alban W.H. Phillips [12,13]. The macrodynamic models of business cycles, where the continuous uniform distribution of delay time is used, were considered by Michal A. Kalecki in 1935 [11], (see also Section 8.4 of [14], (pp. 251–254)). The economic growth models with continuously distributed lag were proposed by Alban W.H. Phillips [12,13] in 1954. In his works, the distribution of delay time has been described by exponential distribution. Operators with continuously distributed lag were considered by Roy G.D. Allen [14] (pp. 23–29), in 1956.

The time delay (lag) is caused by finite speeds of processes, i.e., the change of one variable does not lead to instant changes of another variable. Therefore, the distributed lag (time delay) cannot be considered as a long memory in processes. For example, in physics, the retarded potential of the electromagnetic field is well known. The change in the value of the electromagnetic field at the point of observation r_1 is delayed with respect to the change in the sources of the field located at the point r_2 at the time $t = |r_1 r_2| / c$, where c is the speed of propagation of disturbances. It is known that the processes of propagation of the electromagnetic field in a vacuum do not mean the presence of memory in this process.

Long memory has been considered in economics starting with the works of C.W.J. Granger [15–18]. For the first time, the importance of long-range time dependence in economic data was recognized by C.W.J. Granger [15,16] in 1964, 1966. The long-range time dependencies have empirically observed in economics [19–21]. For these dependencies, the correlations between values of variables decay to zero more slowly than it can be expected from independent variables or variables from classical Markov model and autoregressive moving average (ARMA) model [19–25]. An interpretation of these dependencies between variables is that this process has a long memory. In economics, long memory was first related to fractional differencing and integrating by C.W.J. Granger., R. Joyeux [26] and J.R.M. Hosking [27] by using the discrete time approach. Granger, Joyeux, and Hosking independently proposed the so-called autoregressive fractional integrated moving average models (ARFIMA models). These models use the difference operator $\Delta^d := (1-L)^d$, where L is the lag operator $LX(t) = X(t-1)$, and d is the order of the fractional differencing if $d > 0$ (fractional integrating if $d < 0$), which need not be an integer. In papers [28,29], we noted that the operator Δ^d coincides with the Grunwald–Letnikov fractional difference $\Delta^\alpha_\tau := (1 - T_\tau)^\alpha$ of order $\alpha = d$ and the unit step $\tau = 1$, where T_τ is the translation (shift) operator that is defined [1], (pp. 95–96), by the expression $(T_\tau Y)(t) = Y(t - \tau)$, where $\tau > 0$ is the delay time. The Grunwald–Letnikov fractional differences were proposed over a hundred and fifty years ago. In mathematics these fractional differences are actively used (for example, see [1], (pp. 371–387), and [3], (pp. 43–62) and [4], (pp. 121–123)). Due to the historical circumstances, the description of processes with memory in economics was based on the Granger—Joyeux approach and models with discrete time only. The continuous time form of economic models with memory was practically not considered and advanced mathematical methods of fractional calculus were not applied in mathematical economics.

An application of advanced mathematical methods of fractional calculus in Keynesian economic models with continuous time was proposed by authors [30,31] in 2016 (see also [32–34]). The fractional differential equations of the dynamic Keynesian model with power-law memory and their solutions have been considered in [30–34]. Continuously distributed lag was not discussed in these works.

For macroeconomics, it is important to simultaneously take into account lagging and memory phenomena. In this article, we consider memory with power-law fading and lag with gamma distribution of delay time. The memory is described by the Riemann–Liouville fractional integrals and the Caputo fractional derivatives. The distributed lag is described by the translation T_τ, in which the delay time $\tau > 0$ is considered as a random variable that is distributed by probability law (distribution) on positive semiaxis. The composition of these operators is represented as the Abel-type integral and integro-differential operators with the confluent hypergeometric Kummer function in the kernel. Using these operators, we propose the fractional differential equation for the generalized dynamic Keynesian model that describes the fractional dynamics of national income. We obtain a solution for this equation that describes the macroeconomic growth with power-law fading memory and gamma distribution of delay time.

2. Standard Dynamic Keynesian Model

In macroeconomic growth models, two types of variables are used [14,35,36]. First, exogenous variables are considered as independent quantities that are external to the considered economic model. Secondly, endogenous variables are internal variables that are formed within the model. The endogenous variables are described as functions of exogenous variables. In models with continuous time, all these variables are considered as functions of time t.

Let us consider the standard dynamical Keynesian model with continuous time. In the Keynesian model, the following variables are used to describe the dynamics of the revenue and expenditure parts of the economy: $Y(t)$ is a national income; $G(t)$ is the government expenditure; $C(t)$ describes the consumption expenditure; $I(t)$ describes the investment expenditure, $E(t)$ is a total expenditure, i.e., $E(t)$ is defined as the sum of all expenditures:

$$E(t) = C(t) + I(t) + G(t). \tag{1}$$

In dynamic equilibrium, we have

$$Y(t) = E(t). \tag{2}$$

In this case, the balance equation establishes the equality of the national income to the sum of all expenditures

$$Y(t) = C(t) + I(t) + G(t). \tag{3}$$

In the Keynesian model, it is assumed that the consumption expenditure in period t depends on the income level in the same period. The consumption expenditure $C(t)$ is regarded as an endogenous variable equal to the amount of domestic consumption of some part of the national income and final consumption independent of income. As a result, the consumption expenditure $C(t)$ is described by the linear equation of the economic multiplier

$$C(t) = m(t)Y(t) + b(t), \tag{4}$$

where $m(t)$ is the multiplier factor that describes the marginal propensity to consume ($0 < m(t) < 1$), and the function $b(t) > 0$ describes the autonomous consumption that does not depend on income. The expression $m(t)Y(t)$ describes the part of consumption that depends on income.

In the static model, the investment expenditure and government expenditure are considered as exogenous variables. In the dynamic Keynesian model, the investment expenditure $I(t)$ is treated as endogenous and it is assumed to depend on the level of income [35], (pp. 95–97). The investment expenditure $I(t)$ is determined by the rate of change of the national income. This assumption is described by the equation of the economic accelerator

$$I(t) = v(t)Y^{(1)}(t), \tag{5}$$

where $v(t)$ is the rate of acceleration, which characterizes the level of technology and state infrastructure, and $Y^{(1)}(t) = dY(t)/dt$ is the first-order derivative of the income function $Y(t)$ with respect to the time variable.

In the Keynesian model, government expenditure $G(t)$, the propensity to consume $m(t)$, the rate of acceleration $v(t)$, and the autonomous consumption $b(t)$ are exogenous variables that are specified as external to the model and characterize the functioning and development of the economy. These variables, as functions of time, are assumed to be given.

The purpose of the dynamic Keynesian model is to describe the behavior of the national income. For this, it is necessary to find the national income $Y(t)$, as a function of time t. Substituting the multiplier Equation (4) and the accelerator Equation (5) into the balance Equation (3), we obtain

$$Y(t) = m(t)Y(t) + b(t) + v(t)Y^{(1)}(t) + G(t). \tag{6}$$

This equation can be written in the form

$$\frac{dY(t)}{dt} - \frac{1 - m(t)}{v(t)}Y(t) = -\frac{G(t) + b(t)}{v(t)}. \tag{7}$$

Equation (7) of the dynamic Keynesian model is a non-homogeneous linear differential equation with a first-order derivative.

We see that the functions $G(t)$ and $b(t)$ are included in Equation (7) as a sum. This could be expected since G(t) is an independent expenditure on investment, that is independent of income, and $b(t)$ is an independent expenditure on consumption, also not dependent on national income. From the point of view of the main purpose of the Keynesian model, which is to describe the dynamics of the national income, these two types of expenditure simply complement each other [35], (pp. 95–98). Therefore, it is convenient to use the sum

$$G_b(t) = G(t) + b(t), \tag{8}$$

which describes the independent expenditure. In this case, the consumption function $C(t) = m(t)Y(t)$ is that part of consumption that depends on income. All this allows us to write down the equation of the standard dynamic Keynesian model in the form

$$\frac{dY(t)}{dt} = \frac{1 - m(t)}{v(t)} Y(t) - \frac{G_b(t)}{v(t)}, \tag{9}$$

where $0 < 1 - m(t) < 1$, and $v(t) > 0$. Equation (9) is a non-homogeneous first-order differential equation that describes the standard dynamic Keynesian model, which does not take into account the effects of memory and delay.

3. Dynamic Keynesian Model with Memory

In the standard Keynesian model, Equation (9) implies an instantaneous change in the investment expenditure, when the rate of growth of national income changes. This means that the equation of this model does not take into account the effects of memory and delay. Mathematically, this is due to the fact that the standard model equation is a first-order differential equation. The derivative of the first order, which is used in accelerator Equations (5), implies an instantaneous change of the investment expenditure $I(t)$, when changing the rate of the national income $Y(t)$. Because of this, accelerator Equation (5) does not take into account memory and lag. Multiplier Equation (4) also assumes that the consumption expenditure $C(t)$ changes instantly when the national income changes. As a result, model Equation (9) can describe an economy, in which agents have no memory. This fact greatly limits the applicability of the standard model to describe the real processes in the economy. To expand the scope of the model, we should take into account that economic agents can remember the history of changes of the national income and the investment expenditure, because it affects the behavior of these agents.

Generalization of the standard Keynesian model, in which memory [37,38] is taken into account, was proposed by the authors [30,31]. Let us briefly describe this generalized model with memory. The equation of investment accelerator with memory [37,39] can be written as

$$I(t) = v(t) \int_0^t M(t-\tau)\, Y^{(n)}(\tau) d\tau, \tag{10}$$

where $M(t - \tau)$ is the memory function. Note that Equation (10) can also be used to describe the distributed lag. In this case, $M(t - \tau)$ is called the weighting function, which is interpreted as the probability density function. Equation (5) can be obtained from (10) in the case $M(t - \tau) = \delta(t - \tau)$ and $n = 1$. Substitution of the investment $I(t)$ in the form of expression (10), and the consumption expenditure (4) into balance Equation (3), we get

$$v(t) \int_0^t M(t-\tau) Y^{(n)}(\tau) d\tau = (1 - m(t))Y(t) - G_b(t), \tag{11}$$

where $G_b(t)$ is defined by (8). Equation (9) of the standard Keynesian model without memory and lag can be obtained from (11) by using $M(t-\tau) = \delta(t-\tau)$ and $n = 1$.

Equation (11) describes the fractional dynamics of the national income within the framework of the Keynesian model of growth with memory. If the parameters $m(t)$ and $v(t)$ are given, the growth of national income $Y(t)$ is conditioned by the behavior of the independent expenditure (8).

The memory with one-parameter power-law fading is described [37,39] by the function

$$M(t-\tau) = M_{RL}^{n-\alpha}(t-\tau) = \frac{1}{\Gamma(n-\alpha)}(t-\tau)^{n-\alpha-1}, \tag{12}$$

where $\Gamma(\alpha)$ is the gamma function and $n-1 < \alpha \leq n$. Using (12), the accelerator with memory (10) is represented [37,39] as

$$I(t) = v(t)\,(D_{C,0+}^{\alpha}Y)(t), \tag{13}$$

where $(D_{C,0+}^{\alpha}Y)(t)$ is the Caputo fractional derivative [4,5]. In general, the rate of acceleration $v(t)$ depends on the parameter of memory fading, i.e., $v(t) = v(t,\alpha)$. The parameter $\alpha > 0$ is interpreted as a fading parameter of power-law memory [37]. The concept of an accelerator with memory [39] allows us to get the equation of the Keynesian model with power-law memory in the form of the fractional differential equation

$$(D_{C,0+}^{\alpha}Y)(t) = \frac{1-m(t)}{v(t)}Y(t) - \frac{G_b(t)}{v(t)}, \tag{14}$$

where the Caputo fractional derivative can be represented by the Laplace convolution

$$(D_{C,0+}^{\alpha}Y)(t) = (M_{RL}^{n-\alpha} * Y^{(n)})(t) = \frac{1}{\Gamma(n-\alpha)}\int_0^t (t-\tau)^{n-\alpha-1}Y^{(n)}(\tau)d\tau, \tag{15}$$

where $n = [\alpha]+1$ for $\alpha \notin \mathbb{N}$ and $n = \alpha$ for $\alpha \in \mathbb{N}$, and the function $Y(\tau)$ has integer-order derivatives $Y^{(j)}(\tau)$, $j = 1, \ldots, (n-1)$, that are absolutely continuous.

The solution of Equation (14) with constant values of $m(t) = m$ and $v(t) = v$ has the form

$$Y(t) = \sum_{j=0}^{n-1} Y^{(j)}(0)t^j E_{\alpha,j+1}\left[\frac{1-m}{v}t^{\alpha}\right] - \frac{1}{v}\int_0^t (t-\tau)^{\alpha-1}E_{\alpha,\alpha}\left[\frac{1-m}{v}(t-\tau)^{\alpha}\right]G_b(t)d\tau, \tag{16}$$

where $n-1 < \alpha \leq n$. Solution (16) of the fractional differential Equation (14) and its properties are described in [30,31] (see also [32–34]).

4. Memory and Lag by Abel-Type Integral and Derivative with Kummer Functions

The economic accelerator and multiplier with continuously distributed lag were proposed by Alban W.H. Phillips [12,13] in 1954 (see also Sections 3.4, 3.5 and 8.7 in [14]). The distribution of delay time has been described by the exponential distribution. In 1956, the operators with continuously distributed lag were considered by Roy G.D. Allen in the book [14], (pp. 23–29). In the general case, the distribution of delay time can be described by other probability distributions [40], not just exponential distributions. Note that the time delay is caused by finite speeds of processes. Therefore, the distributed lag (time delay) cannot be interpreted as a memory.

4.1. Fractional Integral with Memory and Lag

The translation operator T_{τ} is defined [1], (pp. 95–96), by the expression $(T_{\tau}Y)(t) = Y(t-\tau)$, where $\tau > 0$ is the delay time. In the general case, the delay time $\tau > 0$ can be considered as a random variable, which is distributed by probability law (distribution) on positive semiaxis [40]. The translation operator with the continuously distributed delay time can be defined [40] by the equation

$$(\boldsymbol{T_M}Y)(t) = \int_0^\infty M_T(\tau)\,(T_\tau Y)(t)\,d\tau = \int_0^\infty M_T(\tau)\,Y(t-\tau)d\tau, \tag{17}$$

where we assume that $Y(t)$ and $M_T(t)$ are piecewise continuous functions on $\mathbb{R}$ and the integral $\int_0^\infty M_T(\tau)\,|Y(t-\tau)|\,d\tau$ converges. In Equation (17), the kernel $M_T(\tau)$ is the weighting function that satisfies the condition

$$M_T(\tau) \geq 0, \; \int_0^\infty M_T(\tau)d\tau = 1. \tag{18}$$

To take into account the distributed time delay and power-law fading memory, we can use a composition of the translation operator (17) and integration of non-integer order. The Riemann–Liouville fractional integral with a distributed time delay can be defined [40] by the equation

$$(\boldsymbol{I^\alpha_{T;RL;t_0}}Y)(t) = (\boldsymbol{T_M}(I^\alpha_{RL,t_0}Y))(t) = \int_0^\infty M_T(\tau)\,(I^\alpha_{RL,t_0}Y)(t-\;\tau)d\tau, \tag{19}$$

where $(I^\alpha_{RL,t_0}Y)(\mathrm{t})$ is the Riemann–Liouville fractional integral of the order $\alpha \in \mathbb{R}_+$, and $M_T(\tau)$ is the weighting function that satisfies the conditions (18). Here we can assume that $(I^\alpha_{RL,t_0}Y)(t)$ and $M_T(t)$ are piecewise continuous functions on $\mathbb{R}$ such that $\int_0^\infty M_T(\tau)\,\left|(I^\alpha_{RL,t_0}Y)(t-\;\tau)\right|\,d\tau$ converges. The Riemann–Liouville fractional integral I^α_{RL,t_0} is defined [4], (p. 92), by the expression

$$(I^\alpha_{RL,t_0}Y)(t) = \frac{1}{\Gamma(\alpha)}\int_{t_0}^t (t-\tau)^{\alpha-1}Y(\tau)d\tau, \tag{20}$$

where $\alpha \geq 0$ is the order of the fractional integral, $\Gamma(\alpha)$ is the gamma function, and $\tau \in [t_0, t]$.

However, in order for definition (19) to be correct, it needs to take into account that the fractional integral (20) is defined for the case $t > t_0$ [4], (p. 69). By virtue of this, we can consider the two following cases.

For $t_0 = -\infty$, Equation (20) defines the Liouville fractional integral $(I^\alpha_{L,+}Y)(\mathrm{t})$, where $\tau \in (-\infty, \mathrm{t}]$. Then the Liouville fractional integration with continuously distributed lag is defined in the form

$$(\boldsymbol{I^\alpha_{T,L}}Y)(t) = (\boldsymbol{T_M}(I^\alpha_{L,+}Y))(t) = \int_0^\infty M_T(\tau)\,(I^\alpha_{L,+}Y)(t-\;\tau)d\tau. \tag{21}$$

Considering that integral (20) is defined for the case $t > t_0 = 0$ [4], (p. 69), we can assume that $(I^\alpha_{RL,0+}Y)(t) = 0$ for $t < t_0 = 0$ in expression (20) with $t_0 = 0$. Here the function $Y(\tau)$ is defined on the finite interval $[t_0, t_1]$ with $0 = t_0 < t_1 < \infty$, [4], (p. 69). This allows us to use the upper limit $\mathrm{t} > 0$ insteat of infinity in Equation (19), such that

$$(\boldsymbol{I^\alpha_{T;RL;0+}}Y)(t) = (\boldsymbol{T_M}(I^\alpha_{RL,0+}Y))(t) = \int_0^t M_T(\tau)\,(I^\alpha_{RL,0+}Y)(t-\tau)d\tau. \tag{22}$$

As a result, we can define the fractional integration with the gamma-distributed lag in the form

$$(\boldsymbol{I^{\lambda,a;\alpha}_{T;C;0+}}Y)(t) = (M_T^{\lambda,a}(\tau) * (I^\alpha_{RL,0+}Y))(t) = \int_0^t M_T^{\lambda,a}(\tau)(I^\alpha_{RL,0+}Y)(t-\tau)\,d\tau, \tag{23}$$

where $M_T(\tau) = M_T^{\lambda,a}(\tau)$ is the probability density function of the gamma distribution

$$M_T^{\lambda,a}(\tau) = \begin{cases} \frac{\lambda^a\,\tau^{a-1}}{\Gamma(a)}\,exp(-\lambda\,\tau) & if\,\tau > 0, \\ 0 & if\,\tau \leq 0 \end{cases} \tag{24}$$

with the shape parameter $a > 0$ and the rate parameter $\lambda > 0$. If $a = 1$, the function (24) describes the exponential distribution.

In economics, the gamma distribution (24) is applied to take into account waiting times, when there is a sharp increase in the average delay time. For example, this distribution is used to describe delays orders in queues, delays in payments, and to take into account the likelihood of risk events. The distribution also describes the time of receipt of the order for the enterprise, the service life of device components, and time between store visits.

Substitution of (24) into (23) gives the Riemann–Liouville fractional integral with gamma distribution of delay time in the form of the Laplace convolution of memory and weighting functions

$$(\boldsymbol{I}^{\lambda,a;\alpha}_{T;C;0+}Y)(t) = \int_0^t M_T^{\lambda,a}(\tau)(I^{\alpha}_{RL,0+}Y)(t-\tau)\,d\tau = (M_T^{\lambda,a} * (M^{\alpha}_{RL} * Y))(t), \tag{25}$$

where $M^{\alpha}_{RL}(t) = (t-\tau)^{\alpha-1}/\Gamma(\alpha)$ is the kernel of the Riemann–Liouville fractional integral (20). The associativity of the Laplace convolution gives

$$(M_T^{\lambda,a} * (M^{\alpha}_{RL} * Y^{(n)}))(t) = (M^{\lambda,a;\alpha}_{TRL} * Y^{(n)})(t), \tag{26}$$

where $M^{\lambda,a;\alpha}_{TRL}(t)$ is the memory-and-lag function

$$M^{\lambda,a;\alpha}_{TRL}(t) = (M_T^{\lambda,a} * M^{\alpha}_{RL})(t). \tag{27}$$

This allows us to represent operator (25) in the form

$$(\boldsymbol{I}^{\lambda,a;\alpha}_{T;C;0+}Y)(t) = (\boldsymbol{T}^{\lambda,a}_{\boldsymbol{M}}(I^{\alpha}_{RL,0+}Y))(t) = \int_0^t M^{\lambda,a;\alpha}_{TRL}(t-\tau)Y(\tau)\,d\tau, \tag{28}$$

where $M^{\lambda,a;\alpha}_{TRL}(t-\tau)$ is defined by Equation (27). Let us obtain an explicit form of the memory-and-lag function $M^{\lambda,a;\alpha}_{TRL}$ (t). For this purpose, we can use Equation 2.3.6.1 of [41], (p. 324), that has the form

$$\int_0^t (t-\tau)^{\alpha-1}\tau^{\beta-1}\,exp(-\lambda\,\tau)d\tau = \frac{\Gamma(\alpha)\Gamma(\beta)}{\Gamma(\alpha+\beta)}\,t^{\alpha+\beta-1}F_{1,1}(\beta;\alpha+\beta;-\lambda t), \tag{29}$$

where $Re(\alpha) > 0$, $Re(\beta) > 0$ and $F_{1,1}(a;b;z)$ is the confluent hypergeometric Kummer function. The function $F_{1,1}(a;b;z)$ can be defined (see [42], (p. 115), and [4], (pp. 29–30)) by the equation

$$F_{1,1}(a;c;z) = \frac{\Gamma(c)}{\Gamma(a)\Gamma(c-a)}\int_0^1 t^{a-1}(1-t)^{c-a-1}\exp(zt)dt = \sum_{k=0}^{\infty}\frac{\Gamma(a+k)\Gamma(c)}{\Gamma(a)\Gamma(c+k)}\frac{z^k}{k!}, \tag{30}$$

where a, $z \in \mathbb{C}$, $Re(c) > Re(a) > 0$ such that $c \neq 0, -1, -2, \ldots$. Series (30) is absolutely convergent for all $z \in \mathbb{C}$.

Using Equation (29), the memory-and-lag function (27) can be written as

$$M^{\lambda,a;\alpha}_{TRL}(t) = \frac{\lambda^a\,\Gamma(a)}{\Gamma(a+n-\alpha)}\,t^{a+\alpha-1}F_{1,1}(a;a+\alpha;-\lambda t) \tag{31}$$

that defines the kernel of operator (28). In general, the function (31) can be interpreted as a new memory function. Note that equality $F_{1,1}(a;c;z) = \Gamma(c)E^{a}_{1,c}(z)$ (see Equation 5.1.18 of [43]) allows us to represent the memory kernel (31) through the three parameter Mittag–Leffler functions [43].

As a result, the Riemann–Liouville fractional integral with gamma distribution of delay time can represented [40] by the equation

$$(\boldsymbol{I}^{\lambda,a;\alpha}_{T;RL;0+}Y)(t) = \frac{\lambda^a\,\Gamma(a)}{\Gamma(a+\alpha)}\int_0^t (t-\tau)^{\alpha+a-1}F_{1,1}(a;a+\alpha;-\lambda(t-\tau))Y(\tau)\,d\tau, \tag{32}$$

where $\alpha > 0$ is the order of integration and the parameters $a > 0$, $\lambda > 0$ describe the shape and rate of the gamma distribution, respectively.

It is known (see Section 37 of [2] and [44]), the integral operators of the form

$$(\boldsymbol{A}_{\alpha} Y)(t) = \int_0^t (t-\tau)^{\alpha-1} K(t,\tau) Y(\tau)\, d\tau \tag{33}$$

are called the Abel-type integral operators. For example, we can consider the confluent hypergeometric Kummer function $F_{1,1}(\beta;\alpha;\lambda(t-\tau))$ as the kernel $K(t,\tau)$ of the operator (33). It is known that the Abel-type (AT) fractional integral operator with Kummer function in the kernel (see equation 37.1 in [1], (p. 731), and [45]) is defined by the equation

$$(\boldsymbol{I}^{\alpha,\beta,\lambda,}_{AT;0+} Y)(t) = \frac{1}{\Gamma(\alpha)} \int_0^t (t-\tau)^{\alpha-1} F_{1,1}(\beta;\alpha;\lambda(t-\tau)) Y(\tau)\, d\tau. \tag{34}$$

In paper [45], the integral operator (34) is denoted as $K_0(\beta,\alpha,\lambda)$.

As a result, the Riemann–Liouville fractional integral with gamma-distributed lag (32) can be expressed through the AT fractional integral (34) by the equation

$$(\boldsymbol{I}^{\lambda,a;\alpha}_{T;RL;0+} Y)(t) = \lambda^a\, \Gamma(a) (\boldsymbol{I}^{a+\alpha,a,-\lambda,}_{AT;0+} Y)(t). \tag{35}$$

The AT fractional integral (34) can be represented as an infinite series of the Riemann–Liouville fractional integrals

$$I^{\alpha,\beta,\lambda}_{AT;0+} = \sum_{k=0}^{\infty} \frac{(\beta)_k}{k!} \lambda^k I^{\alpha+k}_{RL,0+}, \tag{36}$$

where $(\beta)_k = \Gamma(\beta+k)/\Gamma(\beta)$ is the Pochhammer symbol. Expression (36) is called the Neumann generalized series (see Equation 37.10 of [1], (p. 732)), which characterizes the structure of the AT fractional integral operator (34). Using (35) and (36), the Riemann–Liouville fractional integral with gamma-distributed lag (32) can be represented as the series

$$(\boldsymbol{I}^{\lambda,a;\alpha}_{T;RL;0+} Y)(t) = \sum_{k=0}^{\infty} \frac{\Gamma(a+k)}{\Gamma(k+1)} (-1)^k \lambda^{k+a} I^{\alpha+a+k}_{RL,0+}. \tag{37}$$

Since the Riemann–Liouville fractional integrals (20) are bounded in $L_p(t_0,t_1)$, where $p \geq 1$, $t_1 < \infty$ (see the proof of Theorem 2.6 in [1], (pp. 48–51)), then the series (37) may be summed for $|\lambda| < \| I^1_{RL,a+} \|^{-1}_{L_p(t_0,t_1)}$. After evaluating the sums one may remove this restriction on λ and on the sum (37), since the suggested fractional integral operators (32) are analytic functions with respect to λ (see [1], (p. 732)).

Using Equation (37) and the semigroup property of the Riemann–Liouville fractional integrals, we can get the semigroup property for the Riemann–Liouville fractional integral with gamma-distributed lag (32) in the form

$$\boldsymbol{I}^{\lambda,a;\alpha}_{T;RL;0+} \boldsymbol{I}^{\lambda,b;\beta}_{T;RL;0+} = B(\alpha,\beta) \boldsymbol{I}^{\lambda,a+b;\alpha+\beta}_{T;RL;0+}, \tag{38}$$

where

$$B(\alpha,\beta) = \frac{\Gamma(\alpha)\Gamma(\beta)}{\Gamma(\alpha+\beta)} \tag{39}$$

is the beta function. Equality (38) directly follows from Equation 37.14 of [1], (p. 733).

Using Theorem 37.1 of [1], (p. 733), and Equation (35), we can state that the suggested fractional integral with lag (32) has the same range in $L_p(t_0,t_1)$ as the Riemann–Liouville fractional integrals and it is bounden from $L_p(t_0,t_1)$ onto $I^{\alpha}_{RL,0+}[L_p(t_0,t_1)] \subset L_p(t_0,t_1)$.

Using the condition of the invertibility of the AT operators (34), which is described by Theorem 37.2 of [1], (p. 736), and Equation 37.32 of [1], (p. 735), we get that the solution of the fractional integral Equation

$$m(t)(\boldsymbol{I}^{\alpha,\beta,\lambda}_{\boldsymbol{AT;0+}}Y)(t) = X(t) \tag{40}$$

can be represented in the form

$$Y(t) = \frac{m(t)}{\lambda^a\,\Gamma(a)} e^{-\lambda t} D^a_{RL,0+}\,(e^{\lambda t} D^{\alpha}_{RL,0+}\,X)(t)\,, \tag{41}$$

where $D^{\alpha}_{RL,0+}$ and $D^{a}_{RL,0+}$ are the Riemann–Liouville fractional derivatives of orders $\alpha > 0$ and $a > 0$ respectively. These derivatives can be defined by the Laplace convolution as $(D^{\alpha}_{RL,0+}X)(t) = D^n_t(M^{n-\alpha}_{RL} * X)(t)$, where $D^n_t = d^n/dt^n$, $n \in \mathbb{N}$.

Note that Equation (40) can be interpreted as an equation of economic multiplier with power-law memory and distributed lag [39], which is a generalization of the multiplier Equation (4), where $m(t)$ is the multiplier factor. In this case, Equation (41) can be considered as the equation of an economic accelerator with memory [39]. The equation of multiplier with memory is a reversible, such that the dual (inverse) equation describes an accelerator with memory (see Section 4 of [39]).

4.2. Fractional Derivative with Memory and Lag

Using the integral operator (32), we can define the fractional derivatives with continuously distributed lag [40]. For example, the Caputo fractional derivative with gamma-distributed lag is defined by the Laplace convolution in the form

$$(\boldsymbol{D}^{\lambda,a;\alpha}_{\boldsymbol{T;C;0+}}Y)(t) = \int_0^t M^{\lambda,a}_T(\tau)(D^{\alpha}_{C,0+}Y)(t-\tau)\,d\tau = (M^{\lambda,a}_T * (M^{n-\alpha}_{RL} * Y^{(n)}))(t). \tag{42}$$

The convolution is an associative operation that allows us to write

$$(M^{\lambda,a}_T * (M^{n-\alpha}_{RL} * Y^{(n)}))(t) = (M^{\lambda,a;n-\alpha}_{TRL} * Y^{(n)})(t), \tag{43}$$

where $M^{\lambda,a;n-\alpha}_{TRL}(t)$ is defined by Equation (31). This allows us to represent operators (42) in the form

$$(\boldsymbol{D}^{\lambda,a;\alpha}_{\boldsymbol{T;C;0+}}Y)(t) = \int_0^t M^{\lambda,a;n-\alpha}_{TRL}(t-\tau)Y^{(n)}(\tau)\,d\tau, \tag{44}$$

where $n-1 < \alpha \leq n$. Using Equation (31), we have the representation of the kernel $M^{\lambda,a;n-\alpha}_{TRL}(t)$ in the form

$$M^{\lambda,a;n-\alpha}_{TRL}(t) = \frac{\lambda^a\,\Gamma(a)}{\Gamma(a+n-\alpha)}\,t^{a+n-\alpha-1}F_{1,1}(a;a+n-\alpha;-\lambda t). \tag{45}$$

As a result, the Caputo fractional derivative with gamma-distributed lag is represented [40] by the Equation

$$(\boldsymbol{D}^{\lambda,a;\alpha}_{\boldsymbol{T;C;0+}}Y)(t) = \frac{\lambda^a\,\Gamma(a)}{\Gamma(a+n-\alpha)}\int_0^t (t-\tau)^{n-\alpha+a-1}F_{1,1}(a;a+n-\alpha;-\lambda(t-\tau))Y^{(n)}(\tau)\,d\tau, \tag{46}$$

where $n-1 < \alpha \leq n$. Fractional differential operator (46) can be expressed through the Riemann–Liouville fractional integral (32) with gamma-distributed lag in the form

$$(\boldsymbol{D}^{\lambda,a;\alpha}_{\boldsymbol{T;C;0+}}Y)(t) = (\boldsymbol{I}^{\lambda,a;n-\alpha}_{\boldsymbol{T;RL;0+}}Y^{(n)})(t), \tag{47}$$

where $n-1 < \alpha \leq n$.

Using the Laplace transform of the Caputo fractional derivative and the gamma distribution function, we get [40] the Laplace transform of the Caputo fractional derivative with gamma-distributed lag in the form

$$(\mathcal{L}(\mathbf{D}^{\lambda,a;\alpha}_{T;C;0+}Y)(t))(s) = \frac{\lambda^a}{(s+\lambda)^a}\left(s^\alpha(\mathcal{L}Y)(s) - \sum_{j=0}^{n-1} s^{\alpha-j-1}Y^{(j)}(0)\right), \tag{48}$$

where $n-1 < \alpha \leq n$.

Let us note that the operator (42) with $a = 1$ describes the Caputo fractional derivative with exponentially distributed lag [40]. Using the fact that the Caputo fractional derivatives with integer values $\alpha = n \in \mathbb{N}$ are the integer-order derivatives $(D^{\alpha}_{C,0+}Y)(t) = d^nY(t)/dt^n$, the operator (42) with $a = 1$ and $\alpha = n$ describes the integer-order derivatives with the exponential distribution [40]. Note that these operators with exponentially distributed lag were defined in the works of Caputo and Fabrizio [46,47], where they have been misinterpreted as fractional derivatives of non-integer orders. We can state [40] that the derivative of integer order with exponentially distributed lag coincides with the Caputo–Fabrizio operator of the order $\beta = n - 1/(\lambda+1)$, where λ is the rate parameter of the distribution and $n = [\beta] + 1$. As a result, the Caputo–Fabrizio operator can be interpreted as an integer-order derivative with the exponentially distributed lag.

The proposed operator (42) in the form (46) can be interpreted as a new generalized operator with the memory function given by the confluent hypergeometric function. The generalized fractional derivative (46) can be used to simultaneously account of long memory with power-law fading and distributed lag with the gamma distribution of delay time. In the next section, we describe a macroeconomic model with memory and distributed lag by using this proposed operator.

5. Fractional Differential Equation of a Keynesian Model with Memory and Lag

Let us take into account that the relationship between the investment expenditure $I(t)$ and the national income $Y(t)$ depends on memory and lag effects. For the case of the power-law memory and gamma distribution of the delay time, we can use the generalized accelerator equation

$$I(t) = v(t)\,(\mathbf{D}^{\lambda,a;\alpha}_{T;C;0+}Y)(t), \tag{49}$$

where $\mathbf{D}^{\lambda,a;\alpha}_{T;C;0+}$ is the Caputo fractional derivative with distributed lag given by (46). Equation (49) describes the economic accelerator that takes into account the power-law fading memory and the gamma-distributed lag.

Substituting expressions (49) and (4) into balance Equation (3), we obtain the fractional differential equation of the Keynesian model with power-law memory and gamma distribution of delay time in the form

$$(\mathbf{D}^{\lambda,a;\alpha}_{T;C;0+}Y)(t) = \frac{1-m(t)}{v(t)}Y(t) - \frac{G_b(t)}{v(t)}, \tag{50}$$

where α is the parameter of memory fading, $a > 0$ is the shape parameter and $\lambda > 0$ is the rate parameter of the gamma distribution of the delay time.

Let us consider the case where $v(t)$ and $m(t)$ are constant quantities. Then the Keynesian model with one-parameter power-law memory and gamma-distributed lag is described by the fractional differential equation

$$(\mathbf{D}^{\lambda,a;\alpha}_{T;C;0+}Y)(t) = \omega Y(t) + F(t), \tag{51}$$

where $\omega = (1-m)/v$ and $F(t) = -v^{-1}G_b(t)$.

The general solution of Equation (51) can be written as

$$Y(t) = Y_0(t) + Y_F(t), \tag{52}$$

where $Y_0(t)$ is the solution of Equation (51) with $F(t) = 0$, i.e., the homogeneous equation

$$(\mathbf{D}^{\lambda,a;\alpha}_{T;C;0+} Y)(t) = \omega Y(t), \tag{53}$$

and $Y_F(t)$ is particular solution of (51) that can be represented in the form

$$Y_F(t) = \int_0^t G_\alpha[t-\tau]\, F(\tau) d\tau, \tag{54}$$

where $G_\alpha[t-\tau]$ is the generalized Green function [4], (p.281,295). Equation (54) yields the solution $Y_F(t)$ for Equation (51) with initial conditions, $Y^{(j)}(0) = 0$ for all $j = 0, \ldots, (n-1)$.

Theorem 1. *The fractional differential Equation*

$$(\mathbf{D}^{\lambda,a;\alpha}_{T;C;0+} Y)(t) = \omega Y(t) + F(t), \tag{55}$$

where $\mathbf{D}^{\lambda,a;\alpha}_{T;C;0+}$ is the fractional derivative of order $\alpha > 0$ with gamma-distributed lag, in which $a > 0$ and $\lambda > 0$ are the shape and rate parameters of the gamma distribution respectively, has the solution

$$Y(t) = \sum_{j=0}^{n-1} S^{\alpha-j-1}_{\alpha,a}[\omega\,\lambda^{-a}, \lambda|t] Y^{(j)}(0) + \frac{1}{\omega} F(t) - \frac{1}{\omega}\int_0^t S^{\alpha}_{\alpha,a}\,[\omega\,\lambda^{-a}, \lambda|t-\tau]\; F(\tau) d\tau, \tag{56}$$

where $n = [\alpha] + 1$, and $S^{\gamma}_{\alpha,\delta}\,[\mu, \lambda|t]$ is the special function that is defined by the expression

$$S^{\gamma}_{\alpha,\delta}\,[\mu, \lambda|t] = -\sum_{k=0}^{\infty} \frac{t^{\delta(k+1)-\alpha k-\gamma-1}}{\mu^{k+1}\Gamma(\delta(k+1)-\alpha k-\gamma)} F_{1,1}(\delta(k+1); \delta(k+1)-\alpha k-\gamma, -\lambda t), \tag{57}$$

where $F_{1,1}(a;b;z)$ is the confluent hypergeometric Kummer function (30).

Proof. The first step is to find a solution for the homogeneous Equation (53). Using the Laplace transform of Equation (53), we get

$$\frac{\lambda^a}{(s+\lambda)^a}\left(s^{\alpha}(\mathcal{L}Y)(s) - \sum_{j=0}^{n-1} s^{\alpha-j-1} Y^{(j)}(0)\right) = \omega(\mathcal{L}Y)(s). \tag{58}$$

Then we can write

$$(\mathcal{L}\,Y)(s) = \sum_{j=0}^{n-1} \frac{s^{\alpha-j-1}}{s^{\alpha} - \mu(s+\lambda)^a} Y^{(j)}(0), \tag{59}$$

where $\mu = \omega\,\lambda^{-a}$. Using Equation 5.4.9 of [48] in the form

$$\left(\mathcal{L}^{-1}\left(\frac{s^a}{(s+b)^c}\right)\right)(s) = \frac{1}{\Gamma(c-a)} t^{c-a-1} F_{1,1}(c; c-a, -bt), \tag{60}$$

where $Re(\mathrm{c}-\mathrm{a}) > 0$, we get [40] the Laplace transform of the function (57) (**Theorem 1**) as

$$\mathcal{L}\,(S^{\gamma}_{\alpha,\delta}\,[\mu, \lambda\big|t])(s) = \frac{s^{\gamma}}{s^{\alpha} - \mu(s+\lambda)^{\delta}}. \tag{61}$$

Using Equation (61) the solution of the homogenous fractional differential Equation (53) has the form

$$Y_0(t) = \sum_{j=0}^{n-1} S^{\alpha-j-1}_{\alpha,a}[\omega\,\lambda^{-a}, \lambda|t] Y^{(j)}(0), \tag{62}$$

where $S^{\alpha-j-1}_{\alpha,a}[\omega\, \lambda^{-a}, \lambda \big| t]$ is defined by Equation (57).

The second step is to find a particular solution (54) of Equation (55). The Laplace transform of Equation (55) with conditions $Y^{(j)}(0) = 0$ for all $j = 0, \ldots (n-1)$ gives the expression

$$\frac{\lambda^a}{(s+\lambda)^a} s^{\alpha} (\mathcal{L}Y)(s) = \omega (\mathcal{L}Y)(s) + (\mathcal{L}F)(s) \tag{63}$$

that can be rewritten in the from

$$(\mathcal{L}Y)(s) = \frac{(s+\lambda)^a}{\lambda^a\, s^{\alpha} - \omega (s+\lambda)^a} (\mathcal{L}F)(s). \tag{64}$$

The equality

$$\frac{(s+\lambda)^a}{\lambda^a\, s^{\alpha} - \omega (s+\lambda)^a} = -\frac{1}{\omega} + \frac{1}{\omega} \frac{s^{\alpha}}{s^{\alpha} - \mu (s+\lambda)^a}, \tag{65}$$

where $\mu = \omega\, \lambda^{-a}$, gives

$$(\mathcal{L}Y)(s) = -\frac{1}{\omega} (\mathcal{L}F)(s) + \frac{1}{\omega} \frac{s^{\alpha}}{s^{\alpha} - \mu (s+\lambda)^a} (\mathcal{L}F)(s). \tag{66}$$

Using (61) with $\delta = a$, $\gamma = \alpha$, we have

$$G_{\alpha}[t-\tau] = -\frac{1}{\omega} \delta(t-\tau) + \frac{1}{\omega} S^{\alpha}_{\alpha,a}\, [\mu, \lambda | t-\tau]. \tag{67}$$

As a result, we obtain

$$Y_F(t) = \frac{1}{\omega} F(t) - \frac{1}{\omega} \int_0^t S^{\alpha}_{\alpha,a}\, [\mu, \lambda | t-\tau]\, F(\tau) d\tau \tag{68}$$

that describes the particular solution of Equation (55).

Substitution of (62) and (68) into (52) gives (56).

This ends the proof. □

As a result, the solution of Equation (55) of the Keynesian model with one-parameter power-law memory and gamma-distributed lag for constant $v(t) = v$ and $m(t) = m$ is described by the expression

$$Y(t) = \sum_{j=0}^{n-1} S^{\alpha-j-1}_{\alpha,a} [\omega\, \lambda^{-a}, \lambda | t] Y^{(j)}(0) - \frac{1}{1-m} G_b(t) + \frac{1}{1-m} \int_0^t S^{\alpha}_{\alpha,a}\, [\omega\, \lambda^{-a}, \lambda | t-\tau]\, G_b(\tau) d\tau, \tag{69}$$

where $G_b(\tau) = G(\tau) + g(\tau)$ and $\omega = (1-m)/v$.

6. Asymptotic Behavior of National Income Growth with Memory and Lag

In economic theory, the important concept of the Harrod's warranted rate of growth [14] is used, which is also called the technological growth rate. The warranted growth rate describes the growth when the following two conditions are satisfied. The first condition is the constancy of the structure of the economy. This condition means that the parameters of the model do not change over time. In the Keynesian model, we should consider the parameters (t), $b(t)$, and $m(t)$ as constant quantities. The second condition is the absence of external influences. This condition means the absence of exogenous variables. In the Keynesian model, we should consider the case of the absence of the independent expenditure, i.e., $G_b(t) = 0$. Mathematically the warranted growth rate can be obtained from the asymptotic expression of the solution of the homogeneous differential equation of the macroeconomic model. In the standard Keynesian model, the solution of Equation (9) with $G_b(t) = 0$ has the form $Y(t) = Y(0) \exp(\omega\, t)$. Therefore, the warranted growth rate of this model is described by the value $\omega = (1 - \mathrm{m})/v$, when $G_b(t) = 0$.

The Keynesian model with memory was suggested by authors [30,31] in 2016 (see also [32–34]). For this growth model, the fractional differential equation, its solution, and properties have been described. We proved [32,33] that the warranted growth rate with memory is equal to the value $\omega_{\text{eff}}(\alpha) = \omega^{1/\alpha}$, where $\alpha > 0$ is a parameter of power-law memory fading and ω is the rate of growth without memory ($\alpha = 1$). The warranted growth rates of models with memory do not coincide with the growth rates of the standard Keynesian model. The memory effects can significantly change the growth rates of the economy [32–34] and lead to new types of behavior for the same parameters of the economic model. The principles of changing growth rates by memory have been proposed in [32,33]. The memory effects can both increase and decrease the warranted growth rates in comparison with the standard Keynesian model. For the memory fading with $\alpha < 1$, we get a slowdown in the growth and decline of the economy. We can state that memory with $\alpha < 1$ leads to inhibition of economic growth or decline, i.e., we have stagnation of the economy if $\alpha < 1$. For the memory fading with $\alpha > 1$ we have an improvement in the economy. In this case, the memory effect leads either to the slowdown in the decline rate or to the replacement of the decline with growth, or to the increase in the growth rate.

To consider the warranted growth rate of national income for the Keynesian model with memory and distributed lag, we should obtain an asymptotic behavior of the solution (62) of homogenous model Equation (53). This solution is expressed by the function $S^{\gamma}_{\alpha,\delta}[\mu,\lambda|t]$ that is represented as an infinite sum (57) of the confluent hypergeometric Kummer function $F_{1,1}(a;c;z)$. We can use the asymptotic expression of the function $F_{1,1}(a;c;z)$ at infinity $z \to -\infty$ that is given in [4] (p. 29), in the form

$$F_{1,1}(a;c;z) = \frac{\Gamma(c)}{\Gamma(c-a)} e^{-i\pi a} z^{-a} \left(1 + O\left(\frac{1}{z}\right)\right), \tag{70}$$

where $z = -\lambda t < 0$. Therefore the asymptotic expression for $t \to \infty$ is

$$F_{1,1}(\delta(k+1);\delta(k+1)-\alpha k-\gamma,-\lambda t) = \frac{\Gamma(\delta(k+1)-\alpha k-\gamma)}{\Gamma(-\alpha k-\gamma)} (\lambda t)^{-\delta(k+1)} e^{-i\pi\delta(k+1)} \left(1 + O\left(\frac{1}{t}\right)\right). \tag{71}$$

Equation (71) leads to the asymptotic expression at infinity ($t \to \infty$) of the function (57) in the form

$$S^{\gamma}_{\alpha,\delta}[\mu,\lambda|t] = -\sum_{k=0}^{\infty} \frac{\lambda^{-\delta(k+1)} t^{-\alpha k-\gamma-1}}{\mu^{k+1}\Gamma(-\alpha k-\gamma)} e^{-i\pi\delta(k+1)} \left(1 + O\left(\frac{1}{t}\right)\right) = -\frac{\lambda^{-\delta} t^{-\gamma-1}}{\mu^{1}\Gamma(-\gamma)} e^{-i\pi\delta}\left(1 + O(t^{-\alpha-\gamma-1})\right), \tag{72}$$

where $\alpha k + \gamma \neq 0, 1, 2, \ldots$ for all integer k.

Solution (62) is expressed by the function $S^{\alpha-j-1}_{\alpha,a}[\mu,\lambda|t]$, where $\mu = \omega\, \lambda^{-a}$. Then using Equation (72), we get the asymptotic expression

$$S^{\alpha-j-1}_{\alpha,a}[\omega\,\lambda^{-a},\lambda|t] = -\sum_{k=0}^{\infty} \frac{\lambda^{-a} t^{-\alpha(k+1)+j}}{\omega^{k+1}\,\Gamma(-\alpha(k+1)+j+1)} e^{-i\pi a}\left(1 + O\left(\frac{1}{t}\right)\right). \tag{73}$$

For the gamma distribution with the integer shape parameter ($a = m \in \mathbb{N}$), which is called the Erlang distribution, expression (73) gives

$$S^{\alpha-j-1}_{\alpha,m}[\omega\,\lambda^{-a},\lambda|t] = -\sum_{k=0}^{\infty} \frac{(-1)^m \lambda^{-m} t^{-\alpha(k+1)+j}}{\omega^{k+1}\,\Gamma(-\alpha(k+1)+j+1)} \left(1 + O\left(\frac{1}{t}\right)\right), \tag{74}$$

where $e^{-i\pi m} = (-1)^m$ is used.

Using Equation 5.1.18 of [43], (p. 99), in the form $F_{1,1}(a;c;z) = \Gamma(c)E^{a}_{1,c}(z)$, we can get asymptotic expressions (72) and (74) by using asymptotic expressions of the three parameter Mittag–Leffler functions $E^{\delta(k+1)}_{1,\delta(k+1)-\alpha k-\gamma}(-\lambda t)$ (for example, see [49]). Using Equation (74) we can see that the series

(74) can be represented through the two-parameter Mittag–Leffler function with the negative first parameter [50] as

$$-\sum_{k=0}^{\infty} \frac{(-1)^m \lambda^{-m} t^{-\alpha(k+1)+j}}{\omega^{k+1}\, \Gamma(-\alpha(k+1)+j+1)} = \frac{(-1)^m \lambda^{-m} t^j}{\Gamma(j+1)} - (-1)^m \lambda^{-m} t^j E_{-\alpha,j+1}(t^{-\alpha}/\omega). \tag{75}$$

This allows us to state that the function $S_{\alpha,m}^{\alpha-j-1}[\omega\, \lambda^{-a}, \lambda|t]$ in the long time limit leads to a series, which can be interpreted as a two-parameter Mittag–Leffler function, which is analogous to the property of the three-parameter Mittag–Leffler function described in [51].

Equation (74) allows us to give the asymptotic expression

$$S_{\alpha,a}^{\alpha-j-1}[\omega\, \lambda^{-a}, \lambda|t] = -\frac{(-1)^m \lambda^{-m} t^{-\alpha+j}}{\omega^1\, \Gamma(-\alpha+j+1)} \left(1 + O\left(t^{-2\alpha+j}\right)\right). \tag{76}$$

As a result, the asymptotic behavior of solution (69) with $G_b(t) = 0$ can be described by the Equation

$$Y(t) = \sum_{j=0}^{n-1} \frac{(-1)^{m+1} \lambda^{-m} t^{-\alpha+j}}{\omega\, \Gamma(-\alpha+j+1)}\, Y^{(j)}(0) \left(1 + O\left(t^{-2\alpha+j}\right)\right) \tag{77}$$

that characterizes warranted growth of national income, which is represented by solution (69) with $G_b(\tau) = 0$ and $a = m \in \mathbb{N}$. Equation (77) allows us to state that the warranted growth has the power-law form with the power $-\alpha + j$, where $j \in \{0, \ldots, n-1\}$ is the smallest value at which $Y^{(j)}(0) \neq 0$. As a result, the warranted growth of the national income with memory and distributed lag has the power-law type instead of the exponential type of growth with memory without time delay [32,33], where warranted growth rate is $\omega_{\text{eff}}(\alpha) = \omega^{1/\alpha}$. Therefore we can state that the distributed lag (time delay) suppresses the effects of fading memory.

7. Conclusions

The standard Keynesian model [8–10,14] describes the dynamics of national income in the absence of long memory and distributed time delay. The Keynesian model with power-law memory has been suggested by authors [30,31]. The effects of continuously distributed lag are not considered in [30,31]. In this paper, we generalize the Keynesian model with memory by taking into account gamma distribution of delay time. To take into account the distributed lag, we use the operators that are compositions of the translation operator with distributed delay time and the fractional derivatives or integrals. These operators allow us to take into account the memory and lag in the economic accelerator. These operators are the Abel-type integro-differential operators with the confluent hypergeometric Kummer function in the kernel. The solution of the suggested fractional differential Equation, which describes the fractional dynamics of national income, has been suggested. The asymptotic behavior of economic processes with memory and distributed lag demonstrates power-law growth. In the absence of delays, the processes with fading memory demonstrate exponential growth. The warranted growth rate with memory [30–34] is equal to the value $\omega_{\text{eff}}(\alpha) = \omega^{1/\alpha}$, where $\alpha > 0$ is a memory fading parameter. Therefore effects of long memory can significantly accelerate the growth rate of the economy by several orders of magnitude [30–33]. Fading memory can lead to an increase in the growth rate in processes without lag [30–34]. The appearance of distributed lag does not accelerate growth due to the memory effect (see also [52,53]). Moreover, we can state that the lag can suppress the effect of fading memory. The distributed lag leads to slower growth. We assume that the suggested approach and model can be used for economic growth modeling by analogy with the computer simulation of the economy in [54–58]. It should also be noted that fractional differential Equations have been applied to describe power-law memory in continuous-time finance [59–72]. This fact allows us to assume that the proposed approach can be used to take into account the continuously distributed time delay and memory in financial processes with the waiting-time distribution [60,62].

Author Contributions: V.E.T.: Contributed by the ideas, analysis and writing the manuscript in mathematical part. V.V.T.: Contributed by the ideas, analysis and writing the manuscript in economical part.

References

1. Samko, S.G.; Kilbas, A.A.; Marichev, O.I. *Fractional Integrals and Derivatives Theory and Applications*; Gordon and Breach: New York, NY, USA, 1993; 1006p, ISBN 9782881248641.
2. Kiryakova, V. *Generalized Fractional Calculus and Applications*; Longman and J. Wiley: New York, NY, USA, 1994; 360p, ISBN 9780582219779.
3. Podlubny, I. *Fractional Differential Equations*; Academic Press: San Diego, CA, USA, 1998; 340p.
4. Kilbas, A.A.; Srivastava, H.M.; Trujillo, J.J. *Theory and Applications of Fractional Differential Equations*; Elsevier: Amsterdam, The Netherlands, 2006; 540p, ISBN 9780444518323.
5. Diethelm, K. *The Analysis of Fractional Differential Equations: An Application-Oriented Exposition Using Differential Operators of Caputo Type*; Springer: Berlin, Germany, 2010; 247p.
6. Mainardi, F. (Ed.) *Fractional Calculus Theory and Applications*; MDPI: Basel, Switzerland, 2018; 208p, ISBN 978-3-03897-207-5. Available online: https://www.mdpi.com/books/pdfview/book/755 (accessed on 24 January 2019).
7. Tenreiro Machado, J.A. (Ed.) *Handbook of Fractional Calculus with Applications*; De Gruyter: Berlin, Germany, 2019; Volumes 1–8.
8. Keynes, J.M. *The General Theory of Employment, Interest and Money*; Macmillan: London, UK, 1936.
9. Keynes, J.M. *The General Theory of Employment, Interest and Money: With the Economic Consequences of the Peace*; Series: Classics of World Literature; Wordsworth Editions: Hertfordshire, UK, 2017; 576p, ISBN 978-1840227475.
10. Keynes, J.M. *The General Theory of Employment, Interest and Money: With the Economic Consequences of the Peace*; Palgrave Macmillan: London, UK, 2018; 438p.
11. Kalecki, M. A macrodynamic theory of business cycles. *Econometrica* **1935**, *3*, 327–344. [CrossRef]
12. Phillips, A.W. Stabilisation policy in a closed economy. *Econ. J.* **1954**, *64*, 290–323. [CrossRef]
13. Phillips, A.W. *A. W. H. Phillips Collected Works in Contemporary Perspective*; Leeson, R., Ed.; Cambridge University Press: Cambridge, UK, 2000; 515p, ISBN 9780521571357.
14. Allen, R.G.D. *Mathematical Economics*, 2nd ed.; Macmillan: London, UK, 1959; 812p.
15. Granger, C.W.J. *The Typical Spectral Shape of an Economic Variable*; Technical Report No. 11. January 30, 1964; Department of Statistics, Stanford University: Stanford, CA, USA, 1964; 21p, Available online: https://statistics.stanford.edu/research/typical-spectral-shape-economic-variable (accessed on 24 January 2019).
16. Granger, C.W.J. The typical spectral shape of an economic variable. *Econometrica* **1966**, *34*, 150–161. [CrossRef]
17. Granger, C.W.J. *Essays in Econometrics: Collected Papers of Clive W. J. Granger. Volume. I. Spectral Analysis, Seasonality, Nonlinearity, Methodology, and Forecasting*; Ghysels, E., Swanson, N.R., Watson, M.W., Eds.; Cambridge University Press: Cambridge, UK; New York, NY, USA, 2001; 523p.
18. Granger, C.W.J. *Essays in Econometrics Collected Papers of Clive W.J. Granger. Volume II: Causality, Integration and Cointegration, and Long Memory*; Ghysels, E., Swanson, N.R., Watson, M.W., Eds.; Cambridge University Press: Cambridge, UK, 2001; 398p.
19. Bearn, J. *Statistics for Long-Memory Processes*; Capman and Hall: New York, NY, USA, 1994; 315p, ISBN 0-412-04901-5.
20. Palma, W. *Long-Memory Time Series: Theory and Methods*; Wiley-InterScience: Hoboken, NJ, USA, 2007; 304p, ISBN 978-0-470-11402-5.
21. Beran, J.; Feng, Y.; Ghosh, S.; Kulik, R. *Long-Memory Processes: Probabilistic Properties and Statistical Methods*; Springer: Berlin/Heidelberg, Germany; New York, NY, USA, 2013; 884p.
22. Robinson, P.M. (Ed.) *Time Series with Long Memory*; Series: Advanced Texts in Econometrics; Oxford University Press: Oxford, UK, 2003; 392p, ISBN 978-0199257300.
23. Teyssiere, G.; Kirman, A.P. (Eds.) *Long Memory in Economics*; Springer: Berlin/Heidelberg, Germany, 2007; 390p.

24. Baillie, R. Long memory processes and fractional integration in econometrics. *J. Econom.* **1996**, *73*, 5–59. [CrossRef]
25. Graves, T.; Gramacy, R.; Watkins, N.; Franzke, C. A Brief History of Long Memory: Hurst, Mandelbrot and the Road to ARFIMA, 1951–1980. *Entropy* **2017**, *19*, 437. [CrossRef]
26. Granger, C.W.J.; Joyeux, R. An introduction to long memory time series models and fractional differencing. *J. Time Ser. Anal.* **1980**, *1*, 15–39. [CrossRef]
27. Hosking, J.R.M. Fractional differencing. *Biometrika* **1981**, *68*, 165–176. [CrossRef]
28. Tarasov, V.E.; Tarasova, V.V. Long and short memory in economics: Fractional-order difference and differentiation. *IRA Int. J. Manag. Soc. Sci.* **2016**, *5*, 327–334. [CrossRef]
29. Tarasova, V.V.; Tarasov, V.E. Comments on the article «Long and short memory in economics: Fractional-order difference and differentiation». *Probl. Modern Sci. Educ.* **2017**, *113*, 26–28. [CrossRef]
30. Tarasova, V.V.; Tarasov, V.E. Keynesian model of economic growth with memory. *Econ. Manag. Probl. Solut.* **2016**, *58*, 21–29. (In Russian)
31. Tarasova, V.V.; Tarasov, V.E. Memory effects in hereditary Keynes model. *Probl. Modern Sci. Educ.* **2016**, *80*, 56–61. (In Russian) [CrossRef]
32. Tarasov, V.E.; Tarasova, V.V. Macroeconomic models with long dynamic memory: Fractional calculus approach. *Appl. Math. Comput.* **2018**, *338*, 466–486. [CrossRef]
33. Tarasov, V.E. Economic models with power-law memory. In *Handbook of Fractional Calculus with Applications*; Volume 8: Applications in Engineering, Life and Social Sciences, Part B; Chapter 1; De Gruyter: Berlin, Germany, 2019; ISBN 978-3-11-057092-2.
34. Tarasov, V.E. Self-organization with memory. *Commun. Nonlinear Sci. Numer. Simul.* **2019**, *72*, 240–271. [CrossRef]
35. Volgina, O.A.; Golodnaya, N.Y.; Odiako, N.N.; Shuman, G.I. *Mathematical Modeling of Economic Processes and Systems*, 3rd ed.; Kronus: Moscow, Russia, 2014; 200p, ISBN 978-5-406-03252-7.
36. Shone, R. *An Introduction to Economic Dynamics*, 2nd ed.; Cambridge University Press: Cambridge, UK, 2012; 224p, ISBN 9781139165020.
37. Tarasova, V.V.; Tarasov, V.E. Concept of dynamic memory in economics. *Commun. Nonlinear Sci. Numer. Simul.* **2018**, *55*, 127–145. [CrossRef]
38. Tarasov, V.E.; Tarasova, V.V. Criterion of existence of power-law memory for economic processes. *Entropy* **2018**, *20*, 414. [CrossRef]
39. Tarasova, V.V.; Tarasov, V.E. Accelerator and multiplier for macroeconomic processes with memory. *IRA Int. J. Manag. Soc. Sci.* **2017**, *9*, 86–125. [CrossRef]
40. Tarasov, V.E.; Tarasova, S.S. Fractional and integer derivatives with continuously distributed lag. *Commun. Nonlinear Sci. Numer. Simul.* **2019**, *70*, 125–169. [CrossRef]
41. Prudnikov, A.P.; Brychkov, Y.A.; Marichev, O.I. *Integrals and Series. Vol. 1. Elementary Functions*, 5th ed.; Taylor & Francis: London, UK, 2002; 798p, ISBN 2-88124-089-5.
42. Luke, Y.L. *The Special Functions and Their Approximations*; Academic Press: San Diego, CA, USA; New York, NY, USA, 1969; Volume 1, 348p, ISBN 0-12-459901-X.
43. Gorenflo, R.; Kilbas, A.A.; Mainardi, F.; Rogosin, S.V. *Mittag-Leffler Functions, Related Topics and Applications*; Springer: Berlin, Germany, 2014; 443p.
44. Gorenflo, R.; Vessella, S. *Abel Integral Equations: Analysis and Applications*; Springer: Berlin/Heidelberg, Germany, 1991; 215p, ISBN 3-540-53668-X.
45. Prabhakar, T.R. Some integral equations with Kummer's functions in the kernels. *Can. Math. Bull.* **1971**, *4*, 391–404. [CrossRef]
46. Caputo, M.; Fabrizio, M. A new definition of fractional derivative without singular kernel. *Prog. Fract. Differ. Appl.* **2015**, *1*, 73–85. [CrossRef]
47. Caputo, M.; Fabrizio, M. Applications of new time and spatial fractional derivatives with exponential kernels. *Prog. Fract. Differ. Appl.* **2016**, *2*, 1–11. [CrossRef]
48. Bateman, H. *Tables of Integral Transforms*; McGraw-Hill: New York, NY, USA, 1954; Volume I, 391p, ISBN 07-019549-8.
49. Mainardi, F.; Garrappa, R. On complete monotonicity of the Prabhakar function and non-Debye relaxation in dielectrics. *J. Comput. Phys.* **2015**, *293*, 70–80. [CrossRef]

50. Hanneken, J.W.; Narahari Achar, B.N.; Puzio, R.; Vaught, D.M. Properties of the Mittag–Leffler function for negative alpha. *Phys. Scr.* **2009**, *2009*, 014037. [CrossRef]
51. Sandev, T.; Metzler, R.; Tomovski, Z. Correlation functions for the fractional generalized Langevin equation in the presence of internal and external noise. *J. Math. Phys.* **2014**, *55*, 023301. [CrossRef]
52. Tarasov, V.E.; Tarasova, V.V. Phillips model with exponentially distributed lag and power-law memory. *Comput. Appl. Math.* **2019**, *38*, 13. [CrossRef]
53. Tarasov, V.E.; Tarasova, V.V. Harrod-Domar growth model with memory and distributed lag. *Axioms* **2019**, *8*, 9. [CrossRef]
54. Tejado, I.; Valerio, D.; Perez, E.; Valerio, N. Fractional calculus in economic growth modelling: The Spanish and Portuguese cases. *Int. J. Dyn. Control* **2015**, *5*, 208–222. [CrossRef]
55. Tejado, I.; Valerio, D.; Perez, E.; Valerio, N. Fractional calculus in economic growth modelling: The economies of France and Italy. In Proceedings of the International Conference on Fractional Differentiation and its Applications, Novi Sad, Serbia, 18–20 July 2016; Spasic, D.T., Grahovac, N., Zigic, M., Rapaic, M., Atanackovic, T.M., Eds.; pp. 113–123.
56. Tejado, I.; Perez, E.; Valerio, D. Fractional calculus in economic growth modelling of the group of seven. *SSRN Electron. J.* **2018**. [CrossRef]
57. Luo, D.; Wang, J.R.; Feckan, M. Applying fractional calculus to analyze economic growth modelling. *J. Appl. Math. Stat. Inform.* **2018**, *14*, 25–36. [CrossRef]
58. Tejado, I.; Perez, E.; Valerio, D. Economic growth in the European Union modelled with fractional derivatives: First results. *Bull. Pol. Acad. Sci. Tech. Sci.* **2018**, *66*, 455–465. [CrossRef]
59. Scalas, E.; Gorenflo, R.; Mainardi, F. Fractional calculus and continuous-time finance. *Physica A* **2000**, *284*, 376–384. [CrossRef]
60. Mainardi, F.; Raberto, M.; Gorenflo, R.; Scalas, E. Fractional calculus and continuous-time finance II: The waiting-time distribution. *Physica A* **2000**, *287*, 468–481. [CrossRef]
61. Laskin, N. Fractional market dynamics. *Physica A* **2000**, *287*, 482–492. [CrossRef]
62. Raberto, M.; Scalas, E.; Mainardi, F. Waiting-times and returns in high-frequency financial data: An empirical study. *Physica A* **2002**, *314*, 749–755. [CrossRef]
63. West, B.J.; Picozzi, S. Fractional Langevin model of memory in financial time series. *Phys. Rev. E* **2002**, *65*, 037106. [CrossRef] [PubMed]
64. Picozzi, S.; West, B.J. Fractional Langevin model of memory in financial markets. *Phys. Rev. E* **2002**, *66*, 046118. [CrossRef]
65. Scalas, E. The application of continuous-time random walks in finance and economics. *Physica A* **2006**, *362*, 225–239. [CrossRef]
66. Meerschaert, M.M.; Scalas, E. Coupled continuous time random walks in finance. *Physica A* **2006**, *370*, 114–118. [CrossRef]
67. Cartea, A.; Del-Castillo-Negrete, D. Fractional diffusion models of option prices in markets with jumps. *Physica A* **2007**, *374*, 749–763. [CrossRef]
68. Mendes, R.V. A fractional calculus interpretation of the fractional volatility model. *Nonlinear Dyn.* **2009**, *55*, 395–399. [CrossRef]
69. Blackledge, J. Application of the fractional diffusion equation for predicting market behavior. *Int. J. Appl. Math.* **2010**, *40*, 130–158.
70. Tenreiro Machado, J.; Duarte, F.B.; Duarte, G.M. Fractional dynamics in financial indices. *Int. J. Bifurc. Chaos* **2012**, *22*, 1250249. [CrossRef]
71. Kerss, A.; Leonenko, N.; Sikorskii, A. Fractional Skellam processes with applications to finance. *Fract. Calc. Appl. Anal.* **2014**, *17*, 532–551. [CrossRef]
72. Korbel, J.; Luchko, Yu. Modeling of financial processes with a space-time fractional diffusion equation of varying order. *Fract. Calc. Appl. Anal.* **2016**, *19*, 1414–1433. [CrossRef]

3

Dynamic Modelling of Interactions between Microglia and Endogenous Neural Stem Cells in the Brain during a Stroke

Awatif Jahman Alqarni [1,2], Azmin Sham Rambely [1,*] and Ishak Hashim [1]

1 Department of Mathematical Sciences, Faculty of Science and Technology, Universiti Kebangsaan Malaysia, 43600 UKM Bangi, Selangor, Malaysia; aw-aw910@hotmail.com (A.J.A.); ishak_h@ukm.edu.my (I.H.)

2 Department of Mathematics, College of Sciences and Arts in Blqarn, University of Bisha, Bisha 61922, Saudi Arabia

* Correspondence: asr@ukm.edu.my.

Abstract: In this paper, we study the interactions between microglia and neural stem cells and the impact of these interactions on the brain cells during a stroke. Microglia cells, neural stem cells, the damage on brain cells from the stroke and the impacts these interactions have on living brain cells are considered in the design of mathematical models. The models consist of ordinary differential equations describing the effects of microglia on brain cells and the interactions between microglia and neural stem cells in the case of a stroke. Variables considered include: resident microglia, classically activated microglia, alternatively activated microglia, neural stem cells, tissue damage on cells in the brain, and the impacts these interactions have on living brain cells. The first model describes what happens in the brain at the stroke onset during the first three days without the generation of any neural stem cells. The second model studies the dynamic effect of microglia and neural stem cells on the brain cells following the generation of neural stem cells and potential recovery after this stage. We look at the stability and the instability of the models which are both studied analytically. The results show that the immune cells can help the brain by cleaning dead cells and stimulating the generation of neural stem cells; however, excessive activation may cause damage and affect the injured region. Microglia have beneficial and harmful functions after ischemic stroke. The microglia stimulate neural stem cells to generate new cells that substitute dead cells during the recovery stage but sometimes the endogenous neural stem cells are highly sensitive to inflammatory in the brain.

Keywords: chemokines; cytokines; eigenvalue stability analysis; neurogenesis; numerical solution; system of ordinary differential equations

---

1. Introduction

Strokes are diseases that affect the brain, which may cause death and disabilities in mammals [1–3]. When ischemic stroke occurs, the cells surrounding the ischemic area start to die once the outflow of blood declines below 10% of the normal blood influx [1,2]. Cells can die through necrosis or apoptosis [1,4]. Necrosis occurs very quickly after stroke onset; the necrotic cells lead to pollution of the local environment in the brain and damage the surrounding cells through laceration and the release of intracellular contents into the brain [1,4]. On the other hand, apoptosis occurs several hours or days after the stroke onset, where apoptosis occurs in the penumbra and is typically not detrimental to the neighboring cells [1,4]. Due to this, resident microglia are activated by the dead cells causing damage to the surrounding area, which also contributes to an increase of dead brain cells [1,5,6].

Microglia are the assigned immune cells aimed at protecting brain cells from any damage [1,2,7]. Microglia reside within the normal brain as a ubiquitously distributed quiescent cell population which respond to changes within the system's micro-environment, in order to interact swiftly with diseases [1,8,9]. When microglia are activated their functions change, with increased capacity for phagocytosis and production of cytokines and chemokines [1]. The small secreted proteins, cytokines, which are released by the cells produce a specific effect, especially on inter-cellular interactions and communications [1,5,6]. When cells go through the process of phagocytosis, they are able to identify and consume large particles like pathogens, apoptotic cells, and cellular debris [10].

In living systems, free radicals and oxidants have double functions as both toxic and beneficial complex since they can be either harmful or beneficial to the body [11–13]. The activation of microglia has both beneficial and deleterious effects. The positive effects manifest through the prevention of damage extension by phagocytosis and the production of trophic molecules and anti-inflammatory cytokines, which can be conducive to tissue repair as well as neuroprotection. The harmful effects, which are toxic to tissues, occur primarily through the production of free radicals [12,13]. This reaction is important in removing damaged cells from the brain tissue; however, it can also increase the harm to brain cells by producing free radicals that are toxic to healthy cells [2,14,15]. There has been an increasing amount of evidence suggesting that ischemic inflammation may be vital to the issue of pathogens, as ischemic stroke results not only in damage and neuronal loss but also in sustained neuroinflammation after stroke [3,16,17]. Microglial cells have been shown to take part in the neural analysis associated with adult brain function responsible for the harm occurring in the brain when a stroke occurs [3]. After an ischemic stroke, a dynamic microglial polarization occurs within the harm region [18]. The microglia composition M_1 unleashes proinflammatory cytokines and oxidative mediators that prejudice living neurons, whereas the microglia composition M_2 tends to unleash neurotrophic factors, in order to stop neuronal death and promote brain repair [17,18]. The anti-inflammatory phenotype composition M_2 is capable of manufacturing anti-inflammatory cytokines, which are conducive to inhibiting inflammation and tissue restoration [17–19]. This shows that microglia play both positive and negative roles in stroke [20,21].

In adult brains, neural stem cells (which are active throughout life) are capable of generating different cells [22,23]. Neurogenesis continues for the duration of life in the subventricular zone in all mammalian types, generating new neural stem cells during the recovery stage after ischemic stroke [24–27]. Neural stem cells appear to have many important roles, such as self-renewal and multipotency, replacement of the dead cells, and inhibition of inflammation [22,28–30]. However, endogenous neural stem cells may not be able to generate enough cells to repair the neurological damage caused by a major disease such as stroke [31]. Furthermore, the pathologic environment created after an ischemic stroke poses numerous hurdles for new neurons, which make the utilization of endogenous repair mechanisms more difficult; in particular, in directing the migration and differentiation of endogenous neural stem cells needed to repair the tissue [3,31].

Numerous mathematical models have been proposed for the dynamics of brain diseases, the functions of immune cells in terms of inflammation, and neural stem cell generation. Several models of inflammatory processes have been proposed. For example, the dynamics of inflammation from a stroke were modeled by Di Russo et al. [1] where they studied the dynamics of the densities of cells dead by necrosis and apoptosis, activated and inactivated resident microglia and the proportion of neutrophils and macrophages in the tissue. Mathematical models of inflammation proposed by the authors; Reynolds et al. [32] and Kumar et al. [33] ; for blood cells inflammation. Alharbi and Rambely [34] discussed a mathematical model of the ability of the immune system to inhibit and eliminate abnormal cells, as well as the role of dietary habits in boosting the immune system. Mathematical models for brain diseases have also been proposed by several authors [1,35]. For example, Hao and Friedman [35] described a model for Alzheimer's disease which utilized the microglia. Also, mathematical models have been proposed for compartmental adult neurogenesis. Nakata et al. [36] studied models of a hierarchical cell building system controlled by the mature cells. Ziebell et al. [37] introduced a mathematical model that represents different states of the adult hippocampus and the changing dynamics in stem cells during that time. Cacao and Cucinotta [38] developed a mathematical model of radiation-induced changes to neurogenesis, while a different study by Huang and Zhang [39] discussed the knowledge of strategies and mechanisms for neural stem cell-based therapies on brain hypoxic-ischemic injury. These studies were all focused on brain diseases and the roles of the immune system in several diseases in the blood or brain. Furthermore, some of the studies focused on the behavior of neural stem cell generation in the subventricular zone. As a starting point, we considered the model which has been recently proposed by Leah et al. [40]. We modified our models based on the model presented by Leah et al. [40], which was derived by studying the behaviors of several types of cells, microglia, brain cells, and the impact of microglial pro-inflammatory factors in living brain cells. In their model, they studied methods to represent interactions between proinflammatory and anti-inflammatory cytokines, types of microglia, and central nervous system (CNS) tissue damage, using clinical data [40].

In our model, we incorporated variables for several cytokines. Furthermore, we introduced the effect of the proinflammation by microglia on brain cells and focused on the brain during the time of stroke. However, we did not consider many types of cytokines. We studied the cytokines as parameters and their influence on the brain. Furthermore, we developed a second model to study the dynamics and stability of the interaction between the damage from microglia and the endogenous neural stem cells in the brain in the recovery stage, in order to study the ability of neural stem cells to heal the brain after stroke. The purposes of this work are to describe the biological interaction between the inflammatory cytokines from microglia and living brain cells in stroke; to study the damage on brain cells during the early stage of stroke over a 72-h duration; to study the interactions between microglia and neural stem cells and the influence of proinflammatory and anti-inflammatory cytokines on brain cells; to analyze the dynamic effect of microglia by stimulating the generation and the proliferation of neural stem cells during the recovery phase after the stroke; and to draw conclusions about the dynamics of neural stem cells to improve the brain after a stroke within a mathematical framework. Our final aim is to understand the positive and negative aspects of this biological process, which could be helpful for the development of therapeutic methods using endogenous neural stem cells in ischemic stroke, by studying the stability of the models.

In Section 2, we present a model—called stroke-microglia-damage (SMD)—of the effect of microglia on the brain during stroke onset, and another model—called stroke-microglia-neural stem cells-recovery (SMNR)—of the interaction between the microglia and neural stem cells and their impact on the brain during a stroke, including analysis of the equilibrium points of the models and their stability states. Numerical experiments of the modified models are detailed in Section 3. Finally, Section 4 presents the conclusions of the study.

2. Mathematical Models

We modified mathematical models of biological processes to investigate the impact of microglia and neural stem cells on the brain by increasing damage or assisting it in the recovery stage in two steps, illustrated by systems of ordinary differential equations. We developed the SMD model based on a model presented by Leah et al. [40], who proposed a mathematical model for neuroinflammation in traumatic brain injury using mathematical modeling with clinical data. The system of ordinary differential equations they derived described the dynamics of biological processes for multiple interactions post-injury. The first model, called the SMD model, that we shall develop describes the roles of microglia in the first 72 h after stroke onset without any generation of neural stem cells, while the second model, called the SMNR model, describes the state from the third day when neural stem cells start to generate. The resting microglia are the immune cells, M_0, which reside in the brain, and will be activated when a stroke occurs to a classic proinflammatory M_1 or an alternative anti-inflammatory M_2 state, where the resting microglia polarizes into two activation states, M_1 or M_2, in reaction to the dead cells from ischemia [18,19]. The cytokines signal for the activation of resting microglia into two phenotypes, proinflammation M_1, and anti-inflammation M_2, microglia denoted by R_1 and R_2, respectively. When microglia cells are activated in stroke, they become either M_1 or M_2

phenotypes. At the start of the stroke, the microglia are biased towards the M_1 phenotype rather than M_2 phenotype, since $R_1 > R_2$. We assume that in stroke onset with this bias towards M_1, the transition rate from M_1 into M_2 phenotype is very small, approximately equal to zero in the first 60 h after stroke onset [1], and the rate of anti-inflammation cytokines, R_2, is at a smaller rate than the rate of proinflammation cytokines, R_1, in the SMD model. We take $R_1 < R_2$ when the rate of proinflammation cytokines, R_1, is at a lower rate than the rate of anti-inflammation cytokines, R_2 in the recovery stage in the SMNR model. Neural stem cells in the subventricular zone from the adult mammalian brain give rise to neuroblasts, which migrate to the injury region and generate new cells [22,28,41]. Microglia have a positive and negative impact on generating neural stem cells [42]. M_1 can impair neural stem cells [18], while M_2 shows the ability to stimulate the generation and the proliferation of neural stem cells N_{sc} Ref. [19]. We focus our model on the mature neural stem cells only in the SMNR because no generation of neural stem cells occurs in this stage of stroke onset.

2.1. Modelling of the Effect of Microglia on the Brain in a Stroke Onset (SMD)

In this study, we modified the dynamic system of the SMD model based on the model presented in [40], which illustrates the behavior of the impact of activated microglia on brain cells in the 72 h following a stroke. We consider a non-linear system of ordinary differential equations, which describes the effects of microglia $M_0(t)$, $M_1(t)$, $M_2(t)$, and the damage from microglia, $D(t)$, on brain cells, $C(t)$, in the 72 h after stroke onset as follows:

$$\frac{dM_0}{dt} = \alpha - [R_1 + R_2 + \mu] M_0(t).$$

This equation demonstrates that the microglia are activated after an ischemic stroke and are polarized either towards a classic state, M_1, by proinflammatory cytokines, leading to an adaptive immune response and causing additional neuronal damage; or towards an alternative state, M_2, which is the anti-inflammatory phenotype induced by anti-inflammatory cytokines, which is thought to inhibit inflammation and enhance tissue repair [18,19,40]. Although we do not specify the types of cytokines in our models, we focus on the influences of the cytokines which gather in the region of damage; which are determined to either increase M_1 polarization or to change the microglia into the M_2 state. For mathematical modeling purposes, we assume that resting microglia M_0 are generated at a constant rate (α) and die at a constant rate (μ). The function of the activated microglia is to clean up dead cells produced by ischemia and cytokines caused by dead cells [43]. Microglia are activated by the cytokines caused by dead cells: the M_1 phenotype is induced by proinflammation cytokine signals R_1 and the M_2 phenotype is induced by anti-inflammation cytokine signals R_2 [5,17,43]. Furthermore, a shift from M_1 to M_2 may be induced by R_3 signaling [5], where R_3 is the result of a transition from proinflammatory to anti-inflammatory cytokines. This transition only appears in the SMNR model, occurring 60 h after the stroke onset [1]. The following differential equations illustrate the behavior of the activated microglia M_1 and M_2:

$$\begin{aligned} \frac{dM_1}{dt} &= R_1 M_0(t) - [\delta D(t) + \gamma_1] M_1(t), \\ \frac{dM_2}{dt} &= R_2 M_0(t) - \gamma_2 M_2(t), \end{aligned}$$

where R_1 represents the rate of M_1 activation, R_2 is the rate of M_2 activation, δ is the rate of damage by M_1 activation, γ_1 is the death rate of M_1, and γ_2 is the death rate of M_2. The M_1 phenotype is essential for cell recovery, due to their protection mechanisms against the damage which recruits immune cells to the region of injury. The M_2 phenotype microglia also play a role in reducing damage, by clearing the brain of dead cells and assisting in neurogenesis, as well as inhibiting inflammation. On the other hand, the microglia can also play a negative role, through secretion of damaging proinflammatory cytokines which increase inflammation in the area of healthy cells [5,17]. We consider the damage D caused by the activated microglia to represent secondary damage in the case of stroke [2,14]. The following differential equation describes this damage:

$$\frac{dD}{dt} = D(t)\left[M_1(t)(\delta - r_1) - \beta_1 C(t) - r_2 M_2(t)\right],$$

where β_1 denotes the rate of effect of damage on brain cells, r_1 denotes the rate of damage clearance by M_1, and r_2 denotes the rate of damage clearance by M_2. In our model, we use the impact of the damage by microglia on the living brain cells, C. The following ordinary differential equation describes this effect on brain cells:

$$\frac{dC}{dt} = C(t)\left[\beta_1 D(t) - \beta_0\right],$$

where β_0 is the rate of cells dying from ischemic stroke.

Thus, the SMD model is expressed as follows:

$$\frac{dM_0}{dt} = \alpha - \left[R_1 + R_2 + \mu\right] M_0(t), \quad (1)$$

$$\frac{dM_1}{dt} = R_1 M_0(t) - \left[\delta D(t) + \gamma_1\right] M_1(t), \quad (2)$$

$$\frac{dM_2}{dt} = R_2 M_0(t) - \gamma_2 M_2(t), \quad (3)$$

$$\frac{dD}{dt} = D(t)\left[M_1(t)(\delta - r_1) - \beta_1 C(t) - r_2 M_2(t)\right], \quad (4)$$

$$\frac{dC}{dt} = C(t)\left[\beta_1 D(t) - \beta_0\right], \quad (5)$$

with initial values $M_0(0) = 1$ [1], $M_1(0) = 0.1514$ [1], $M_2(0) = 0.02$ [35], $D(0) = 0.4$ [1], and $C(0) = 0.28$ [35]. Furthermore, we assumed that, in steady state, $M_1 > M_2$ for 72 h and that neural stem cells are not generated.

2.1.1. Equilibrium Points for the SMD Model

In this section, we calculate the equilibrium points of the system and determine the parameters for the existence of different types of biological states. We now determine the steady-state solutions as follows:

$$\frac{dM_0}{dt} = \frac{dM_1}{dt} = \frac{dM_2}{dt} = \frac{dD}{dt} = \frac{dC}{dt} = 0.$$

Equilibrium points are stable if they remain constant over time or continually balance change in one direction by that in another. We classify three steady states as follows:

Definition 1 (State of the activation of microglia). *We define the microglia that reside in a healthy brain as the absence of a high activation for these cells in the brain when any damage occurs in the region. The steady-state of the form $M_0; M_1; M_2 > 0$ and $D, C = 0$ indicates that the function of immune cells is normal and the microglia do not cause any damage in the brain.*

Definition 2 (State of the beginning of damage from activated microglia M_1). *We define the beginning of the damage by the increased rate of proinflammation cytokines where the existence of high activation of microglia will cause damage to the living brain cells. The steady-state of the form $M_0; M_1; M_2; D > 0, C = 0$ indicates the beginning activation of proinflammation from M_1 microglia.*

Definition 3 (State of the damage from activated microglia in brain cells). *We define the impact of the damage caused by microglia on the living brain cells. The steady state of the form* $M_0; M_1; M_2; D; C > 0$ *indicates damage of the living brain cells by active microglia.*

Populations of microglia cells (M_0,M_1,M_2), the damage on the tissue by proinflammation, D and the impact of this damage on the living brain cells in the CNS, C are positive or equal to zero for all equations. Thus, we obtain the equilibrium points by solving system (1)–(5) to determine the positive equilibrium points if and only if M_0, M_1, M_2, D and C exemplify the positive solutions.

Proposition 1 (Nonnegative Equilibrium for the SMD model). *We assume the equilibrium points for SMD system,* $M_0; M_1; M_2; D; C > 0$ *satisfy the following conditions:*

- $\delta > r_1$
- $R_1\gamma_2(-r_1+\delta) > r_2R_2\gamma_1$
- $\alpha R_1\beta_1\gamma_2(-r_1+\delta) > r_2R_2(\beta_1\gamma_1+\beta_0\delta)$.

Then and only then can there exist nonnegative real steady states.

Proof. The proof can be clearly found from Definition 1, 2 and 3. □

Hence, the equilibrium point is given as follows:

The equilibrium point for the activation of microglia, E_r

Corollary 1. *The equilibrium point* $E_r = (M_0, M_1, M_2, 0, 0)$ *exists in* $\Re^5_+$ *if and only if* M_0, M_1, M_2 *are the nonnegative roots and* $D = 0, C = 0$. *This point is given by*

$$E_r = (M_0, M_1, M_2, 0, 0) = \left(\frac{\alpha}{x}, \frac{\alpha R_1}{\gamma_1 x}, \frac{R_2\alpha}{\gamma_2 x}, 0, 0\right). \tag{6}$$

where,

$$x = R_1 + R_2 + \mu.$$

Thus, in the normal state in mammalian brains, we obtain a normal situation when there is no activation of microglia on the brain where $D, C = 0$ given that there is no damage on the brain cells, where the equilibrium point E_r indicated the activation of microglia steady state.

The equilibrium point for high activation of microglia, E_p

Corollary 2. *The equilibrium point* E_p *exists in* $\Re^5_+$ *if and only if,*M_0, M_1, M_2 *and* D *are the nonnegative roots and* $C = 0$. *Then, the equilibrium point of proinflammation microglia through 72 h after stroke is given as follows:*

$$E_p = \left(\frac{\alpha}{x}, \frac{r_2R_2\alpha}{\gamma_2(-r_1+\delta)x}, \frac{R_2\alpha}{\gamma_2 x}, \frac{-r_2R_2\gamma_1 + R_1\gamma_2(-r_1+\delta)}{r_2R_2\delta}, 0\right). \tag{7}$$

The model has an equilibrium point for the activation of microglia. During this stage, M_1 has a significant impact. The equilibrium point E_p describes the activation by cytokines which obtains higher proinflammation from R_1 than the rate of the second kind of cytokines R_2 if and only if the rate of damage on the living brain cells by M_1 is more than r_1, which is the clearance damage by the immune cells microglia M_1.

The equilibrium points for the impact of proinflammation microglia M_1 **on brain cells,** E_d

Corollary 3. *The equilibrium point E_d exists in $\Re^5_+$ if and only if M_0, M_1, M_2, D and C are the nonnegative roots. Then, the equilibrium point E_d has an impact on the damage to the living brain cells through 72 h after stroke onset given as follows:*

$$E_d = \left(\frac{\alpha}{x}, \frac{R_1\beta_1\alpha}{\kappa x}, \frac{R_2\alpha}{\gamma_2 x}, \frac{\beta_0}{\beta_1}, \frac{(R_1\beta_1\gamma_2(-r_1+\delta) - r_2R_2\kappa)\alpha}{\beta_1\gamma_2\kappa x}\right), \tag{8}$$

where,

$$\kappa = (\beta_1\gamma_1 + \beta_0\delta).$$

2.1.2. Stability of Equilibrium Points

We now study the stability of equilibrium points for the effect of microglia on brain cells for 72 h without neural stem cells by using definitions 1–3. For the eigenvalues associated without neural stem cell equilibrium in stroke onset, the (5×5) Jacobian matrix of the system (1)–(5) is given by

$$J_1 = \begin{bmatrix} -x & 0 & 0 & 0 & 0 \\ R_1 & -\gamma_1 - D\delta & 0 & -M_1\delta & 0 \\ R_2 & 0 & -\gamma_2 & 0 & 0 \\ 0 & D(-r_1+\delta) & -Dr_2 & -M_2r_2 - c\beta_1 + M_1(-r_1+\delta) & -D\beta_1 \\ 0 & 0 & 0 & c\beta_1 & D\beta_1 - \beta_0 \end{bmatrix}.$$

1. Stability analysis of equilibrium point, E_r :

Theorem 1. *Suppose that the function $f : \Gamma \to \Re^5_+$ where Γ is a domain in $\Re^5_+$, and suppose that $E_r = (M_0, M_1, M_2, 0, 0) \in \Gamma$ is an equilibrium point at which at least one eigenvalue of the Jacobian matrix has a positive real part. Then, E_r is an unstable equilibrium point of f.*

Proof. The Jacobian J_1 at equilibrium point E_r is calculated as follows:

$$J[E_r] = \begin{bmatrix} -x & 0 & 0 & 0 & 0 \\ R_1 & -\gamma_1 & 0 & a_{24} & 0 \\ R_2 & 0 & -\gamma_2 & 0 & 0 \\ 0 & 0 & 0 & a_{44} & 0 \\ 0 & 0 & 0 & 0 & -\beta_0 \end{bmatrix},$$

where,

$$a_{24} = \frac{-R_1\alpha\delta}{\gamma_1 x}, \quad a_{44} = \frac{-r_2R_2\alpha}{\gamma_2 x} + \frac{R_1\alpha(-r_1+\delta)}{\gamma_1 x}.$$

The characteristic equation for the Jacobian $J[E_r]$ is given by

$$(\beta_0 + \lambda)(\gamma_2 + \lambda)(\gamma_1 + \lambda)(x + \lambda)(\lambda + y) = 0. \tag{9}$$

We assume that $x = R_1 + R_2 + \mu$, $y = \frac{r_2R_2\alpha}{\gamma_2 x} + \frac{R_1\alpha(r_1-\delta)}{\gamma_1 x}$.

Then, the eigenvalues of Jacobian matrix $J[E_r]$ are given by:

$$\begin{aligned} \lambda_1 &= -x < 0, \ \lambda_2 = -\beta_0 < 0, \\ \lambda_3 &= -\gamma_1 < 0, \ \lambda_4 = -\gamma_2 < 0, \ \lambda_5 = -y = \frac{\alpha(-r_2R_2\gamma_1 + R_1\gamma_2(-r_1+\delta))}{(\gamma_1\gamma_2 x)} > 0. \end{aligned}$$

The eigenvalues $\lambda_1, \ldots, \lambda_4$ are negative, but λ_5 is positive. Therefore, E_r is an unstable equilibrium point. □

2. Stability Analysis of Equilibrium point, E_p:

Theorem 2. *Suppose that the function $f : \Gamma \rightarrow \Re^5_+$ where Γ is a domain in $\Re^5_+$, and suppose that $E_p = (M_0, M_1, M_2, D, 0) \in \Gamma$ is an equilibrium point at which at least one eigenvalue of the Jacobian matrix has a positive real part and $D > 0$, $C = 0$. Then, E_p is an unstable equilibrium point of f.*

Proof. The Jacobian J_1, at E_p calculated as:

$$J[E_p] = \begin{bmatrix} -x & 0 & 0 & 0 & 0 \\ R_1 & b_{22} & 0 & b_{24} & 0 \\ R_2 & 0 & -\gamma_2 & 0 & 0 \\ 0 & b_{42} & b_{43} & b_{44} & b_{45} \\ 0 & 0 & 0 & 0 & b_{55} \end{bmatrix},$$

where,

$$\begin{aligned} b_{22} &= -\gamma_1 - \frac{(-r_2R_2\gamma_1 + R_1\gamma_2(-r_1+\delta))}{r_2R_2} < 0, \; b_{24} = \frac{r_2R_2\alpha\delta}{\gamma_2(r_1-\delta)x} < 0, \\ b_{42} &= \frac{(-r_1+\delta)(-r_2R_2\gamma_1 + R_1\gamma_2(-r_1+\delta))}{r_2R_2\delta} > 0, \; b_{43} = \frac{-(-r_2R_2\gamma_1 + R_1\gamma_2(-r_1+\delta))}{R_2\delta}, \\ b_{44} &= \frac{-r_2R_2\alpha}{\gamma_2 x} - \frac{r_2R_2\alpha(-r_1+\delta)}{\gamma_2(r_1-\delta)x} < 0, \; b_{45} = -\frac{\beta_1(-r_2R_2\gamma_1 + R_1\gamma_2(-r_1+\delta))}{r_2R_2\delta}, \\ b_{55} &= \frac{R_1\beta_1\gamma_2(-r_1+\delta) - r_2R_2(\beta_1\gamma_1 + \beta_0\delta)}{r_2R_2\delta} > 0. \end{aligned}$$

From the Jacobian $J[E_p]$, the characteristic equation is given by

$$(\gamma_2 + \lambda)(b_{55} - \lambda)(x + \lambda)(q_2\lambda^2 + q_1\lambda^1 + q_0) = 0,$$

where,

$$\begin{aligned} q_2 &= r_2R_2\gamma_2(r_1-\delta)x, \; q_1 = -R_1\gamma_2^2(r_1-\delta)^2x, \\ q_0 &= -r_2R_2\alpha(r_2R_2\gamma_1 + R_1\gamma_2(r_1-\delta))(r_1-\delta), \\ x &= R_1 + R_2 + \mu. \end{aligned}$$

By Proposition 1 one of the eigenvalues is positive. So, $J(E_p)$ has one at least positive root, which indicates that the equilibrium point E_p is unstable [44]. □

3. Stability analysis of equilibrium point, E_d:

Theorem 3. *Suppose that the function $f : \Gamma \rightarrow \Re^5_+$ where Γ is a domain in $\Re^5_+$, and suppose that $E_d = (M_0, M_1, M_2, D, C) \in \Gamma$ is an equilibrium point where all the eigenvalues of the Jacobian matrix have negative real parts at the equilibrium point E_d and $D > 0$, $C > 0$. Then, E_d is a stable equilibrium point of f.*

Proof. The Jacobian J_1 at E_d is calculated as follows:

$$J[E_d] = \begin{bmatrix} -x & 0 & 0 & 0 & 0 \\ R_1 & c_{22} & 0 & c_{24} & 0 \\ R_2 & 0 & -\gamma_2 & 0 & 0 \\ 0 & c_{42} & c_{43} & c_{44} & -\beta_0 \\ 0 & 0 & 0 & c_{54} & 0 \end{bmatrix},$$

where,

$$\begin{aligned} c_{22} &= -\gamma_1 - \frac{\beta_0\delta}{\beta_1} < 0, \ c_{24} = -\frac{R_1\alpha\beta_1\delta}{(\beta_1\gamma_1 + \beta_0\delta)x} < 0, \\ c_{42} &= \frac{\beta_0(-r_1+\delta)}{\beta_1} > 0, \ c_{43} = -\frac{r_2\beta_0}{\beta_1} < 0, \\ c_{44} &= -\frac{r_2R_2\alpha}{\gamma_2 x} + \frac{R_1\alpha\beta_1(-r_1+\delta))}{(\beta_1\gamma_1+\beta_0\delta)x} \\ &+ \frac{\alpha(R_1\beta_1\gamma_2(r_1-\delta) + r_2R_2(\beta_1\gamma_1+\beta_0\delta))}{\gamma_2(\beta_1\gamma_1+\beta_0\delta)x} > 0, \ c_{45} = -\beta_0 < 0 \\ c_{54} &= -\frac{\alpha(R_1\beta_1\gamma_2(r_1-\delta) + r_2R_2(\beta_1\gamma_1+\beta_0\delta))}{\gamma_2(\beta_1\gamma_1+\beta_0\delta)x} > 0. \end{aligned}$$

The characteristic equation is given by

$$(\gamma_2+\lambda)(x+\lambda)(N_3\lambda^3 + N_2\lambda^2 + N_1\lambda + N_0) = 0. \tag{10}$$

Thus, we can find the first two eigenvalues directly:

$$\lambda_1 = -\gamma_2, \ \lambda_2 = -x.$$

Here, we can apply the Routh–Hurwitz Criterion if and only if [45]:

- $N_2 > 0$,
- $N_0 > 0$,
- $N_2N_1 > N_0N_3$.

where

$$\begin{aligned} N_0 &= \alpha\beta_1\beta_0(\beta_1\gamma_1+\beta_0\delta)(R_1\beta_1\gamma_2(-r_1+\delta) - r_2R_2(\beta_1\gamma_1+\beta_0\delta)) > 0, \\ N_1 &= \alpha\beta_1^2\beta_0(R_1\gamma_2(-r_1+\delta)(\beta_1+\delta) - r_2R_2(\beta_1\gamma_1+\beta_0\delta)) > 0, \\ N_2 &= \beta_1\gamma_2(\beta_1\gamma_1+\beta_0\delta)^2 x > 0, N_3 = \beta_1^2\gamma_2(\beta_1\gamma_1+\beta_0\delta)x > 0. \end{aligned}$$

Since from Proposition 1,

$$N_2N_1 - N_0N_3 = R_1\alpha\beta_1^3\beta_0\gamma_2^2\delta(-r_1+\delta)(\beta_1\gamma_1+\beta_0\delta)^2 x > 0.$$

Then, $N_2, N_0 > 0$ and $N_2N_1 - N_0N_3 > 0$,

Now, we apply the Routh–Hurwitz theorem for $N_3\lambda^3 + N_2\lambda^2 + N_1\lambda + N_0 = 0$, giving

$$\begin{vmatrix} \lambda^3 & N_3 & N_1 \\ \lambda^2 & N_2 & N_0 \\ \lambda^1 & N^* & 0 \\ \lambda^0 & N_0 & 0 \end{vmatrix},$$

where,

$$N^* = \frac{N_2 N_1 - N_3 N_0}{N_2},$$

thus,

$$N^* = R_1 \alpha \beta_1^2 \beta_0 \gamma_2 \delta(-r_1 + \delta) > 0.$$

Since all the coefficients in the first column have positive signs; the Equation (10) has no roots with positive real parts and two of the eigenvalues are negative; thus, the equilibrium point E_d is stable. Activated microglia are capable of cleaning dead cells; however, they produce free radicals from brain cells, which increases the damage in brain cells during a stroke. This lead to further death of brain cells [17,32]. □

Remark 1. *The effect of microglia on brain cells in a stroke (which includes activated microglia at stroke onset) on the dynamic system of SMD model can be deduced, as follows:*

- *As a result of Theorem 1 and Definition 1, the damage, D, can invade the SMD system if $\lambda_5 > 0$.*
- *As a result of Theorem 2 and Definition 2, this means that the damage, $D > 0$, invades C.*
- *As a result of Theorem 3 and Definition 3, this means that the damage, $D > 0$, causes the death of C.*
- *The SMD model is stable when the brain cells are affected by the proinflammatory cytokines of microglia; however, when the rate of production of proinflammatory cytokines leads to an increase in damage, the possibility of death of the brain cells is introduced.*

2.2. Modeling the Interaction between Microglia and Neural Stem Cells and Impact on the Brain in Stroke (SMNR)

We formulate the following a non-linear system of ordinary differential equations to describe the interaction between microglial cells and neural stem cells and to investigate the damage to brain cells in the recovery stage after a stroke. This system uses the same equations as the SMD model, but also includes the transition from M_1 to M_2 and the behavior of neural stem cells N_{SC}. The system is as follows:

$$\frac{dM_0}{dt} = \alpha - \left[R_1 + R_2 + \mu\right] M_0(t) \quad (11)$$

$$\frac{dM_1}{dt} = R_1 M_0(t) - \left[R_3 + \delta D(t) + k_1 N_{SC}(t) + \gamma_1\right] M_1(t) \quad (12)$$

$$\frac{dM_2}{dt} = R_2 M_0(t) + R_3 M_1(t) - \left[k_2 N_{SC}(t) + \gamma_2\right] M_2(t) \quad (13)$$

$$\frac{dN_{SC}}{dt} = N_{SC}(t)\left[k_2 M_2(t) + k_1 M_1(t) - \beta_2 C(t) - \sigma\right] \quad (14)$$

$$\frac{dD}{dt} = D(t)\left[\delta M_1(t) - \beta_1 C(t) - r_1 M_1(t) - r_2 M_2(t)\right] \quad (15)$$

$$\frac{dC}{dt} = C(t)\left[\beta_1 D(t) + \beta_2 N_{SC}(t) - \beta_0\right], \quad (16)$$

with initial values $M_0[0] = 1$ [1], $M_1[0] = 0.02$ [35], $M_2[0] = 0.1514$ [1], $N_{SC}[0] = 0.85$ [38], $D[0] = 0.4$ Ref. [1], and $C[0] = 0.28$ [35].

The parameters are real and positive, where R_3 denotes the rate of transition from M_1 to M_2, k_1 is the rate of interaction between neural stem cells and M_1 microglia, k_2 is the rate of interaction between neural stem cells and M_2 microglia, β_2 is the rate of the effect of neural stem cells on brain cells, and σ is the death rate of neural stem cells. The other parameters are the same as in the SMD model. In the SMNR model, $M_1 < M_2$ occurs during the recovery stage, where neural stem cells also start to generate and help the brain. Therefore, microglial activation is important for directing the replacement of damaged or lost cells in the brain.

2.2.1. Equilibrium Points

In this section, we calculate the equilibrium points of the system (11)–(16) and determine the parameter conditions for the existence of the different types of biological states. We find the steady-state solutions as follows:

$$\frac{dM_0}{dt} = \frac{dM_1}{dt} = \frac{dM_2}{dt} = \frac{dN_{SC}}{dt} = \frac{dD}{dt} = \frac{dC}{dt} = 0.$$

We classify three states of steady states:

Definition 4 (The transformation state). *We define the transition cytokines from proinflammation microglia to anti-inflammation (M_1 to M_2). The steady state of the form $M_0; M_1; M_2 > 0$ and $N_{SC}, D, C = 0$ indicates the beginning of recovery stage.*

Definition 5 (The interaction between microglia and neural stem cells state). *We define the interaction between microglia and neural stem cells microglia in which the proinflammation phenotype (M_1) affects negatively on neural stem cells generation and the anti-inflammation phenotype (M_2) stimulates its generation. The steady state of the form $M_0; M_1; M_2; N_{SC} > 0$ and $C, D = 0$, indicates the interaction between brain cells in the recovery stage.*

Definition 6 (The impact of neural stem cells on brain cells state). *We define the impact of generating neural stem cells on brain cells, the steady state of the form $M_0; M_1; M_2; N_{SC}; C > 0$ and $D = 0$ indicates the generation of neural stem cells on brain cells.*

Populations of microglia cells (M_0, M_1, M_2), D, N_{SC} and C are positive or equal to zero. Thus, we obtain the equilibrium points by solving the system (11)–(16) to determine the positive equilibrium points if and only if, M_0,M_1, M_2, D, C exemplify the positive solution. Subpopulation of microglia cells M_0, M_1, M_2, N_{SC}, D and C are positive for all $t > 0$.

Proposition 2 (Nonnegative equilibriums for the SMNR model). *We assume the equilibrium points for the system SMNR is nonnegative for the following conditions given by:*

- $\delta > r_1$
- $\alpha\beta_2 > \beta_0\sigma$
- $R_2\alpha\beta_2 > \beta_0\sigma(R_2 + \mu)$
- $k_1\alpha\beta_2 > k_1\beta_0\sigma + \beta_2\gamma_1\sigma$
- $(k_1\beta_0(R_2 + \mu)\sigma + \beta 2(R_3 + \gamma_1)x\sigma < k_1R_1(\alpha\beta_2 - \beta_0\sigma)$
- $k_2\gamma_1 < k_1\gamma_2$
- $k_1k_2\alpha + k_2R_3\sigma + k_2\gamma_1\sigma > k_1\gamma_2\sigma$
- $k_1k_2(R_1 + R_2)\alpha + k_2(R_3 + \gamma_1)x\sigma < k_1\gamma_2x\sigma + \nu$
- $k_1k_2(R_1 + R_2)\alpha + k_2R_3x\sigma + k_1\gamma_2x\sigma > k_2\gamma_1x\sigma + \nu$
- $k_1k_2(R_1 + R_2)\alpha + \nu > k_1\gamma_2x\sigma + k_2(R_3 + \gamma_1)x\sigma.$

Then and only then there exist non-negative real steady states.

Proof. The proof is clearly found from the steady states imposing $(M_0; M_1; M_2; N_{SC}; D; C) > 0$. Thus the models (11)–(16) obtain three points of equilibrium in the recovery stage by using MATHEMATICA. □

The first equilibrium point: the transformation state of microglia from proinflammation to anti-inflammation, E_t

Corollary 4. *The equilibrium point E_t has transformation microglia from proinflammation to anti-inflammation point recovery state from stroke given as follows:*

The equilibrium point $E_t = (M_0, M_1, M_2, N_{SC}, D, C)$ exists in $\Re^6_+$, in the beginning of activation of microglia that inhibits and transfers part of M_1 phenomena to the second M_2 phenomena. Then, $C = 0$, $D = 0$, $N_{SC} = 0$. This point E_t is given by

$$E_t = \left(\frac{\alpha}{x}, \frac{(R_1\alpha)}{(R_3+\gamma_1)x}, \frac{\alpha(R_1R_3+R_2(R_3+\gamma_1))}{(R_3+\gamma 1)\gamma_2 x}, 0, 0, 0\right), \tag{17}$$

where

$$x = R_1 + R_2 + \mu.$$

The second equilibrium point: the interaction between microglia and neural stem cells on brain cells, E_I

Corollary 5. *The equilibrium point $E_I = (M_0, M_1, M_2, N_{SC}, D, C)$ exists in $\Re^6_+$. At the beginning of the neural stem cells generation, the damage decreases in this state. Then, $C = 0$, $D = 0$. This point is given by*

$$\begin{aligned} E_I = \Bigg(&\frac{\alpha}{x}, \frac{1}{2k_1(-k_2\gamma_1+k_1\gamma_2)x}[(R_1+R_2)(-k_1k_2\alpha - k_2R_3\sigma - k_2\gamma_1\sigma + k_1\gamma_2\sigma) \\ &-\mu\sigma[k_2(R_3+\gamma_1) - k_1\gamma_2] + v], \frac{1}{2k_2(-k_2\gamma_1+k_1\gamma_2)x}[(R_1+R_2)(k_1k_2\alpha \\ &+k_2R_3\sigma - k_2\gamma_1\sigma + k_1\gamma_2\sigma) + \mu\sigma(k_2(R_3-\gamma_1)+k_1\gamma_2)] - v \\ &, \frac{1}{2k_1k_2x\sigma}[(R_1+R_2)(k_1k_2\alpha - k_2R_3\sigma - k_2\gamma_1\sigma \\ &-k_1\gamma_2\sigma) - k_2\mu\sigma(R_3+\gamma_1) - k_1\gamma_2\mu\sigma + v), 0, 0\Bigg) \end{aligned} \tag{18}$$

where,

$$v = \sqrt{(4k_1k_2R_1\alpha(-k_2\gamma_1+k_1\gamma_2)x\sigma + (k_1k_2(R_1+R_2)\alpha + k_2(R_3+\gamma_1)x\sigma - k_1\gamma_2x\sigma)^2)}.$$

The third equilibrium point: the effect of generating neural stem cells on brain cells, E_R

Corollary 6. *The equilibrium point $E_R = (M_0, M_1, M_2, N_{SC}, D, C)$ exists in of $\Re^6_+$. When the neural stem cells are generated, the damage fades away in this state. Then, $D = 0$. This point is given by:*

$$\begin{aligned} E_R = \Bigg(&\frac{\alpha}{x}, \frac{R_1\alpha\beta_2}{\epsilon_1 x}, \frac{\alpha\beta_2(R_1R_3\beta_2+R_2\epsilon_1)}{\epsilon_1\epsilon_2 x}, \frac{\beta_0}{\beta_2}, 0 \\ &, (\frac{\beta_2\gamma_2(-k_1\beta_0(R_2+\mu)\sigma - \beta_2(R_3+\gamma_1)x\sigma + k_1R_1\epsilon_3) + k_2\epsilon_7}{\beta_2\epsilon_1\epsilon_2 x}\Bigg), \end{aligned} \tag{19}$$

where,

$$\begin{aligned}\epsilon_1 &= (R_3\beta_2 + k_1\beta_0 + \beta_2\gamma_1),\ \epsilon_2 = k_2\beta_0 + \beta_2\gamma_2,\\ \epsilon_3 &= (\alpha\beta_2 - \beta_0\sigma),\ \epsilon_4 = (-k_1\alpha\beta_2 + k_1\beta_0\sigma + \beta_2\gamma_1\sigma),\\ \epsilon_5 &= R_3\beta_2 + k_1\beta_0 + \beta_2\gamma_1,\ \epsilon_6 = (R_2\alpha\beta_2 - R_2\beta_0\sigma - \beta_0\mu\sigma),\\ \epsilon_7 &= R_1R_3\beta_2\epsilon_3 - R_1\beta_0\epsilon_4 + \epsilon_5\epsilon_6.\end{aligned}$$

2.2.2. Stability of Equilibrium Points for SMNR Model

We study the stability of the equilibrium points for the second model SMNR by using Definitions 4–6. The first step requires linearization of the system equations that describes the interaction between microglia M_1, M_2, neural stem cells N_{SC}, the damage D in this stage and the impact it has on the brain cells C. The (6×6) Jacobian matrix of the system (11)–(16)is given by:

$$J_2 = \begin{bmatrix} j_{11} & 0 & 0 & 0 & 0 & 0\\ R_1 & j_{22} & 0 & -k_1M_1 & -M_1\delta & 0\\ R_2 & R_3 & j_{33} & -k_2M_2 & 0 & 0\\ 0 & k_1N_{SC} & k_2N_{SC} & j_{54} & 0 & -N_{SC}\beta_2\\ 0 & D(-r_1+\delta) & -Dr_2 & 0 & j55 & -D\beta_1\\ 0 & 0 & 0 & c\beta_2 & c\beta_1 & j_{66}\end{bmatrix},$$

where,

$$\begin{aligned} j_{11} &= -x, x = R_1 + R_2 + \mu,\ j_{22} = -R_3 - k_1N_{SC} - \gamma_1 - D\delta,\\ j_{33} &= -k_2N_{SC} - \gamma_2,\ j_{54} = k_1M_1 + k_2M_2 - C\beta_2 - \sigma,\\ j_{55} &= -M_2r_2 - C\beta_1 + M_1(-r_1+\delta),\ j_{66} = D\beta_1 + N_{SC}\beta_2 - \beta_0.\end{aligned}$$

1. Stability analysis of equilibrium point, E_t:

Theorem 4. *Suppose that the function $f : \Gamma \to \Re^6_+$ where Γ is a domain in $\Re^6_+$, and suppose that $E_t = (M_0, M_1, M_2, 0, 0, 0) \in \Gamma$ is an equilibrium point at which at least one eigenvalue of the Jacobian matrix has a positive real part. Then, E_t is an unstable equilibrium point of f.*

Proof. The Jacobian J_2 corresponding to the equilibrium point E_t is given by

$$J[E_t] = \begin{bmatrix} -x & 0 & 0 & 0 & 0 & 0\\ R_1 & -R_3-\gamma_1 & 0 & d_{24} & d_{25} & 0\\ R_2 & R_3 & -\gamma_2 & d_{34} & 0 & 0\\ 0 & 0 & 0 & d_{44} & 0 & 0\\ 0 & 0 & 0 & 0 & d_{55} & 0\\ 0 & 0 & 0 & 0 & 0 & -\beta_0\end{bmatrix},$$

where,

$$\begin{aligned} d_{24} &= -\frac{k_1R_1\alpha}{(R_3+\gamma_1)x},\ d_{25} = -\frac{R_1\alpha\delta}{(R3+\gamma_1)x},\\ d_{34} &= -\frac{k_2\alpha(R_1R_3 + R_2(R_3+\gamma_1))}{(R_3+\gamma_1)\gamma_2x},\ d_{44} = \frac{k_1R_1\alpha}{(R_3+\gamma_1)x} + \frac{k_2\alpha(R_1R_3+R_2(R_3+\gamma_1))}{(R_3+\gamma_1)\gamma_2x} - \sigma,\\ d_{55} &= -\frac{r_2\alpha(R_1R_3+R_2(R_3+\gamma_1))}{(R_3+\gamma_1)\gamma_2x} + \frac{R_1\alpha(-r_1+\delta)}{(R_3+\gamma_1)x}.\end{aligned}$$

From the Jacobian matrix $J[E_t]$, the characteristic equation is given by

$$(\beta_0+\lambda)(R_3+\gamma_1+\lambda)(\gamma_2+\lambda)(x+\lambda)(-\lambda+t_{55})(-\lambda+t_{44})=0. \tag{20}$$

Then, the eigenvalues corresponding to E_t are given by

$$\begin{aligned}
\lambda_1 &= -\beta_0<0,\ \lambda_2=-R_3-\gamma_1<0,\ \lambda_3=-x<0,\ \lambda_4=-\gamma_2<0,\\
\lambda_5 &= \frac{k_1R_1\alpha}{(R_3+\gamma_1)x}+\frac{k_2\alpha(R_1R_3+R_2(R_3+\gamma_1))}{(R_3+\gamma_1\gamma_2)x}-\sigma>0,\\
\lambda_6 &= -\frac{r_2\alpha(R_1R_3+R_2(R_3+\gamma_1))}{(R_3-\gamma_1\gamma_2)x}+\frac{(R_1\alpha(r_1-\delta))}{(R_3+\gamma_1)x}<0.
\end{aligned}$$

One of the eigenvalues, $\lambda_5>0$, then E_t is an unstable point. □

2. Stability analysis of equilibrium point, E_I:

Theorem 5. *Suppose that the function $f:\Gamma\rightarrow\Re^6_+$ where Γ is a domain in $\Re^6_+$, and suppose that $E_I=(M_0,M_1,M_2,N_{SC},0,0)\in\Gamma$ is an equilibrium point at which at least one eigenvalue of the Jacobian matrix has a positive real part and $N_{SC}>0$. Then, E_I is an unstable equilibrium point of f.*

Proof. The Jacobian matrix J_2 corresponding to the equilibrium point E_I is given by:

$$J[E_I]=\begin{bmatrix} -x & 0 & 0 & 0 & 0 & 0\\ R_1 & e_{22} & 0 & e_{24} & e_{25} & 0\\ R_2 & R_3 & e_{33} & e_{34} & 0 & 0\\ 0 & e_{42} & e_{43} & 0 & 0 & e_{46}\\ 0 & 0 & 0 & 0 & e_{55} & 0\\ 0 & 0 & 0 & 0 & 0 & e_{66}\end{bmatrix},$$

where,

$$\begin{aligned}
e_{22} &= -\frac{\Lambda_1+\upsilon}{2k_2x\sigma}<0,\ e_{42}=\frac{\Lambda_2}{2k_2x\sigma}>0,\ e_{33}=-\frac{\Lambda_2}{2k_1x\sigma}<0,\ e_{43}=\frac{\Lambda_2}{2k_1x\sigma}>0,\\
e_{24} &= \frac{-\Lambda_1+\upsilon}{2qx}<0, e_{34}=\frac{\Lambda_3}{2qx}<0,\ e_{25}=-\frac{\delta\Lambda_4}{2k_1qx}<0,\\
e_{46} &= \frac{\Lambda_7}{2k_1k_2x\sigma}<0, e_{55}=-\frac{\Lambda_5}{2k_1k_2qx}>0,\ e_{66}=\frac{\Lambda_6}{2k_1k_2x\sigma}>0,\\
\upsilon &= \sqrt{4k_1k_2R_1\alpha(-k_2\gamma_1+k_1\gamma_2)x\sigma+(k_1k_2(R_1+R_2)\alpha+k_2(R_3+\gamma_1)x\sigma-k_1\gamma_2x\sigma)^2}>0,\\
\Psi &= x(k_2(R_3+\gamma_1)+k_1\gamma_2)\sigma>0, \Lambda_1=k_1k_2(R_1+R_2)\alpha+k_2x(R_3+\gamma_1)\sigma-k_1x\gamma_2\sigma>0,\\
\Lambda_2 &= \upsilon-\Psi+k_1k_2(R_1+R_2)\alpha>0,\\
\Lambda_3 &= -\upsilon+k_1k_2(R_1+R_2)\alpha+k_2x(R_3-\gamma_1)\sigma+k_1x\gamma_2\sigma>0,\\
\Lambda_4 &= -\upsilon+k_1k_2(R_1+R_2)\alpha+(k_2(R_3+\gamma_1)-k_1\gamma_2)x\sigma<0, \Lambda_5=k_2^2(r_1-\delta)(k_1(R_1+R_2)\alpha\\
&\quad +x(R_3+\gamma_1)\sigma)+k_1r_2(\upsilon-k_1x\gamma_2\sigma)+k_2(-k_1r_2(k_1(R_1+R_2)\alpha+x(R_3-\gamma_1)\sigma)\\
&\quad -(r_1-\delta)(\upsilon+k_1x\gamma_2\sigma))>0,\\
\Lambda_6 &= \upsilon\beta_2+k_1k_2(R_1+R_2)\alpha\beta_2-x(k_2\beta_2(R_3+\gamma_1)+k_1(2k_2\beta_0+\beta_2\gamma_2))\sigma>0,\\
\Lambda_7 &= \beta_2(-\upsilon-k_1k_2(R_1+R_2)\alpha+k_2x(R_3+\gamma_1)\sigma+k_1x\gamma_2\sigma)<0,\\
q &= k_2\gamma_1-k_1\gamma_2<0, x=R_1+R_2+\mu>0.
\end{aligned}$$

From the Jacobian $J[E_I]$, the characteristic equation is given by

$$(e_{66}-\lambda)(\lambda+x)(e_{55}-\lambda)[\lambda^3+T_2\lambda^2+T_1\lambda+T_0]=0, \quad (21)$$

where,

$$\begin{aligned}
T_3 &= -8k_1^3k_2^2q^2x^4\sigma^3,\ T_2=-4k_1^2k_2q^2(k_1(v+\Lambda_1)+k_2\Lambda_2)x^3\sigma^2,\\
T_1 &= 2k_1^2k_2qx^2\Lambda_2\sigma(-q(v+\Lambda_1)+\Lambda_2(k_1(v-\Lambda_1)+k_2\Lambda_3)\sigma),\\
T_0 &= k_1^2k_2q\Lambda_2x\sigma(\Lambda_1(-\Lambda_2+\Lambda_3-2k_2R_3x\sigma)+v(\Lambda_2+\Lambda_3+2k_2R_3x\sigma)).
\end{aligned}$$

In Equation (21), the eigenvalue $\lambda=e_{66}$ is distinctly positive, by Proposition 2. Thus, $J(E_I)$ has at least one positive root. Thus, the equilibrium point E_I is unstable [44]. □

3. Stability analysis of equilibrium point, E_R:

Theorem 6. *Suppose that the function $f:\Gamma\rightarrow\Re_+^6$ where Γ is a domain in $\Re_+^6$, and suppose that $E_R=(M_0,M_1,M_2,N_{SC},0,C)\in\Gamma$ is an equilibrium point at which at least one eigenvalue of the Jacobian matrix has a positive real part and $N_{SC}>0$, $C>0$. Then, E_R is an unstable equilibrium point of f.*

Proof. We now study the stability of the equilibrium point E_R, calculated as:
The Jacobian matrix J_2 estimated at E_R is

$$J[E_R]=\begin{bmatrix} -x & 0 & 0 & 0 & 0 & 0\\ R_1 & f_{22} & 0 & f_{24} & f_{25} & 0\\ R_2 & R_3 & f_{33} & f_{34} & 0 & 0\\ 0 & \frac{k_1\beta_0}{\beta_2} & \frac{k_2\beta_0}{\beta_2} & f_{44} & 0 & -\beta_0\\ 0 & f_{52} & f_{53} & 0 & f_{55} & f_{56}\\ 0 & 0 & 0 & 0 & 0 & f_{66}\end{bmatrix},$$

where

$$\begin{aligned}
f_{22} &= -R_3-\frac{k_1\beta_0}{\beta_2}-\gamma_1-\frac{\delta\rho_1}{\beta_2v_1v_2x}<0,\ f_{24}=-\frac{k_1R_1\alpha\beta_2}{v_1x}<0,\ f_{25}=-\frac{(R_1\alpha\beta_2\delta)}{v_1x}<0,\\
f_{55} &= -\frac{(r_2\alpha\beta_2)\rho_2}{v_1v_2x}+\frac{R_1\alpha\beta_2(-r_1+\delta)}{v_1x}>0,\ f_{33}=-\frac{k_2\beta_0}{\beta_2}-\gamma_2,\ f34=-\frac{k_2\alpha\beta_2\rho_2}{v_1v_2x}>0,\\
f_{43} &= \frac{k_2\beta_0}{\beta_2},\ f_{44}=\frac{k_1R_1\alpha\beta_2}{v_1x}+\frac{k_2\alpha\beta_2\rho_2}{v_1v_2x}-\sigma>0,\ f_{42}=\frac{k_1\beta_0}{\beta_2},\ f_{53}=-\frac{r_2\rho_1}{\beta_2v_1v_2x}<0,\\
f_{56} &= -\frac{\beta_1\rho_1}{\beta_2v_1v_2x}<0,\ f_{52}=\frac{(-r_1+\delta)\rho_1}{\beta_2v_1v_2x}<0,\ f_{66}=\frac{\beta_1\rho_1}{\beta_2v_1v_2x}>0,\\
\rho_1 &= (\beta_2\gamma_2(-k_1\beta_0(R_2+\mu)\sigma-\beta_2(R_3+\gamma_1)x\sigma+k_1R_1(\alpha\beta_2-\beta_0\sigma)),\\
&\quad +k_2(R_1R_3\beta_2(\alpha\beta_2-\beta_0\sigma))-R_1\beta_0(-k_1\alpha\beta_2+k_1\beta_0\sigma+\beta_2\gamma_1\sigma)\\
&\quad +(R_3\beta_2+k_1\beta_0+\beta_2\gamma_1)(R_2\alpha\beta_2-R_2\beta_0\sigma-\beta_0\mu\sigma)))\\
\rho_2 &= (R_1R_3\beta_2+R_2(R_3\beta_2+k_1\beta_0+\beta_2\gamma_1))>0,\\
v_1 &= (R_3\beta_2+k_1\beta_0+\beta_2\gamma_1)>0,\ v_2=(k_2\beta_0+\beta_2\gamma_2)>0.
\end{aligned}$$

From the Jacobian $J[E_R]$, the characteristic equation is given by

$$\begin{aligned}
&(f_{66}-\lambda)(\lambda+x)(f_{25}f_{34}(-f_{53}k_1+f_{52}k_2)\beta_0+f_{25}\beta_2(f_{44}-\lambda)(-f_{33}f_{52}+f_{53}R_3+f_{52}\lambda)+(f_{55}-\lambda)\\
&\quad\times((-f_{34}k_2\beta_0+\beta_2(f_{33}-\lambda)(f_{44}-\lambda))(f_{22}-\lambda)+f_{24}\beta_0(-f_{33}k_1+k_2R_3+k_1\lambda)))=0. \quad (22)
\end{aligned}$$

In Equation (22), the eigenvalue $\lambda = f_{66}$ is distinctly positive, by Proposition 2. Thus, $J(E_R)$ has at least one positive root and, so, the equilibrium point E_R is unstable [44]. □

Remark 2. *The effects of microglia and neural stem cells on brain cells in the recovery stage after a stroke, according to analysis of the dynamic system of SMNR, can be deduced as follows:*

- *As a result of Theorem 4 and Definition 4, the neural stem cells, Nsc, can invade the SMNR system if* $\lambda_5 > 0$.
- *As a result of Theorem 5 and Definition 5, this means that the neural stem cells,* $Nsc > 0$, *can invade C and D.*
- *As a result of Theorem 6 and Definition 6, this means that the neural stem cells,* $Nsc > 0$, *can eliminate the damage D.*
- *The SMNR model is unstable, given that the model uses mature neural stem cells, where the neural stem cells can help the brain to inhibit inflammation during a stroke; however, it is not always enough to substitute all dead cells with new neural stem cells [3,27].*

3. Numerical Experiments

The aim of this section is to study the parameters of the systems (1)–(5) and (11)–(16), in order to determine those that affect the behavior of the modified models, by using numerical simulations. One of the main problems in modeling and simulating the interactions between microglia and endogenous neural stem cells is that few parameter values are known. Therefore, we listed the parameters which could be determined from the stroke model of [1]. The other parameters were determined from experimental data of other brain injuries which involve similar biological processes. Furthermore, we obtained some parameter values by simulations using the Mathematica software (11.2, Wolfram, Champaign, IL, USA) with the command NDSolve, in order to study the influence of the interactions during the stroke and the possible therapeutic value of neural stem cells. Tables 1 and 2 show the reference sets of parameter values, along with the corresponding simulation results.

Table 1. Parameters values for the stroke-microglia-damage (SMD) model.

The SMD Model			
Parameters	**Values**	**Meaning**	**Sources**
α	0.38	Source of resting microglia	[1]
R_1	0.12	Rate the activation of resting microglia into M_1	[1]
R_2	0.017	Rate the activation of resting microglia into M_2	[35]
δ	0.2854	Rate of the damage produced by M_1	[1]
β_1	0.1	Rate of effect the damage on $C(t)$	[1]
β_0	0.05	Rate of dying C through stroke	[1]
μ	0.003	Death rate of M_0	simulation
γ_1	0.05	Death rate of M_1	simulation
γ_2	0.06	Death rate of M_2	simulation
r_1	0.05	Damage clearance by M_1	[1]
r_2	0.0125	Damage clearance by M_2	[1]

The numerical simulations in Tables 1 and 2 appeared in the SMD and the SMNR models. The microglial cells at stroke onset had high proinflammatory activation (i.e., high level of M_1), which caused an increase of damage in the region, affecting brain cells as well as neural stem cells during the 72 h period. On the other hand, when the rate of the anti-inflammatory phenotype microglia (M_2) became higher than that of the M_1, the brain was assisted in recovery and the neural stem cells were stimulated to generate new cells to compensate for lost brain cells. These dynamics started approximately on the third day after stroke onset. The parameters of the SMD model, in the simulation, were fixed as $\mu = 0.003$, $\gamma_1 = 0.05$, and $\gamma_2 = 0.06$. The parameters of the SMNR model, in the simulation, were fixed as $R_3 = 0.11$, $K_1 = 0.75$, $K_2 = 0.91$, $\beta_2 = 0.2$, and $\mu = 0.053$. In the SMD model, $M_1 > M_2$ and we set $R_1 = 0.12$ [1] and $R_2 = 0.017$ [35]. In the SMNR model, $M_2 > M_1$, the source of resting microglia was $\alpha = 0.38$, and the rate of the anti-inflammatory cytokines (R_2) was greater than that of the proinflammatory cytokines (R_1) in the recovery stage. Accordingly, we set $R_1 = 0.12$ [1] and estimated $R_2 = 0.26$. The dynamics of the systems were given by solving the systems numerically and then plotting the time series of the solutions of the system (1)–(5) and (11)–(16) for the parameters, which were solved using the fourth-order Runge–Kutta (RK4) method to obtain more stable and convergent solutions. The simulations of SMD and SMNR were carried out for the time spans of three days (72 h) and thirty days, respectively with a step size of 10^{-4} in the RK4 method. The SMD model described the inflammatory process where activation of the microglia cells peaks around three days (72 h) from stroke onset [1,2]. Neural stem cells increase the proliferation and the migration from the subventricular zone to the brain in thirty days and peaking around seven days after stroke [3,27]. Therefore, the SMNR model studied thirty days after stroke to investigate the interactions between microglia and neurogenesis after stroke. The accuracy of solutions by using the RK4 method of the SMD model and the best numerical solution of the model was obtained at 38 steps. While the accuracy of solutions by using the RK4 method of the SMNR model and the best numerical solution of the model was obtained at 35 steps. The reliability and accuracy of the proposed numerical method can be seen from the residual error which is shown in Figures 1 and 2. Figure 1 shows the residual error for the SMD model at the number of steps and the residual error at the time. While Figure 2 shows the residual error for the SMNR model at the number of steps and the residual errors at the time. By comparing the results in our study of microglial activation and the damage during the first three days after stroke onset and the recovery stage within the first thirty days after stroke onset with the effects of neural stem cells and those of the study in [40], we found that the studies agreed, in that the extended proinflammatory activation of microglia may inhibit neurogenesis and contribute to additional neuronal loss. Hence, the behavioural responses of microglia can lead to an increase in damage in the brain or its capability to recover. In this study, we focused on the importance and contribution of neural stem cells in transitioning to the recovery stage, as well as the interaction between microglia and neural stem cells. The generation of neural stem cells was considered to lead to a decrease in damage to the brain, compared to their absence. The behaviors of the solutions of the SMD and SMNR models are shown in Figure 3. It can be seen, in the SMD model, that the population of microglial cells M_1 had a shallow increase, asides from shifting to a steep curve at approximately 15–30 h, and peaked at around 72 h. On the other hand, we can observe that M_2 had a low level after stroke onset and, after three days, reached a stable level. The population of M_1 began to shift into the steep curve (the second stage) with an increase of M_2. Furthermore, the damage increased sharply in the initial stage and decreased in the second stage. On the other hand, in the SMNR model, the response of the neural stem cells began after the initial stage, helping the brain cells to repair damaged cells and lost cells, depending on whether the number of living cells was greater than that of the dead cells. As depicted in Figure 3, the curve of neural stem cells increased and then decreased during the 15 days following stroke onset; this can be explained by the ability of neural stem cells to inhibit inflammation, as has been studied in [3], where the survival of neural stem cells seemed to be prevented, with up to 80% of the new neurons dying within two weeks after their generation in vivo. Furthermore, the curve of neural stem cells in Figure 3 reached its peak during the first week, in agreement with studies in rats and mice, where the increase in proliferation of neural stem cells peaked at around 7 days after ischemic injury [27].

Table 2. Parameters values for the stroke-microglia-neural stem cells-recovery (SMNR) model.

The SMNR Model			
Parameters	**Values**	**Biological Meaning**	**Sources**
α	0.38	Source of resting microglia	[1]
R_1	0.12	Rate the activation of resting microglia intoM_1	[1]
R_2	0. 26	Rate the activation of resting microglia into M_2	estimated
R_3	0.11	The transition rate of $M_1 \rightarrow M_2$	simulation
K_1	0.91	Rate of interaction between M_1 and N_{SC}	simulation
K_2	0.75	Rate of interaction between M_2 and N_{SC}	simulation
δ	0.2854	Rate of the damage produced by M_1	[1]
β_1	0.1	Rate of effect the damage on $C(t)$	[1]
β_2	0.2	Rate of effectN_{SC} on $C(t)$	simulation
β_0	0.05	Rate of dying C through stroke	[1]
μ	0.053	Death rate of M_0	simulation
γ_1	0.015	Death rate of M_1	[35]
γ_2	0.015	Death rate of M_2	[35]
σ	0.015	Death rate of N_{SC}	[38]
r_1	0.05	Damage clearance by M_1	[1]
r_2	0.0125	Damage clearance by M_2	[1]

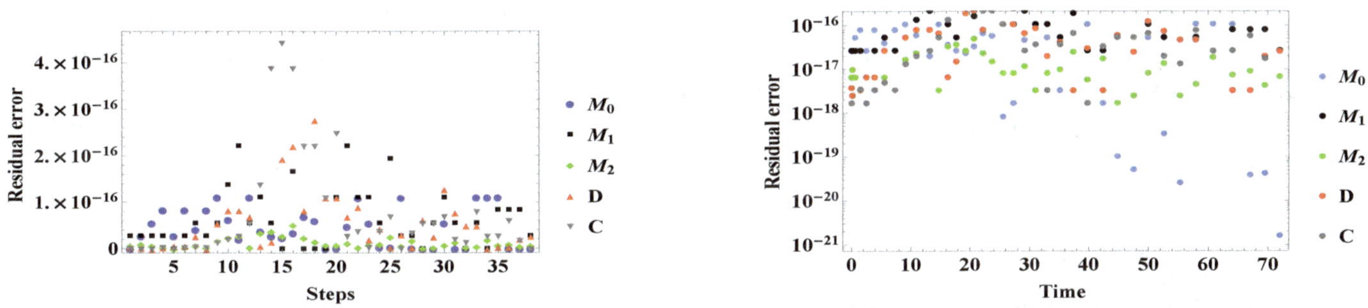

Figure 1. Residual error for the step and time of the numerical method in the SMD model.

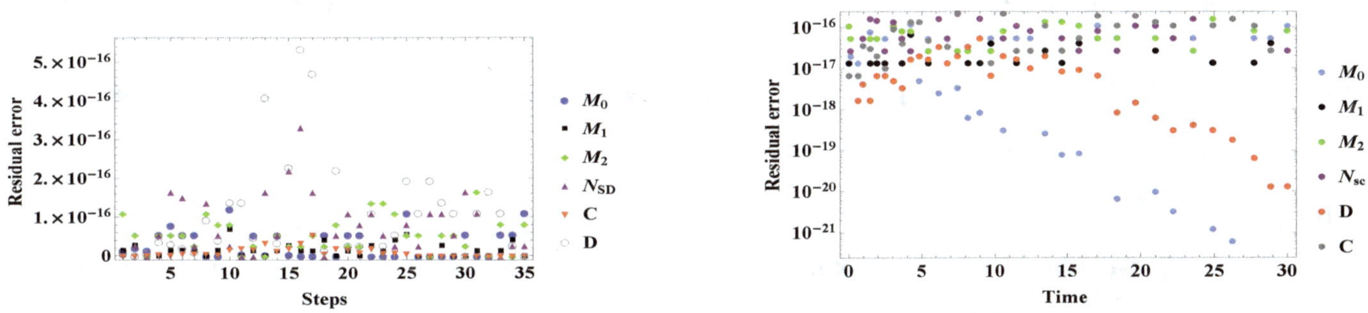

Figure 2. Residual error for the step and time of the numerical method in the SMNR model.

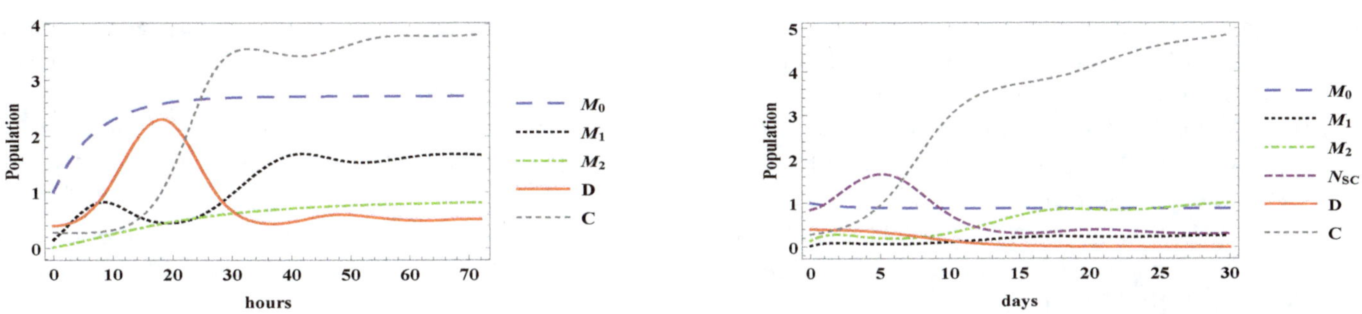

Figure 3. The behaviour of the SMD model within 72 h from stroke onset and the SMNR model within 30 days after stroke onset.

4. Conclusions

We modified two models—SMD ((1)–(5)) and SMNR ((11)–(16))—based on the model of Leah et al. [40], in order to study the impact of the microglial activation stage and the interaction between microglia and neural stem cells during a stroke. The dynamic models of the effects of microglia and the interaction between neural stem cells and microglia in stroke over time were studied both analytically and numerically. From the results of the analysis and simulation of both models, two states of microglial cells (M_1 to M_2) emerged from the resting state. In the first disease stage, the activation of microglial cells went from resting-state microglia to either of the states M_1 or M_2. The results of this interaction led to increased damage in the brain after stroke onset when M_2 was very low, there was no shift from M_1 to M_2, and no generation of neural stem cells occurred in the first 72 h after the stroke. Subsequently, in the recovery stage, the rate of M_1 decreased and the rate of M_2 increased and the neural stem cells began to generate. Subsequent to that, the rate of damage decreased. The instability of the SMNR model can be explained: in the nervous system, the number of endogenous neural stem cells is very low under normal physiological conditions, showing a very limited capacity for cell replacement under normal physiological conditions [20,21]. Understanding the biology in vivo of neural stem cells could lead to new therapeutic strategies for brain repair by endogenous neural stem cells [20,21], encouraging inflammatory inhibition. In conclusion, our modified models can lead to lead to an understanding of the effectiveness risks of the inflammatory responses associated with strokes and their positive and negative effects on the brain in stroke patients, as well as the general dynamics of microglial effects on neural stem cells, both during stroke and in the recovery stage. In the future, we will expand this work to study the mechanisms that improve, stimulate, and generate the neural stem cells in the early stage, and where that information could contribute to understanding the effects of therapeutic strategies. Additionally, it could be interesting to incorporate the dynamics of anti-inflammatory and proinflammatory cytokines from microglia and the cytokines of endogenous neural stem cells into the SMNR model, in order to describe the interaction processes of the different types of cytokines in ischemic stroke.

Author Contributions: Conceptualization, A.J.A.; funding acquisition, A.S.R.; methodology, A.J.A.; Project administration, A.S.R.; supervision, A.S.R. and I.H.; validation, A.S.R.; writing—original draft, A.J.A.; writing—review and editing, I.H. All authors have read and agreed to the published version of the manuscript.

Acknowledgments: This study was supported by a research grant from the Universiti Kebangsaan Malaysia with code GUP-2017-112.

References

1. Di Russo, C. A mathematical model of inflammation during ischemic stroke. *Math. Model. Med.* **2010**, *30*, 15–33. [CrossRef]
2. Guruswamy, R.; ElAli, A. Complex roles of microglial cells in ischemic stroke pathobiology: New insights and future directions. *Int. J. Mol. Sci.* **2017**, *18*, 496. [CrossRef]
3. Tobin, M.K.; Bonds, J.A.; Minshall, R.D.; Pelligrino, D.A.; Testai, F.D.; Lazarov, O. Neurogenesis and inflammation after ischemic stroke: What is known and where we go from here. *J. Cereb. Blood Flow Metab.* **2014**, *10*, 1573–1584. [CrossRef] [PubMed]
4. Radak, D.; Katsiki, N.; Resanovic, I.; Jovanovic, A.; Sudar-Milovanovic, E., Zafirovic, S.; Mousad, S.A.; Isenovic, E.R. Apoptosis and acute brain ischemia in ischemic stroke. *Curr. Vasc. Pharmacol.* **2017**, *15*, 115–122. [CrossRef] [PubMed]
5. Anttila, J.E.; Whitaker, K.W.; Wires, E.S.; Harvey, B.K.; Airavaara, M. Role of microglia in ischemic focal stroke and recovery: Focus on toll-like receptors. *Neuro-Psychopharmacol. Biol. Psychiatry* **2016**, *79*, 3–14. [CrossRef] [PubMed]

6. Stonesifer, C.; Corey, S.; Ghanekar, S.; Diamandis, Z.; Acosta, S.A.; Borlongan, C.V. Stem cell therapy for abrogating stroke-induced neuroinflammation and relevant secondary cell death mechanisms. *Prog. Neurobiol. Prog. Neurobiol.* **2017**, *158*, 94–131. [CrossRef]
7. Banjara, M.; Ghosh, C. Sterile neuroinflammation and strategies for therapeutic intervention. *Int. J. Inflamm.* **2017**, *10*, 8385961. [CrossRef]
8. Rajab, N.F.; Musa, S.M.; Ahmad Munawar, M.; Leong, L.M.; Heng, K.Y.; Ibrahim, F.W.; Chan, K.M. Antineuroinflammatory effects of hibiscus sabdariffa Linn.(Roselle) on Lipopolysaccharides-induced Microglia and Neuroblastoma Cells. *Malays. J. Health Sci.* **2016**, *14*, 111–117.
9. Rune, L.; Martin, W.; Daniel, G.; Christina, F.; Lasse, D.; Ishar, D.; Bente, F. Microglial cell population dynamics in the injured adult central nervous system. *Brain Res. Rev.* **2005**, *48*, 196–206.
10. Galloway, D.A.; Phillips, A.E.; Owen, D.R.; Moore, C.S. Phagocytosis in the Brain: Homeostasis and Disease. *Front. Immunol.* **2019**, *10*, 790. [CrossRef]
11. Pham-Huy, L.A.; He, H.; Pham-Huy, C. Free radicals, antioxidants in disease and health. *Int. J. Biomed. Sci. IJBS* **2008**, *4*, 89–96. [PubMed]
12. Di Meo, S.; Reed, T.T.; Venditti, P.; Victor, V.M. Role of ROS and RNS Sources in Physiological and Pathological Conditions. Oxid. Med. *Cell. Longev.* **2016**, *2016*, 1245049. [CrossRef]
13. Fresta, C.G.; Chakraborty, A.; Wijesinghe, M.B.; Amorini, A.M.; Lazzarino, G.; Lazzarino, G.; Tavazzi, B.; Lunte, S.M.; Caraci, F.; Dhar, G.; et al. Non-toxic engineered carbon nanodiamond concentrations induce oxidative/nitrosative stress, imbalance of energy metabolism, and mitochondrial dysfunction in microglial and alveolar basal epithelial cells. *Cell Death Dis.* **2018**, *9*, 245. [CrossRef] [PubMed]
14. Vay, S.U.; Flitsch, L.J.; Rabenstein, M.; Rogall, R.; Blaschke, S.; Kleinhaus, J.; Bach, A.; Fink, G.R.; Schroeter, M.; Rueger, A.M. The plasticity of primary microglia and their multifaceted effects on endogenous neural stem cells in vitro and in vivo. *J. Neuroinflamm.* **2018**, *15*, 226. [CrossRef]
15. Hui-Yin, Y.; Ahmad, N.; Azmi, N.; Makmor-Bakry, M. Curcumin: The molecular mechanisms of action in inflammation and cell death during kainate-induced epileptogenesis. *Indian J. Pharm. Edu. Res.* **2018**, *52*, 32–41. [CrossRef]
16. Hake, I.; Schönenberger, S.; Neumann, J.; Franke, K.; Paulsen-Merker, K.; Reymann, K.; Ismail, G.; Bin Din, L.; Said, I.M.; Latiff, A.; et al. Neuroprotection and enhanced neurogenesis by extract from the tropical plant Knema laurina after inflammatory damage in living brain tissue. *J. Neuroimmunol.* **2009**, *206*, 91–99. [CrossRef]
17. Cherry, J.D.; Olschowka, J.A.; O'Banion, M.K. Neuroinflammation and M_2 microglia: The good, the bad, and the inflamed. *J. Neuroinflamm.* **2014**, *11*, 98. [CrossRef]
18. Ma, Y.; Wang, J.; Wang, Y.; Yang, G.Y. The biphasic function of microglia in ischemic stroke. *Prog. Neurobiol.* **2016**, *157*, 247–272. [CrossRef]
19. Choi, J.Y.; Kim, J.Y.; Kim, J.Y.; Park, J.; Lee, W.T.; Lee, J.E. M_2 phenotype microglia-derived cytokine stimulates proliferation and neuronal differentiation of endogenous stem cells in ischemic brain. *Exp. Neurobiol.* **2017**, *26*, 33–41. [CrossRef]
20. Ahn, S.; Joyner, A. In vivo analysis of quiescent adult neural stem cells responding to Sonic hedgehog. *Nature* **2005**, *437*, 894–897. [CrossRef]
21. Bauer, S . Cytokine control of adult neural stem cells. *Ann. N. Y. Acad. Sci.* **2009**, *1153*, 48–56. [CrossRef] [PubMed]
22. Vieira, M.S.; Santos, A.K.; Vasconcellos, R.; Goulart, V.A.; Parreira, R.C.; Kihara, A.H.; Ulrich, H.; Resende, R.R. Neural stem cell differentiation into mature neurons: Mechanisms of regulation and biotechnological applications. *Biotechnol. Adv.* **2018**, *7*, 1946–1970. [CrossRef] [PubMed]
23. Martínez-Garza, D.M.; Cantú-Rodríguez, O.G.; Jaime-Pérez, J.C.; Gutiérrez-Aguirre, C.H.; Góngora-Rivera, J.F.; Gómez-Almaguer, D. Current state and perspectives of stem cell therapy for stroke. *Med. Univ.* **2016**, *72*, 169–180. [CrossRef]
24. Matarredona, E.R.; Talaveron, R.; Pastor, A.M. Interactions between neural progenitor cells and microglia in the subventricular zone: Physiological implications in the neurogenic niche and after implantation in the injured brain. *Front. Cell. Neurosci.* **2018**, *12*, 268. [CrossRef]
25. Zhang, Z.; Chopp, M. Neural stem cells and ischemic brain. *J. Stroke* **2016**, *18*, 267–272. [CrossRef] [PubMed]

26. Faiz, M.; Sachewsky, N.; Gascón, S.; Bang, K.A.; Morshead, C.M.; Nagy, A. Adult neural stem cells from the subventricular zone give rise to reactive astrocytes in the cortex after stroke. *Cell Stem Cell* **2015**, *17*, 624–634. [CrossRef] [PubMed]
27. Choi, Y.S.; Lee, M.Y.; Sung, K.W.; Jeong, S.W.; Choi, J.S.; Park, H.J.; Kim, O.N.; Lee, S.B.; Kim, S.Y. Regional differences in enhanced neurogenesis in the dentate gyrus of adult rats after transient forebrain ischemia. *Mol. Cells* **2003**, *16*, 232–238.
28. Barkho, B.Z.; Zhao, X. Adult neural stem cells: Response to stroke injury and potential for therapeutic applications. *Curr. Stem Cell Res. Ther.* **2011**, *6*, 327–338. [CrossRef]
29. Ariffin, S.H.Z.; Wahab, R.M.A.; Ismail, I.; Mahadi, N.M.; Ariffin, Z.Z. Stem cells, cytokines and their receptors. *Asia-Pac. J. Mol. Biol. Biotechnol.* **2005**, *13*, 1–13.
30. Lira-Diaz, E; Gonzalez-Perez, O. Emerging roles of microglia cells in the regulation of adult neural stem cells. *Neuroimmunol. Neuroinflamm.* **2008**, *3*, 204–206.
31. Boese, A.C.; Le, Q.S.E.; Pham, D.; Hamblin, M.H.; Lee, J.P. Neural stem cell therapy for subacute and chronic ischemic stroke. *Stem Cell Res. Ther.* **2018**, *9*. [CrossRef] [PubMed]
32. Reynolds, A.; Rubin, J.; Clermont, G.; Day, J.; Vodovotz, Y.; Bard Ermentrout, G. A reduced mathematical model of the acute inflammatory response: I. derivation of model and analysis of anti-inflammation. *J. Theor. Biol.* **2006**, *242*, 220–236. [CrossRef] [PubMed]
33. Kumar, R.; Clermont, G.; Vodovotz, Y.; Chow, C.C. The dynamics of acute inflammation. *J. Theor. Biol.* **2004**, *230*, 145–155. [CrossRef] [PubMed]
34. Alharbi, S.; Rambely, A. A dynamic simulation of the immune system response to inhibit and eliminate abnormal cells. *Symmetry* **2019**, *11*, 572. [CrossRef]
35. Hao, W.; Friedman, A. Mathematical model on Alzheimer'sdisease. *BMC Syst. Biol.* **2016**, *10*, 108. [CrossRef]
36. Nakata, Y.; Getto, P.; Marciniak-Czochra, A.; Alarcón, T. Stability analysis of multi-compartment models for cell production systems. *J. Biol. Dyn.* **2012**, *6*, 2–18. [CrossRef]
37. Ziebell, F.; Martin-Villalba, A.; Marciniak-Czochra, A. Mathematical modelling of adult hippocampal neurogenesis: Effects of altered stem cell dynamics on cell counts and bromodeoxyuridine-labelled cells. *J. R. Soc. Interface* **2014**, *11*, 20140144. [CrossRef]
38. Cacao, E.; Cucinotta, F.A. Modeling impaired hippocampal neurogenesis after radiation exposure. *Radiat. Res* **2016**, *185*, 319–331. [CrossRef]
39. Huang, L.; Zhang, L. Neural stem cell therapies and hypoxic-ischemic brain injury. *Prog. Neurobiol.* **2018**, *173*. [CrossRef]
40. Leah, E.V.; Prerna, R.R.; Raj, G.K.; Amy, K.; Jonathan, E.R. A mathematical model of neuroinflammation in severe clinical traumatic brain injury. *J. Neuroinflamm.* **2018**, *15*, 345. [CrossRef]
41. Zhao, L.; Zhang, J. *Cellular Therapy for Stroke and CNS Injuries*; Springer Series in Translational Stroke Research; Springer: Berlin, Germany, 2015; pp. 34–39.
42. Loane, D.J.; Kumar, A. Microglia in the TBI brain: The good, the bad, and the dysregulated. *Exp. Neurol.* **2016**, *275*, 316–327. [CrossRef] [PubMed]
43. Kim, J.Y.; Kim, N.; Yenari, M.A. Mechanisms and potential therapeutic applications of microglial activation after brain injury. *CNS Neurosci. Ther.* **2015**, *21*, 309–319. [CrossRef] [PubMed]
44. Saha Ray, S.; Sahoo, S. *Generalized Fractional Order Differential Equations Arising in Physical Models*; CRC Press: Boca Raton, FL, USA, 2018.
45. Gantmacher, F.R. *The Theory of Matrices*; American Mathematical Society: Providence, RI, USA, 1959; Volume 2.

4

Probabilistic Interpretation of Solutions of Linear Ultraparabolic Equations

Michael D. Marcozzi

Department of Mathematical Sciences, University of Nevada Las Vegas, Las Vegas, NV 89154-4020, USA; michael.marcozzi@unlv.edu

Abstract: We demonstrate the existence, uniqueness and Galerkin approximatation of linear ultraparabolic terminal value/infinite-horizon problems on unbounded spatial domains. Furthermore, we provide a probabilistic interpretation of the solution in terms of the expectation of an associated ultradiffusion process.

Keywords: ultraparabolic equation; ultradiffusion process; probabilistic representation; mathematical finance

1. Introduction

The connection between parabolic equations and diffusion processes is well understood; the same cannot be said for ultraparabolic equations and ultradiffusion processes. Until recently, theoretical results have been fairly limited relative to the existence and uniqueness of solutions to ultraparabolic equations, deriving from two methodologies. In one, the analysis is affected along the characteristic of the first-order temporal operator, requiring that the speed of propagation varies only spatially. Such an approach was developed by Piskunov [1] in the classical case and extended by Lions [2] to the generalized sense. The second approach is based on the method of fundamental solutions and was implemented by Il'in [3] for the classical Cauchy problem and extended to more general domains via convolution by Vladimirov and Drožžinov [4], albeit at the expense of necessitating constant coefficients in the operator. Recently, however, using energic techniques Marcozzi [5] has established the well-posedness and Galerkin approximation of the generalized solution (strong and weak) to the terminal value problem for square integrable data on bounded temporal and spatial domains. We extend here the results of [5] to linear ultraparabolic terminal value/infinite-horizon temporal problems posed on unbounded spatial domains. We then provide a probabilistic interpretation of the solution in terms of the expectation of an associated ultradiffusion process.

Historically, the connection between the expectation of ultradiffusion processess and the solution to ultradiffusion equations arose from the work of Kolmogorov [6,7] and Uhlenbeck and Ornstein [8] in relation to Brownian motion in phase space—the same with respect to Chandrasekhar [9] in the context of boundary layers and Marshak [10] relative to the Bolzmann equation. A contemporary example may be found in the formulation of so-called Asian options from mathematical finance (cf. [11]), which obtains theoretical context with the present results. The paper is organized as follows. In Section 2, we consider deterministic aspects of the problem, while, in Section 3, the probabilistic interpretation is presented. Appendix A introduces certain regularity results, which, while essential for the analysis, are too extensive to prove in full. In Appendix B, we show formally that the ultraparabolic/ultradiffusion association is locally that of a parameterized parabolic/diffusion.

2. Approximation Solvability

We consider here the existence, uniqueness and approximation of the terminal value/infinite horizon problem on unbounded spatial domains for the linear ultraparabolic equations. To this end, let $\widetilde{\mathcal{O}}_{T,\vartheta} = (0,T) \times (0,\infty)$, $\widetilde{\mathcal{O}}_{\vartheta,x} = (0,\infty) \times (-\infty,\infty)$, $\widetilde{\mathcal{O}}_{T,x} = (0,T) \times (-\infty,\infty)$, and finally $\widetilde{\mathcal{Q}} = (0,T) \times (0,\infty) \times (-\infty,\infty)$, for some finite $T > 0$. The functional setting will be the weighted Sobolev spaces defined as follows. Spatially, we let

$$m_\mu(x) = e^{-\mu|x|},$$

such that

$$H_\mu = \{v(x) \mid m_\mu(x)\, v(x) \in L_2(\mathbb{R})\ \}$$

and

$$V_\mu = \left\{v \in H_\mu \,\middle|\, \partial v(x)/\partial x \in H_\mu\ \right\},$$

with their respective norms

$$|v|_\mu = \left\{\int_{\mathbb{R}} [m_\mu(x)\, v(x)]^2 dx\right\}^{1/2},$$

for all $v \in H_\mu$, and

$$\|v\|_\mu = \left\{|v|_\mu^2 + |\partial v/\partial x|_\mu^2\right\}^{1/2},$$

for all $v \in V_\mu$. The relation "$V_\mu \subseteq H_\mu \subset V_\mu^*$" constitutes an *evolution triple*.

Temporally, let $\boldsymbol{t} = (t,\vartheta)$ and

$$n_\gamma(\boldsymbol{t}) = e^{-\gamma\vartheta},$$

such that

$$\mathcal{X}_{\gamma,\mu} = L_\gamma^2(\widetilde{\mathcal{O}}_{T,\vartheta}; V_\mu) = \left\{u(\boldsymbol{t}) \mid n_\gamma(\boldsymbol{t})\, u(\boldsymbol{t}) \in L_2(\widetilde{\mathcal{O}}_{T,\vartheta}; V_\mu)\ \right\},$$

which we equip with the norm

$$\|u\|_{\gamma,\mu} = \|u\|_{\mathcal{X}_{\gamma,\mu}} = \left\{\int_{\widetilde{\mathcal{O}}_{T,\vartheta}} \|n_\gamma(\boldsymbol{t})\, u(\boldsymbol{t})\|_\mu^2\, d\mathcal{O}\right\}^{1/2},$$

for all $u \in \mathcal{X}_{\gamma,\mu}$. We associate with $\mathcal{X}_{\gamma,\mu}$ the dual space

$$\mathcal{X}_{\gamma,\mu}^* = L_\gamma^2(\mathcal{O}_{t,\vartheta}; V_\mu^*)$$

and the norm $\|u^*\|_{\mathcal{X}_{\gamma,\mu}^*}$, for all $u^* \in \mathcal{X}_{\gamma,\mu}$. In addition, let

$$W_{\gamma,\mu} = W_\gamma^1(\mathcal{O}_{t,\vartheta}; V_\mu, H_\mu) = \left\{u \in \mathcal{X}_{\gamma,\mu}\ :\ \nabla_{\boldsymbol{t}}(u) \in \mathcal{X}_{\gamma,\mu}^* \times \mathcal{X}_{\gamma,\mu}^*\right\},$$

where $\nabla_{\boldsymbol{t}}(u) = (\partial u/\partial t, \partial u/\partial\vartheta)$, which we associate with the norm

$$\|u\|_{W_{\gamma,\mu}} = \left(\|u\|_{\mathcal{X}_{\gamma,\mu}}^2 + \|\partial u/\partial t\|_{\mathcal{X}_{\gamma,\mu}^*}^2 + \|\partial u/\partial\vartheta\|_{\mathcal{X}_{\gamma,\mu}^*}^2\right)^2,$$

for all $u \in W_{\gamma,\mu}$. Finally, we define

$$L_\gamma^2((0,\infty); H_\mu) = \left\{u(\vartheta) \mid n_\gamma(\boldsymbol{t})\, u(\boldsymbol{t}) \in L_2(\widetilde{\mathcal{O}}_{T,\vartheta}; H_\mu)\ \right\}.$$

We consider the *ultraparabolic* t-terminal value/infinite ϑ-horizon problem for $u \in W_{\gamma,\mu}$ satisfying the evolutionary equation

$$-\frac{\partial u}{\partial t} - \frac{\partial(bu)}{\partial\vartheta} + A(\boldsymbol{t})\, u = f\ \ a.e.\text{ on } \widetilde{\mathcal{Q}}, \tag{1}$$

subject to the terminal condition

$$u(T,\vartheta,x) = v(\vartheta,x) \ \ a.e. \text{ on } \widetilde{\mathcal{O}}_{\vartheta,x}, \tag{2}$$

where

$$A(t)\,u = -\frac{\partial}{\partial x}\left(a_2\frac{\partial u}{\partial x}\right) + a_1\frac{\partial u}{\partial x} + a_0\,u\,,$$

for given

$$a_0\,,\ a_1\,,\ a_2 \in L_\infty(\widetilde{\mathcal{Q}})\,,\ b \in C^1\left(\overline{\widetilde{\mathcal{Q}}}\right)\,, \tag{3}$$

$$0 < \underline{b} \le b(t,x) \le 2\,\overline{b}\,,\ \partial b/\partial\vartheta \text{ bounded}\,, \tag{4}$$

$$v \in L^2_\gamma((0,\infty);H_\mu)\,, \tag{5}$$

$$f \in \mathcal{X}^*_{\gamma,\mu}\,, \tag{6}$$

$$0 < \alpha \le a_2 \text{ and } 0 < \beta \le a_0\,, \tag{7}$$

for some sufficiently large β.

The *generalized problem* associated with (1)–(2) is: supposing (3)–(7), find $u \in W_{\gamma,\mu}$ satisfying

$$-\frac{\partial}{\partial t}\,(u(t)|v)_{H_\mu} - \frac{\partial}{\partial\vartheta}\,(b(t)\,u(t)|v)_{H_\mu} + a_\mu\,(t;u(t),v) = \langle f(t),v\rangle_{V_\mu}\,, \tag{8}$$

for almost all $t \in \widetilde{\mathcal{O}}_{T,\vartheta}$, such that

$$u(T_\vartheta) = v(\vartheta) \text{ on } \widetilde{\mathcal{O}}_{\vartheta,x}\,, \tag{9}$$

where $(u|v)_X$ is the scalar product canonically defined on the Hilbert space X, $\langle f(t),v\rangle_{V_\mu}$ denotes the value of the linear functional $f(t) \in V^*_\mu$ at $v \in V_\mu$, $T_\vartheta = (T,\vartheta)$ and

$$a_\mu\,(t;u,v) = \int_{\mathbb{R}} a_2\,\frac{\partial u}{\partial x}\,\frac{\partial v}{\partial x}\,m^2_\mu\,dx + \int_{\mathbb{R}} a_1\,\frac{\partial u}{\partial x}\,v\,m^2_\mu\,dx + \int_{\mathbb{R}} a_0\,u\,v\,m^2_\mu\,dx\,,$$

for all $u,v \in V_\mu$ and $t \in \widetilde{\mathcal{O}}_{T,\vartheta}$. In Equation (8), the expressions $\partial/\partial t$ and $\partial/\partial\vartheta$ denote generalized derivatives on $\widetilde{\mathcal{O}}_{T,\vartheta}$; that is, Equation (8) means explicitly

$$\int_{\widetilde{\mathcal{O}}_{T,\vartheta}} (u(t)|v)_{H_\mu}\,\frac{\partial}{\partial t}\varphi(t)\,n^2_\gamma\,d\mathcal{O} + \int_{\widetilde{\mathcal{O}}_{T,\vartheta}} (b(t)\,u(t)|v)_{H_\mu}\,\frac{\partial}{\partial\vartheta}\varphi(t)\,n^2_\gamma\,d\mathcal{O} \tag{10}$$

$$+\int_{\widetilde{\mathcal{O}}_{T,\vartheta}} a\,(t;u(t),v)\,\varphi(t)\,d\mathcal{O} = \int_{\widetilde{\mathcal{O}}_{T,\vartheta}} \langle f(t),v\rangle_{V_\mu}\,\varphi(t)\,n^2_\gamma\,d\mathcal{O}\,,$$

for all test functions $\varphi \in C^\infty_0(\widetilde{\mathcal{O}}_{T,\vartheta})$.

For $t \in \widetilde{\mathcal{O}}_{T,\vartheta}$, the mapping $a_\mu(t) : V_\mu \times V_\mu$ is bilinear and bounded; we likewise assume that $a_\mu(t)$ is strongly positive;

$$c\|u(t)\|^2_\mu \le a_\mu(t;u(t),u(t))\,. \tag{11}$$

Remark 1. *We note that Equations (1)–(2) is an* infinite horizon *problem in ϑ. That is, the far-field behavior of ϑ is implicitly defined relative to the weight γ.*

Remark 2. *In general, the validity of (11) will be problem dependent, predicated upon the spatial asymptotic behavior of u.*

For $t \in \widetilde{\mathcal{O}}_{T,\vartheta}$, we define the operator $A_\mu(t) : V_\mu \to V^*_\mu$ such that

$$\langle A_\mu(t)\, u, v\rangle_{V_\mu} = a_\mu(t; u, v)\,, \tag{12}$$

from which it follows that $A_\mu(t)$ is linear, continuous, and strongly monotone by (11). In particular, we have

$$\|A(t)\, u\|_{\mathcal{X}^*_{\gamma,\mu}} \leq C \|u\|_{\mathcal{X}_{\gamma,\mu}} \tag{13}$$

and

$$c\|u\|^2_\mu \leq \langle A(t)\, u, u\rangle_{V_\mu}\,, \tag{14}$$

for all $u \in V_\mu$ and $t \in \widetilde{\mathcal{O}}_{T,\vartheta}$.

Lemma 1. *Given (3)–(7), the formulations (1)–(2) and (8)–(9) are equivalent.*

Proof of Lemma 1. By integration by parts and the density of test functions in V_μ, we have

$$\frac{\partial}{\partial t}(u(t)|v) = \left\langle \frac{\partial u(t)}{\partial t}, v\right\rangle_{V_\mu} \quad \text{and} \quad \frac{\partial}{\partial \vartheta}(u(t)|v) = \left\langle \frac{\partial u(t)}{\partial \vartheta}, v\right\rangle_{V_\mu}\,, \tag{15}$$

for all $v \in V_\mu$ and almost all $t \in \widetilde{\mathcal{O}}_{T,\vartheta}$. From (8), (12) and (15), we deduce that

$$\left\langle -\frac{\partial u(t)}{\partial t} - \frac{\partial\,(b(t)u(t))}{\partial \vartheta} + A(t)u(t) - f(t), v\right\rangle_{V_\mu} = 0\,,$$

for all $v \in V_\mu$ and almost all $t \in \mathcal{O}_{T,\Theta}$, in which case (1) follows. The converse derives from (1) and $u \in W_{\gamma,\mu}$, which imply (8). □

Proposition 1. *Uniqueness. We suppose (3)–(7) and (11); let $0 < \gamma < c$. Then, there exists at most one solution to (8)–(9).*

Proof of Proposition 1. We consider (8)–(9) with $f = 0$ and $v = 0$; setting $v = u$ in (8), we obtain

$$-\frac{1}{2}\frac{\partial}{\partial t}\,|u(t)|^2_\mu - \frac{1}{2}\frac{\partial}{\partial \vartheta}\left|\sqrt{b(t)}\, u(t)\right|^2_\mu + a_\mu\left(t; u(t), u(t)\right) = 0$$

or

$$2c\|u(t)\|^2_\mu \leq \frac{\partial}{\partial t}\,|u(t)|^2_\mu + \frac{\partial}{\partial \vartheta}\left|\sqrt{b(t)}\, u(t)\right|^2_\mu\,,$$

from (11), in which case

$$0 \leq \frac{\partial}{\partial t}\left(e^{-2c\vartheta}\,|u(t)|^2_\mu\right) + \frac{\partial}{\partial \vartheta}\left(e^{-2c\vartheta}\left|\sqrt{b(t)}\, u(t)\right|^2_\mu\right).$$

Integrating over the domain $(0, T) \times (0, \Theta)$, for some $\Theta > 0$, and applying Green's Theorem, it follows that

$$\int_0^\Theta e^{-2c\vartheta}\,|u(0,\vartheta)|^2_\mu\, d\vartheta + \int_0^T \left|\sqrt{b(t,0)}\, u(t,0)\right|^2_\mu dt \leq \int_0^T e^{-2c\Theta}\left|\sqrt{b(t,\Theta)}\, u(t,\Theta)\right|^2_\mu dt$$

and so

$$0 < \tilde{c} \leq e^{-2c\Theta}\,|u(t,\Theta)|^2_\mu\,.$$

However,

$$\tilde{c}e^{-2(c-\gamma)\Theta} \leq e^{-2c\Theta}\,|u(t,\Theta)|^2_\mu$$

is not summable on $(0,\infty)$, which contradicts the condition $u \in \mathcal{X}_{\gamma,\mu}$, from which it follows that $u = 0$. $\square$

We consider the regularization of (1)–(2) to domains of finite extent. To this end, it suffices for v to have an extension to, or to be of compact support in, $\widetilde{\mathcal{Q}}$. Without loss of generality, we may assume that $v = 0$. For $m \in \mathbb{N}$, let

$$\mathcal{Q}_m = (0,T) \times (0,m) \times (-m,m)\,,$$
$$f_m = f \ \ on \ \ \mathcal{Q}_m\,,$$
$$V_m = H_0^1(-m,m)\,,$$
$$H_m = L_2(-m,m)\,,$$
$$\mathcal{X}_m = L_2((0,T) \times (0,m); V_m)\,,$$
$$\mathcal{X}_m^* = L_2((0,T) \times (0,m); V_m^*),$$

and

$$W_m = \{u_m \in \mathcal{X}_m \mid \nabla_t(u_m) \in \mathcal{X}_m^* \times \mathcal{X}_m^* \}\,.$$

There exists a unique $u_m \in W_m$ satisfying the ultraparabolic terminal value problem (cf. [5])

$$-\frac{\partial u_m}{\partial t} - \frac{\partial (b u_m)}{\partial \vartheta} + A(t,\vartheta)\, u_m = f_m \ \ a.e. \text{ on } \ \mathcal{Q}_m \tag{16}$$

subject to the terminal conditions

$$u(T,\vartheta,x) = 0 \ \ a.e. \text{ on } \ (0,m) \times (-m,m)\,, \tag{17}$$

$$u(t,m,x) = 0 \ \ a.e. \text{ on } \ (0,T) \times (-m,m)\,, \tag{18}$$

and boundary conditions

$$u(t,\vartheta,-m) = u(t,\vartheta,-m) = 0 \ \ a.e. \text{ on } \ (0,T) \times (0,m)\,. \tag{19}$$

We denote by $\widetilde{u}_m$ the extension of u_m by zero to the compliment of $\mathcal{Q}_m$.

Lemma 2. *We suppose (3)–(7), (11),*

$$\nabla_t(a_i) \in L_\infty(\widetilde{\mathcal{Q}}) \times L_\infty(\widetilde{\mathcal{Q}})\,, \tag{20}$$

$0 < \gamma < c\,/\,\overline{b}$ and $f \in L_\gamma^2(\widetilde{\mathcal{O}}_{t,\vartheta}; H_\mu)$*; then,*

$$\|\widetilde{u}_m\|_{\gamma,\mu} \leq C\,, \tag{21}$$

for all $m \in \mathbb{N}$.

Proof of Lemma 2. Taking the inner product of (16) with $\widetilde{u}_m$, we have

$$-\left(\frac{\partial \widetilde{u}_m}{\partial t}, \widetilde{u}_m\right)_\mu - \left(\frac{\partial b \widetilde{u}_m}{\partial \vartheta}, \widetilde{u}_m\right)_\mu + a_\mu(\boldsymbol{t}; \widetilde{u}_m, \widetilde{u}_m) = (f_m, \widetilde{u}_m)_\mu$$

or

$$-e^{-2\gamma\vartheta}\frac{1}{2}\frac{\partial}{\partial t}\,|\widetilde{u}_m|_\mu^2 - e^{-2\gamma\vartheta}\frac{1}{2}\frac{\partial}{\partial \vartheta}\left|\sqrt{b}\widetilde{u}_m\right|_\mu^2 + ce^{-2\gamma\vartheta}\|\widetilde{u}_m\|_\mu^2 \leq e^{-2\gamma\vartheta}(f_m, \widetilde{u}_m)_\mu\,.$$

Integrating the above over $(0,T)\times(\vartheta,m)$, it follows that

$$-\int_0^T\int_\vartheta^m e^{-2\gamma\vartheta}\frac{1}{2}\frac{\partial}{\partial t}\left|\widetilde{u}_m\right|_\mu^2-\int_0^T\int_\vartheta^m e^{-2\gamma\vartheta}\frac{1}{2}\frac{\partial}{\partial\vartheta}\left|\sqrt{b}\widetilde{u}_m\right|_\mu^2$$
$$+c\int_0^T\int_\vartheta^m e^{-2\gamma\vartheta}\|\widetilde{u}_m\|_\mu^2\leq\int_0^T\int_\vartheta^m e^{-2\gamma\vartheta}(f_m,\widetilde{u}_m)_\mu$$

or

$$-\left\{\int_0^T\int_\vartheta^m\frac{\partial}{\partial t}\left[e^{-2\gamma\vartheta}\left|\widetilde{u}_m\right|_\mu^2\right]+\int_0^T\int_\vartheta^m\frac{\partial}{\partial\vartheta}\left[e^{-2\gamma\vartheta}\left|\sqrt{b}\widetilde{u}_m\right|_\mu^2\right]\right\}$$
$$-\int_0^T\int_\vartheta^m 2\gamma e^{-2\gamma\vartheta}\left|\sqrt{b}\widetilde{u}_m\right|_\mu^2+2c\int_0^T\int_\vartheta^m e^{-2\gamma\vartheta}\|\widetilde{u}_m\|_\mu^2$$
$$\leq 2\int_0^T\int_\vartheta^m e^{-2\gamma\vartheta}(f_m,\widetilde{u}_m)_\mu$$

and so

$$\int_\vartheta^m e^{-2\gamma\vartheta}\left|\widetilde{u}_m(0,\vartheta)\right|_\mu^2+\int_0^T e^{-2\gamma\vartheta}\left|\widetilde{u}_m(t,\vartheta)\right|_\mu^2-\int_0^T\int_\vartheta^m 2\gamma e^{-2\gamma\vartheta}\left|\sqrt{b}\widetilde{u}_m\right|_\mu^2$$
$$+2c\int_0^T\int_\vartheta^m e^{-2\gamma\vartheta}\|\widetilde{u}_m\|_\mu^2\leq 2\int_0^T\int_\vartheta^m e^{-2\gamma\vartheta}(f_m,\widetilde{u}_m)_\mu$$

in which case

$$2c\int_0^T\int_\vartheta^m e^{-2\gamma\vartheta}\|\widetilde{u}_m\|_\mu^2-\int_0^T\int_\vartheta^m 2\gamma e^{-2\gamma\vartheta}\left|\sqrt{b}\widetilde{u}_m\right|_\mu^2$$
$$\leq 2\int_0^T\int_\vartheta^m e^{-2\gamma\vartheta}(f_m,\widetilde{u}_m)_\mu$$

or

$$\left(c-\overline{b}\gamma\right)\int_0^T\int_\vartheta^m e^{-2\gamma\vartheta}\|\widetilde{u}_m\|_\mu^2\leq\int_0^T\int_\vartheta^m e^{-2\gamma\vartheta}(f_m,\widetilde{u}_m)_\mu$$
$$\leq\left(\int_0^T\int_\vartheta^m e^{-2\gamma\vartheta}\|\widetilde{f}_m\|_\mu^2\right)^{1/2}\left(\int_0^T\int_\vartheta^m e^{-2\gamma\vartheta}\|\widetilde{u}_m\|_\mu^2\right)^{1/2}$$

such that

$$\int_0^T\int_\vartheta^m e^{-2\gamma\vartheta}\|\widetilde{u}_m\|_\mu^2\leq C;$$

therefore,

$$\|\widetilde{u}_m\|_{\gamma,\mu}\leq C$$

for all $m\in\mathbb{N}$. □

We obtain a supplementary estimate on $\nabla_t(\widetilde{u}_m)$.

Lemma 3. *We suppose (3)–(7), (11), (20), $0<\gamma<c\,/\,\overline{b}$ and $f\in L^2_\gamma(\mathcal{O}_{t,\vartheta};H_\mu)$, then*

$$\left\|\frac{\partial\widetilde{u}_m}{\partial t}\right\|_{\gamma,\mu}\leq C\ \text{ and }\ \left\|\frac{\partial\widetilde{u}_m}{\partial\vartheta}\right\|_{\gamma,\mu}\leq C. \tag{22}$$

Proof of Lemma 3. We consider the parabolic regularization with respect to ϑ of (16)–(19). To this end, let $\mathcal{H}_m(0,\Theta)=L_2(0,\Theta;H_m)$; we define the space of test functions on $(0,m)\times(-m,m)$ such that

$$\mathcal{V}_m(0,m)=\left\{v\mid v,\frac{\partial v}{\partial x},\frac{\partial v}{\partial\vartheta}\in L_2(0,m;H_m),\ v(t,\vartheta,-m)=v(t,\vartheta,m)=v(t,m,x)=0\right\},$$

in which case we obtain the evolution triple "$\mathcal{V}_m(0,m) \subset \mathcal{H}_m(0,m) \subset \mathcal{V}_m^*(0,m)$". We denote $\mathcal{H}_m = \mathcal{H}_m(0,m)$ and $\mathcal{V}_m = \mathcal{V}_m(0,m)$ for brevity and equip $\mathcal{V}_m$ with the norm

$$\|v\|^2_{\mathcal{V}_m} = \int_{(0,m)} \|v(\theta)\|^2_{H_m}\, d\theta + \int_{(0,m)} \|\partial v(\theta)/\partial x\|^2_{H_m}\, d\theta + \int_{(0,m)} \|\partial v(\theta)/\partial \theta\|^2_{H_m}\, d\theta$$
$$= \|v\|^2_{\mathcal{H}_m} + \|\partial v/\partial x\|^2_{\mathcal{H}_m} + \|\partial v/\partial \vartheta\|^2_{\mathcal{H}_m}.$$

Let

$$\mathcal{W}_m = \mathcal{W}_m(0,T;\mathcal{V}_m,\mathcal{H}_m) = \left\{ v \mid v \in \mathcal{X}_m\,,\ \frac{\partial v}{\partial t} \in \mathcal{X}_m^*\,,\ \frac{\partial v}{\partial \vartheta} \in \mathcal{X}_m^*\,,\ v(t,m,x) = 0 \right\},$$

where $\mathcal{X}_m = L_2((0,m)\times(-m.m);V_m)$ and $\mathcal{X}_m^* = L_2((0,m)\times(-m.m);V_m^*)$, which we equip with the norm

$$\|v\|^2_{\mathcal{W}_m} = \int_0^T \|v(\tau)\|^2_{\mathcal{V}_m}\, d\tau + \int_0^T \|\partial v(\tau)/\partial\tau\|^2_{\mathcal{X}_m^*}\, d\tau\,.$$

The perturbation problem associated with (16)–(19) is: for any $\epsilon > 0$, we seek u_m^ϵ satisfying the parabolic equation

$$-\frac{\partial u_m^\epsilon}{\partial t} - \epsilon\frac{\partial^2 u_m^\epsilon}{\partial \vartheta^2} - \frac{\partial(b\, u_m^\epsilon)}{\partial\vartheta} + A(\boldsymbol{t})\, u_m^\epsilon = f_m \ \ a.e. \text{ on } \ \mathcal{Q}\,, \tag{23}$$

where $A(\boldsymbol{t}) = A(t,\vartheta)$, subject to the terminal condition

$$u_m^\epsilon(T,\vartheta,x) = v(\vartheta,x) \ \ a.e. \text{ on } \ (0,m)\times(-m,m)\,, \tag{24}$$

and boundary conditions

$$u_m^\epsilon(t,\Theta,x) = 0 \ \ a.e. \text{ on } \ (0,T)\times(-m,m)\,, \tag{25}$$

$$u^\epsilon(t,\vartheta,-m) = u^\epsilon(t,\vartheta,m) = 0 \ \ a.e. \text{ on } \ (0,t)\times(0,m)\,, \tag{26}$$

$$\frac{\partial u_m^\epsilon}{\partial\vartheta}(t,m,x) = 0 \ \ a.e. \text{ on } \ (0,T)\times(-m,m)\,. \tag{27}$$

The problem (23)–(27) is well-posed, noting in particular the necessity of the auxiliary boundary condition (27).

We denote by $\widetilde{u}_m^\epsilon$ the extension of u_m^ϵ by zero to the compliment of $\mathcal{Q}_m$. Taking the inner product of (23) with $-n_\gamma^2\, \partial\widetilde{u}_m^\epsilon/\partial t$, it follows that

$$\int_{(0,m)} \left(\frac{\partial\widetilde{u}_m^\epsilon}{\partial t}\middle| n_\gamma^2 \frac{\partial\widetilde{u}_m^\epsilon}{\partial t}\right)_\mu d\theta + a^\epsilon_{0,\mu,\gamma}\left(t;\widetilde{u}_m^\epsilon, -\frac{\partial\widetilde{u}_m^\epsilon}{\partial t}\right) = \int_{(0,m)} \left(f - A_1\widetilde{u}_m^\epsilon\middle| - n_\gamma^2\frac{\partial\widetilde{u}_m^\epsilon}{\partial t}\right)_\mu d\theta\,,$$

where

$$a^\epsilon_{0,\mu,\gamma}(t;u,v) = \epsilon\int_{\widetilde{\mathcal{O}}_{\vartheta,x}} \frac{\partial u}{\partial\vartheta}\frac{\partial v}{\partial\vartheta} m_\mu^2 n_\gamma^2\, d\mathcal{O} + \int_{\widetilde{\mathcal{O}}_{\vartheta,x}} a_2(t)\frac{\partial u}{\partial x}\frac{\partial v}{\partial x} m_\mu^2 n_\gamma^2\, d\mathcal{O}$$

and

$$A_1 u = \frac{\partial(bu)}{\partial\vartheta} - a_1\frac{\partial u}{\partial x} - a_0 u\,,$$

in which case

$$\int_{(0,m)} n_\gamma^2 \left|\frac{\partial\widetilde{u}_m^\epsilon}{\partial t}\right|^2_\mu d\theta - \frac{1}{2}\frac{d}{dt}a^\epsilon_{0,\mu,\gamma}(t;\widetilde{u}_m^\epsilon,\widetilde{u}_m^\epsilon) + \frac{1}{2}\,\dot{a}^\epsilon_{0,\mu,\gamma}(t;\widetilde{u}_m^\epsilon,\widetilde{u}_m^\epsilon)$$
$$= \int_{(0,m)} n_\gamma^2\left(f - A_1\widetilde{u}_m^\epsilon, -\frac{\partial\widetilde{u}_m^\epsilon}{\partial\vartheta}\right)_\mu d\theta\,,$$

where

$$\overset{\bullet}{a}{}^{\epsilon}_{0,\mu,\gamma}(\vartheta;u,v)=\int_{\tilde{\mathcal{O}}_{\vartheta,x}}\frac{da_2(t)}{dt}\frac{\partial u}{\partial x}\frac{\partial v}{\partial x}m_\mu^2 n_\gamma^2\,d\mathcal{O}\,.$$

Integrating the above in time, we have that

$$\int_{\tilde{\mathcal{O}}_{T,\vartheta}} n_\gamma^2\left|\frac{\partial \tilde{u}_m^\epsilon}{\partial t}\right|_\mu^2 d\mathcal{O}+\frac{1}{2}a^{\epsilon}_{0,\mu,\gamma}(0;\tilde{u}_m^\epsilon,\tilde{u}_m^\epsilon)=\frac{1}{2}a^{\epsilon}_{0,\mu,\gamma}(T;\tilde{u}_m^\epsilon,\tilde{u}_m^\epsilon)$$

$$-\frac{1}{2}\int_0^T\overset{\bullet}{a}{}^{\epsilon}_{0}(t;\tilde{u}_m^\epsilon,\tilde{u}_m^\epsilon)\,dt+\int_{\tilde{\mathcal{O}}_{T,\vartheta}} n_\gamma^2\left(f-A_1\tilde{u}_m^\epsilon,-\frac{\partial \tilde{u}_m^\epsilon}{\partial t}\right)_\mu d\mathcal{O}\,.$$

With (21), we proceed as per Lemma 2 to obtain

$$\int_{\tilde{\mathcal{O}}_{T,\vartheta}} n_\gamma^2\left|\frac{\partial \tilde{u}_m^\epsilon}{\partial t}\right|_\mu^2 d\mathcal{O}\le C\,,$$

which is valid for all $\tilde{u}_m^\epsilon$; passing to the limit, we obtain

$$\left\|\frac{\partial \tilde{u}_m}{\partial t}\right\|_{\gamma,\mu}\le C\,,$$

which holds for all m. We determine the estimate in ϑ analogously. □

In the following result, we establish the existence of the solution to (8)–(9) as well as its approximation by the regularization (16)–(19).

Proposition 2. *Existence. We suppose (3)–(7), (11), (20), $0<\gamma<c\,/\,\overline{b}$ and $f\in L^2_\gamma(\mathcal{O}_{t,\vartheta};H_\mu)$; then, there exists a $u\in W_{\gamma,\mu}$ satisfying (8)–(9). Moreover, the sequence $\{\tilde{u}_m\}$ converges such that $\tilde{u}_m\to u$ in $\mathcal{X}_{\gamma,\mu}$ and*

$$\max_{t\in\overline{\mathcal{O}}}\left\{\int_t^T|\sqrt{b(\tau,\vartheta)}(\tilde{u}_m(\tau,\vartheta)-u(\tau,\vartheta)|_\mu^2\,d\tau+\int_\vartheta^\Theta|\tilde{u}_m(t,\theta)-u(t,\theta)|_\mu^2\,d\theta\right\}\to 0\,,$$

as $m\to\infty$, where $\mathcal{O}=(0,T)\times(0,\Theta)$, for any (fixed) $\Theta>0$.

Proof of Proposition 2. From the estimates (21) and (22), it follows that, possibly after extracting a subsequence, $\tilde{u}_m\rightharpoonup u$ in $\mathcal{X}_{\gamma,\mu}$, $\partial\tilde{u}_m/\partial t\rightharpoonup\partial u/\partial t$ in $L^2_\gamma(\mathcal{O},H_\mu))$ and $\partial\tilde{u}_m/\partial\vartheta\rightharpoonup\partial u/\partial\vartheta$ in $L^2_\gamma(\mathcal{O},H_\mu))$, where u satisfies (8)–(9).

In order to show convergence of the regularizations $\tilde{u}_m$, we have from (8) that

$$-\frac{1}{2}\frac{\partial}{\partial t}|u(t)-\tilde{u}(t)|_\mu^2-\frac{1}{2}\frac{\partial}{\partial\vartheta}\left|\sqrt{b(t)}\,(u(t)-\tilde{u}(t))\right|_\mu^2$$

$$+a_\mu(t;u(t)-\tilde{u}(t),u(t)-\tilde{u}(t))=(f(t)|u(t)-\tilde{u}(t))_\mu\,.$$

Multiplying the above by n_γ^2 and applying the Green's formula over $\mathcal{O}$, we obtain

$$\int_t^T|\sqrt{b(\tau,\vartheta)}\,u_n(\tau,\vartheta)-u(\tau,\vartheta)|_\mu^2\,d\tau+\int_\vartheta^\Theta|u_n(t,\theta)-u(t,\theta)|_\mu^2\,d\theta$$

$$+2\int_{\mathcal{O}}a_\mu(t;u(t)-\tilde{u}(t),u(t)-\tilde{u}(t))\,d\mathcal{O}=2\int_{\mathcal{O}}(f(t)|u(t)-\tilde{u}(t))_\mu\,d\mathcal{O}\,,$$

and the result follows from (11) and $\tilde{u}_m\rightharpoonup u$ in $\mathcal{X}_{\gamma,\mu}$. □

Let $\{w_1, w_2, \ldots\}$ denote a basis in V_m. We set

$$u_{m,n}(t,\vartheta,x) = \sum_{k=1}^{n} c_{kn}(t,\vartheta)\, w_k(x)\,, \tag{28}$$

$$v_n(\vartheta,x) = \sum_{k=1}^{n} \alpha_{kn}(\vartheta) w_k(x)\,, \tag{29}$$

where $c_{kn}(t,\vartheta) \in L_2((0,T)\times(0,m))$ and $\alpha_{kn}(\vartheta) \in L_2(0,m)$, such that $\alpha_{kn}(m) = 0$, $v_n \in L_2(0,m;V_m)$ and

$$v_{m,n} \to v \text{ in } L_2(0,m;H_m)\,, \tag{30}$$

as $n \to \infty$. The *Galerkin* equations associated with (16)-(19) are defined

$$-\sum_{k=1}^{n} \frac{\partial}{\partial t} c_{kn}(\boldsymbol{t})(w_k|w_j)_{H_m} - \sum_{k=1}^{n} \frac{\partial}{\partial \vartheta} c_{kn}(\boldsymbol{t})(\sqrt{b(\boldsymbol{t})}\, w_k|\sqrt{b(\boldsymbol{t})}\, w_j)_{H_m} \tag{31}$$

$$+\sum_{k=1}^{n} c_{kn}(\boldsymbol{t})\, a(\boldsymbol{t}; w_k|w_j) = \langle f(\boldsymbol{t}), w_j\rangle_{V_m} \text{ on } (0,T)\times(0,m)$$

for $j = 1,\ldots,n$, such that

$$c_{kn}(T,\vartheta) = \alpha_{kn}(\vartheta) \;\; a.e. \text{ on } (0,m), \tag{32}$$

$$c_{kn}(t,m) = 0 \;\; a.e. \text{ on } (0,T), \tag{33}$$

for $k = 1,\ldots,n$, where

$$a(\boldsymbol{t};u,v) = \int_{(0,X)} a_2 \frac{\partial u}{\partial x}\frac{\partial v}{\partial x}\, dx + \int_{(0,X)} a_1 \frac{\partial u}{\partial x}\, v\, dx + \int_{(0,X)} a_0\, u\, v\, dx\,,$$

for all $u, v \in V$ and $\boldsymbol{t} \in (0,T)\times(0,m)$, $(u|v)_{H_m}$ is the inner product on the Hilbert space H_m, and $\langle f, v\rangle_{V_m}$ is the value of the linear functional $f \in V_m^*$ at $v \in V_m$.

We immediately obtain the constructive approximation of (8)–(9)) by the Galerkin procedure (28)–(33) from ([5], Propositions 4 and 5) and Proposition 2.

Proposition 3. *Galerkin Approximation. We suppose (3)–(7), (11), (20), $0 < \gamma < c/2\bar{b}$, and $f \in L^2_\gamma(\mathcal{O}_{t,\vartheta}; H_\mu)$. Let $u_{m,m}$ be the mth-Galerkin approximation to u_m defined by (28)–(33)* and *u the solution to (8)–(9), then $u_{m,m} \to u$ in $\mathcal{X}$ and*

$$\max_{\boldsymbol{t}\in\overline{\mathcal{O}}}\left\{\int_t^T |\sqrt{b(\tau,\vartheta)}\, u_{m,m}(\tau,\vartheta) - u(\tau,\vartheta)|_\mu^2\, d\tau + \int_\vartheta^\Theta |u_{m,m}(t,\theta) - u(t,\theta)|_\mu^2\, d\theta\right\} \to 0$$

as $m \to \infty$, where $\mathcal{O} = (0,T)\times(0,\Theta)$, for any (fixed) $\Theta > 0$.

Remark 3. *Propositions 2 and 3 likewise hold with $f \in \mathcal{X}^*_{\gamma,\mu}$, where we imply the Galerkin approximation per the proof of Lemma 3.*

3. Probabilistic Interpretation

In order to provide a probabilistic interpretation of the solution to (1)–(2), we make the additional assumptions that

$$a_2\,,\; a_1\,,\; a_0\,,\; b \in C^1\left(\overline{\tilde{\mathcal{Q}}}\right);\; a_2 \text{ bounded}, \tag{34}$$

$$\frac{\partial^2 a_2}{\partial x^2}\,,\; \frac{\partial^2 a_2}{\partial x \partial \vartheta}\,,\; \frac{\partial^2 a_2}{\partial \vartheta^2} \in C^0\left(\overline{\tilde{\mathcal{Q}}}\right),\; \text{ bounded}, \tag{35}$$

$$\frac{\partial a_0}{\partial x}, \frac{\partial a_1}{\partial x}, \frac{\partial a_2}{\partial x}, \frac{\partial b}{\partial x} \quad \text{bounded}, \tag{36}$$

$$\frac{\partial a_0}{\partial \vartheta}, \frac{\partial a_1}{\partial \vartheta}, \frac{\partial a_2}{\partial \vartheta}, \frac{\partial b}{\partial \vartheta} \quad \text{bounded}, \tag{37}$$

$$\frac{\partial^2 a_2}{\partial t \partial x}, \frac{\partial^2 a_2}{\partial t \partial \vartheta}, \frac{\partial a_2}{\partial x^2} \in L^p_{loc}(\tilde{\mathcal{Q}}), \tag{38}$$

$$f, \frac{\partial f}{\partial x} \in C^0\left(\overline{\tilde{\mathcal{Q}}}\right), \tag{39}$$

$$|f(t,\vartheta,x)| \le \left[1 + (\vartheta^2 + x^2)^{m/2}\right], \tag{40}$$

$$\left|\frac{\partial f}{\partial x}\right| \le C\left[1 + (\vartheta^2 + x^2)^{m/2}\right], \tag{41}$$

$$\frac{\partial f}{\partial t} \in L^p_{loc}(\tilde{\mathcal{Q}}). \tag{42}$$

We likewise suppose the existence of a function

$$\Psi \in C^{2,1}\left(\tilde{\mathcal{Q}}\right) \cap C^0\left(\overline{\tilde{\mathcal{Q}}}\right), \tag{43}$$

$$\left|\frac{\partial \Psi}{\partial x}\right| \le C\left[1 + (\vartheta^2 + x^2)^{m/2}\right], \tag{44}$$

such that

$$v(\vartheta, x) = \Psi(T, \vartheta, x) \tag{45}$$

and

$$g = -\frac{\partial \Psi}{\partial t} - \frac{\partial (b\Psi)}{\partial \vartheta} + A(t,\vartheta)\,\Psi \tag{46}$$

satisfies the same assumptions as f. From Proposition 2 and Appendix A, we allow that there exists a unique solution $u \in C^{2,1}(\tilde{\mathcal{Q}}) \cap C^0\left(\overline{\tilde{\mathcal{Q}}}\right)$ to the problem (1)–(2).

We let $\sigma = \sqrt{2a_2}$ (or $a_2 = \sigma^2/2$) and define

$$a(t,\vartheta,x) = \frac{\partial a_2}{\partial x} - a_1 \tag{47}$$

and

$$\alpha_0 = a_0 - \frac{\partial b}{\partial \vartheta}, \tag{48}$$

in which case $\alpha_0 > 0$ for β sufficiently large (cf. (7)). In particular, σ, a and b are elements of $C^1\left(\overline{\tilde{\mathcal{Q}}}\right)$. Moreover, by extending the functions a, b, and σ outside of $\overline{\mathcal{Q}}$, we may assume that

$$|\sigma(t,\vartheta_2,x_2) - \sigma(t,\vartheta_1,x_1)| + |a(t,\vartheta_2,x_2) - a(t,\vartheta_2,x_1)|$$
$$+|b(t,\vartheta_2,x_2) - b(t,\vartheta_1,x_1)| \le K\left(|\vartheta_2 - \vartheta_1|^2 + |x_2 - x_1|^2\right)^{1/2} \tag{49}$$

as well as

$$|\sigma| \le K_o \ \text{ and } \ |a(t,\vartheta,x)|^2 + |b(t,\vartheta,x)|^2 \le K_1^2(1 + |\vartheta|^2 + |x|^2), \tag{50}$$

for all $\boldsymbol{t} \in \overline{\tilde{\mathcal{O}}_{T,\vartheta}}$.

We now seek a probabilistic interpretation of the function u satisfying (1)–(2) by constructing a stochastic differential equation for which the trajectories $(\Theta(t), X(t))$ are the characteristics of $-\partial(b\ \cdot)/\partial\vartheta + A$. To this end, we take a probability space $(\Omega, \mathcal{A}, P)$, an increasing family of sub-σ-algebras $\mathcal{F}_t$ of $\mathcal{A}$, and a $\mathbb{R}$-valued standardized Wiener process $w(t)$, which is an $\mathcal{F}_t$ martingale. We can then consider, on an arbitrary finite interval, the stochastic differential equation

$$dX(t) = a(t,\Theta(t),X(t))\,dt + \sigma(t,y)\,dw(t)\,, \tag{51}$$

$$d\Theta(t) = b(t,\Theta(t),X(t))\,dt, \tag{52}$$

$$X(0) = x \in \mathbb{R}\,, \tag{53}$$

$$\Theta(0) = \vartheta \in \mathbb{R}_+\,, \tag{54}$$

where x and ϑ are fixed and non-random; the solution of (51)–(54) is unique.

Proposition 4. *The assumptions of Proposition 2, as well as (34) through (46); the solution of (1)–(2) is given by*

$$u(t,\vartheta,x) = \mathbb{E}\left\{\int_t^T f(s,\Theta(s),X(s))\exp\left[-\int_t^s \alpha_0(\varsigma,\Theta(\varsigma),X(\varsigma))\,d\varsigma\right] ds\right\} \tag{55}$$
$$+\mathbb{E}\left\{v(\Theta(T),X(T))\exp\left[-\int_t^T \alpha_0(s,\Theta(s),X(s))\,ds\right]\right\}.$$

Proof of Proposition 4. The proof relies on the existence and uniqueness of the regular solution to the ultraparabolic terminal value problem (1)–(2). With this exception, the result is standard such that we will provide only a brief exposition, deferring to e.g., ([12], Chapter 2, Theorem 7.4). For the process $(\Theta(t), X(t))$, we have that

$$\mathbb{E}\left[\left(|\Theta(s)|^2 + |X(s)|^2\right)^k\right] \le C\left[1 + \left(|\vartheta|^2 + |x|^2\right)^k\right], \tag{56}$$

for all $s \in [t,T]$ and all $k \in \mathbb{N}$, in which case the right-hand side of (55) is well-defined.

We shall now prove (55) in the case $v = 0$. We set

$$Z(s) = \exp\left\{-\int_t^s \left[a_0(\varsigma,\Theta(\varsigma),X(\varsigma)) - \frac{\partial b}{\partial\vartheta}\right] d\varsigma\right\}. \tag{57}$$

Then, $Z(s)$ satisfies

$$\frac{dZ(s)}{ds} = \frac{\partial b}{\partial \vartheta} - a_0(s,\Theta(s),X(s))\,Z(s)\,,\ Z(t) = 1\,. \tag{58}$$

Differentiating the functional $\Psi \cdot Z$, applying Ito's formula to Ψ, and integrating from t to T, we obtain

$$\Psi(T,\Theta(T),X(T)) = \Psi(t,\vartheta,x)$$
$$+\int_t^T \left[\frac{\partial\Psi}{\partial s} + \frac{\partial(b\Psi)}{\partial\vartheta} - A(s,\vartheta)\,\Psi\right](s,\Theta(s),X(s))\,Z(s)\,ds$$
$$+\int_t^T Z(s)\left[\frac{\partial\Psi}{\partial x}\sigma\right](s,\Theta(s),X(s))\,dw\,.$$

From (56) with $k = m$ and the assumptions (44) on the growth of $\partial\Psi/\partial x$, we have that the expectation of the stochastic integral is defined and is equal to zero. We therefore have that

$$\mathbb{E}\left\{v(\Theta(T),X(T))\,Z(T)\right\} = \Psi(t,\vartheta,x) - \mathbb{E}\left\{\int_t^T g(s,\Theta(s),X(s))\,Z(s)\right\},$$

in which case (55) is identical to

$$u(\boldsymbol{t},x) - \Psi(t,\vartheta,x) = \mathbb{E}\left\{ \int_t^T [f(s,\Theta(s),X(s)) - g(s,\Theta(s),X(s))]\, Z(s) \right\}$$

and so the problem reduces to proving (55) with $v = 0$, with f replaced by $f - g$, and with u replaced by $u - \Psi$ and a solution to (1)–(2) corresponding to data $f - g$ and 0.

We therefore assume $v = 0$; we prove that

$$u(t,\vartheta,x) = \mathbb{E}\left\{ \int_t^T f(s,\Theta(s),X(s))\, Z(s)\, ds \right\} . \tag{59}$$

We start by considering the bounded case. We approximate f by f_{NM} defined by

$$f_{NM} = \begin{cases} N & \text{if} \quad N \le f, \\ f & \text{if} \quad -M \le f \le N, \\ -M & \text{if} \quad f \le -M. \end{cases}$$

Since $f_{NM} \in C^0\left(\overline{\widetilde{\mathcal{O}}}\right)$ and $\partial f_{NM}/\partial x, \partial f_{NM}/\partial t, \partial f_{NM}/\partial\vartheta \in L^p_{\text{loc}}(\widetilde{\mathcal{O}})$, we can uniquely define u_{NM} as the solution of

$$u_{NM} \in L^2(\widetilde{\mathcal{O}}_{T,\vartheta}; H^1_\mu) \cap C^0\left(\overline{\widetilde{\mathcal{Q}}}\right) \cap C^{2,1}(\mathcal{Q})$$

such that

$$-\frac{\partial u_{NM}}{\partial t} - \frac{\partial b\, u_{NM}}{\partial \vartheta} + A(\boldsymbol{t})u_{NM} = f_{NM} \tag{60}$$

and

$$u_{NM}(T,\vartheta,x) = 0\,. \tag{61}$$

We note that u_{MN} is bounded. This follows as per ([5], Prop. 2').

We show:

$$u_{NM}(t,\vartheta,x) = \mathbb{E}\left\{ \int_t^T f_{NM}(s,\Theta(s),X(s))\, Z(s)\, ds \right\} . \tag{62}$$

To this end, let $\mathcal{O}_R = \{\xi \in \mathbb{R}^2 \mid |\xi| \le R\}$ and τ_R be the exit time form $\mathcal{O}_R$ of the process $(\Theta(t), X(t))$. We can suppose that the (fixed) initial data (x,ϑ) of (51)–(54) belongs to $\mathcal{O}_R$, for R that is sufficiently large. That is, we have, from the continuity of the process a.s. $\tau_R \ge T$ for some $R_0(\omega)$ with $R \ge R_0(\omega)$, in which case

$$a.s. \;\; \tau_R \wedge T = T\,, \tag{63}$$

for $R \ge R_0(\omega)$. As above, with the use of Ito's formula applied to u_{NM} between the instants t and $\tau_R \wedge T - \epsilon$, taking $\epsilon \to 0$, and using the continuity of u_{NM} on $\overline{\mathcal{Q}}$, we have

$$u_{NM}(t,\vartheta,x) = \mathbb{E}\left\{ \int_t^{\tau_R \wedge T} f_{NM}(s,\Theta(s),X(s))\, Z(s)\, ds \right\} \tag{64}$$
$$+\mathbb{E}\left\{ u_{NM}(\tau_R,\Theta(\tau_R),X(\tau_R))\, Z(\tau_R)\, \chi_T(\tau_R) \right\} ,$$

where

$$\chi_T(t) = \begin{cases} 1, & \text{if} \quad t < T, \\ 0, & \text{if} \quad t \ge T\,. \end{cases}$$

However, from (63), we have

$$u_{NM}(\tau_R,\Theta(\tau_R),X(\tau_R))\, z(\tau_R)\, \chi_T(\tau_R) = 0\,,$$

for $R \geq R_0(\omega)$, and so

$$\mathbb{E}\left\{u_{NM}(\tau_R, \Theta(\tau_R), X(\tau_R))\, Z(\tau_R)\, \chi_T(\tau_R)\right\} \to 0 \text{ a.s.},$$

as $R \to \infty$. Application of Lesbesque's theorem then provides the result (62).

From the estimates

$$|u_{NM}| \leq C\left[1 + \left(\vartheta^2 + x^2\right)^{m/2}\right]$$

and

$$\left|\frac{\partial u_{NM}}{\partial x}\right| \leq C\left[1 + \left(\vartheta^2 + x^2\right)^{m/2}\right],$$

it follows that u_{NM} lies in a bounded subset of $L^2(\widetilde{\mathcal{O}}_{T,\vartheta}; H^1_\mu)$ and we obtain (59) by proceeding to the limit successively in M and N. □

4. Conclusions

We have demonstrated the existence and uniqueness of the solution to linear ultraparabolic equations on unbounded domains, both spatial and temporal, as well as the strong convergence of the regularized problem, providing a basis for the subsequent application of a Galerkin approximation. Furthermore, we present a probabilistic interpretation of the solution in terms of the expectation of an associative ultradiffusion process. In practice, the usefulness of this result often stems from the converse formulation; that is, one often wishes to obtain the discounted expectation associated with an ultradiffusion process, e.g., the valuation of an Asian option in mathematical finance (cf. [11]). To this end, the regularity assumptions of Section 3 are necessary for the existence of the solution to the ultradiffusion process (3.5). With respect to a simple regular transformation, the associated ultraparabolic problem maintains the approximation solvability of Section 2, for which efficient and general numerical procedures are readily available (cf. [13–15]).

Appendix A. Regularity

There exist two approaches to obtaining regularity; for parabolic differential equations of the second order, we note that Wloka [16] has provided regularity theorems based on raising the differentiability assumptions of the data while Ladyženskaja et al. [17] have taken the approach of increasing the p-power summability of the data. The results stated below follow the latter approach, the demonstration of which lies outside the scope of this manuscript. As regularity theorems are local, they do not require assumptions on the boundary or the boundedness of the domain .

We let $\mathcal{W}^{1,2,p}(\widetilde{\mathcal{Q}})$ denote the space of functions u such that

$$u, \frac{\partial u}{\partial t}, \frac{\partial u}{\partial \vartheta}, \frac{\partial u}{\partial x}, \frac{\partial^2 u}{\partial x^2} \in L^p(\widetilde{\mathcal{Q}}),$$

for $1 \leq p \leq \infty$. Here, the "1" refers to the order of temporal derivatives, and "2" refers to the number of spatial derivatives. If $p = 2$, we write $\mathcal{W}^{1,2}(\widetilde{\mathcal{Q}})$, which we equip with the natural Banach- and Hilbert-space norm. We denote by $\mathcal{W}^{1,2,p}_{\text{loc}}(\widetilde{\mathcal{Q}})$ the space of functions u such that, for all test functions $\varphi \in \mathcal{D}(\widetilde{\mathcal{Q}})$, the set of infinitely differentiable functions with compact support in $\widetilde{\mathcal{Q}}$, we have $\varphi\, u \in \mathcal{W}^{1,2,p}_{\text{loc}}(\widetilde{\mathcal{Q}})$.

We suppose that

$$a_2, a_1, a_0, b \in C^1(\widetilde{\mathcal{Q}}). \tag{65}$$

Moreover, for $v \in L^p_{\text{loc}}(\widetilde{\mathcal{Q}})$, we denote by Lv the following distribution on $\widetilde{\mathcal{Q}}$:

$$\langle Lv, \psi \rangle = \int_{\widetilde{\mathcal{Q}}} v \left[\frac{\partial \psi}{\partial t} + \frac{\partial (b\,\psi)}{\partial \vartheta} - \frac{\partial}{\partial x}\left(a_2 \frac{\partial \psi}{\partial x} \right) - \frac{\partial}{\partial x}(a_1 \psi) + a_0 \psi \right] d\widetilde{\mathcal{Q}},$$

for all $\psi \in \mathcal{D}(\widetilde{\mathcal{Q}})$.

Proposition A1. *Local Regularity. The assumptions of Proposition 2, as well as (65). Let $u \in L^p_{\text{loc}}(\widetilde{\mathcal{Q}})$ be such that*

$$Lu = -\frac{\partial u}{\partial t} - \frac{\partial (b\,u)}{\partial \vartheta} + A(t,\vartheta)\,u = f \in L^p_{\text{loc}}(\widetilde{\mathcal{Q}}),$$

then $u \in \mathcal{W}^{1,2,p}_{\text{loc}}(\widetilde{\mathcal{Q}})$, for $p > 1$.

Proof of Proposition A1. The case for $p = 2$ follows with a slight modification from ([12], Chapter 2, Theorem 5.5). □

In order to obtain results on the boundary, we set

$$\tilde{f} = \begin{cases} f, & \text{on } (0,T), \\ 0, & \text{on } (-T,0) \text{ and } (T,2T), \end{cases}$$

and extend the operator $A(t,\vartheta)$ in such a way that it is defined over $(-T,2T)$, all the while retaining the properties of the coefficients. Finally, we consider the solution of $\tilde{u}$ of

$$-\frac{\partial \tilde{u}}{\partial t} - \frac{\partial (b\tilde{u})}{\partial \vartheta} + A(t,\vartheta)\,\tilde{u} = \tilde{f} \;\; a.e. \text{ on } \widetilde{\mathcal{Q}} \tag{66}$$

subject to the terminal condition

$$\tilde{u}(2T,\vartheta,x) = 0 \;\; a.e. \text{ on } \widetilde{\mathcal{O}}_{\vartheta,x}\,. \tag{67}$$

Corollary A1. *The assumptions of Proposition A1; we have*

$$u \in C^0\left(\overline{\widetilde{\mathcal{Q}}}\right),$$

for $p > 2$.

Proof of Corollary A1. From Proposition A1 applied to $\tilde{u}$ in (66)–(67), we obtain $\tilde{u} \in \mathcal{W}^{1,2,p}_{\text{loc}}(\widetilde{\mathcal{Q}})$ from which we derive the result. □

Increased smoothness of the data may then be translated into smoothness of the solution.

Proposition A2. *The assumptions of Proposition A1, as well as*

$$\frac{\partial^2 a_2}{\partial t \partial x}, \frac{\partial^2 a_2}{\partial \vartheta \partial x}, \frac{\partial^2 a_2}{\partial x^2} \in L^p_{\text{loc}}(\widetilde{\mathcal{Q}}), \tag{68}$$

$$f, \frac{\partial f}{\partial t}, \frac{\partial f}{\partial \vartheta}, \frac{\partial f}{\partial x} \in L^p_{\text{loc}}(\widetilde{\mathcal{Q}}), \tag{69}$$

$$u \in L^p_{\text{loc}}(\widetilde{\mathcal{Q}}),$$

and $Lu = f$, then

$$u \in \mathcal{W}^{1,3,p}_{\text{loc}}(\widetilde{\mathcal{Q}})$$

and

$$\frac{\partial u}{\partial t}, \frac{\partial u}{\partial \vartheta} \in \mathcal{W}^{1,2,p}_{\text{loc}}(\widetilde{\mathcal{Q}}).$$

In particular, $u \in C^{1,2}(\widetilde{\mathcal{Q}})$.

Proof of Proposition A2. We consider the differential quotients technique per Section 4 of [5] such that, if $f \in \mathcal{W}^{1,1,p}_{\text{loc}}(\widetilde{\mathcal{Q}})$, then $u \in \mathcal{W}^{1,3,p}_{\text{loc}}(\widetilde{\mathcal{Q}})$. □

Finally, the key regularity result is:

Proposition A3. *The assumptions of Proposition A2. If* $f \in L^p(\widetilde{\mathcal{Q}})$ *and* $v = 0$*, then the unique solution to (1)–(2) also satisfies* $u \in L^p_{\text{loc}}(\widetilde{\mathcal{Q}})$.

Appendix B. Localization

In order to highlight the temporal nature of the variable ϑ in the probabilistic framework, we examine formally the localization of the ultradiffusion process (51)–(54) and its relation to (1)–(2). To this end, we freeze the drift and volitility in a neighborhood of $t = 0$ and consider the ultradiffusion process

$$dX(t) = a\,dt + \sigma\,dw(t)\,, \tag{70}$$

$$d\Theta(t) = b\,dt\,, \tag{71}$$

$$X(0) = x \in \mathbb{R}\,, \tag{72}$$

$$\Theta(0) = \vartheta \in \mathbb{R}_+\,, \tag{73}$$

where x and ϑ are again fixed and non-random. It follows then that the solution to ultraparabolic infinite horizon/ terminal boundary value problem

$$-\frac{\partial u}{\partial t} - \frac{\partial (bu)}{\partial \vartheta} + A\,u = 0 \;\; a.e. \text{ on } \widetilde{\mathcal{Q}}, \tag{74}$$

$$u(T, \vartheta, x) = v(\vartheta, x) \;\; a.e. \text{ on } \widetilde{\mathcal{O}}_{\vartheta,x} \tag{75}$$

may be characterized as

$$u(t, \vartheta, x) = \mathbb{E}\left\{v(\Theta(T), X(T)) \exp\left[-a_0(T-t)\right]\right\}\,. \tag{76}$$

In particular, we consider the temporal characteristic transformation $t = t(\tau)$ and $\vartheta = \vartheta(\tau)$ such that

$$\frac{d\,t(\tau)}{d\,\tau} = 1\,;\;\; t(0) = 0\,, \tag{77}$$

$$\frac{d\,\vartheta(\tau)}{d\,\tau} = b\,;\;\; \vartheta(0) = \vartheta_0\,, \tag{78}$$

where $\vartheta_0 > 0$, in which case $t(\tau) = \tau$ and $\vartheta(\tau) = b\,\tau + \vartheta_0$ or, more simply, $t = \tau$ and $\vartheta(t) = b\,t + \vartheta_0$. Note then that, along the characteristic line $(t(\tau), \vartheta(\tau))$, the Formulations (70)–(76) may be restated in terms of the characteristic parameterized diffusion

$$d\mathcal{X}(\tau; \vartheta_0) = a\,d\tau + \sigma\,d\varpi(\tau)\,, \tag{79}$$

$$\mathcal{X}(0; \vartheta_0) = x \in \mathbb{R}\,, \tag{80}$$

such that the solution $v(\tau, x; \vartheta_0) = u(t(\tau), \vartheta(\tau), x)$ to the characteristic parameterized parabolic terminal value problem

$$-\frac{\partial v}{\partial \tau} + A\, v = 0 \;\; a.e. \text{ on } \;(0,T) \times \mathbb{R}, \tag{81}$$

$$v(T,x;\vartheta_0) = v(\vartheta_0,x) \;\; a.e. \text{ on } \;\widetilde{\mathcal{O}}_{\vartheta,x}, \tag{82}$$

satisfies

$$v(\tau,x) = \mathbb{E}' \left\{ v(\mathcal{X}(T)) \exp\left[-a_0(T-\tau)\right] \right\} . \tag{83}$$

The temporal nature of ϑ then follows from the characteristic problem (77)–(83). Due to (77)–(78), the vector $(1,b)$ (resp. b) is known as the *velocity* (resp. *speed*) of propagation.

References

1. Piskunov, N.S. Problémes Limits pour les Équations du Type Elliptic–Parabolique. *Mat. Sb.* **1940**, *7*, 385–424.
2. Lions, J.L. Sur Certaines Équations aux Dérivativées Partielles, à Coefficients Opérateurs non Bornés. *J. Anal. Math.* **1958**, *6*, 333–355. [CrossRef]
3. Il'in, A.M. On a Class of Ultraparabolic Equations. *Soviet Math. Dokl.* **1964**, *5*, 1673–1676.
4. Vladimirov, V.S.; Drožžinov, J.N. Generalized Cauchy Problem for an Ultraparabolic Equation. *Math. USSR Izv.* **1967**, *1*, 1285–1303.
5. Marcozzi, M.D. Well-Posedness of Linear Ultraparabolic Equations on Bounded Domains. *J. Evol. Equ.* **2018**, *18*, 75–104. [CrossRef]
6. Kolmogorov, A.N. Zur Theorie der Stetigen Zufalligen Progresse. *Math. Ann.* **1933**, *108*, 149–160. [CrossRef]
7. Kolmogorov, A.N. Zufällige Bewegungen. *Ann. Math.* **1934**, *35*, 116–117.
8. Uhlenbeck, G.E.; Ornstein, L.S. On the Theory of the Brownian Motion. *Phys. Rev.* **1930**, *36*, 823–841. [CrossRef]
9. Chandrasekhar, S. Stochastic Problems in Physics and Astronomy. In *Selected Papers on Noise and Stochastic Processes*; Wax, N., Ed.; Dover: New York, NY, USA, 2013; Reprint.
10. Marshak, R.E. Theory of the Slowing Down of Neutrons by Elastic Collisions with Atomic Nuclei. *Rev. Mod. Phys.* **1947**, *19*, 185–238. [CrossRef]
11. Hull, J.C. *Options, Futures, and Other Derivatives*, 10th ed.; Pearson: New York, NY, USA, 2017.
12. Bensoussan, A.; Lions, J.L. *Applications of Variational Inequalities in Stochastic Control*; North Holland: Amsterdam, The Netherlands, 1982.
13. Akrivis, G.M.; Crouzeix, M.; Thomee, V. Numerical methods for ultraparabolic equations. *Calcolo* **1996**, *31*, 179–190. [CrossRef]
14. Marcozzi, M.D. On the valuation of Asian options by variational methods. *SIAM J. Sci. Comput.* **2003**, *24*, 1124–1140. [CrossRef]
15. Marcozzi, M.D. Extrapolation discontinuous Galerkin method for ultraparabolic equations. *J. Comput. Appl. Math.* **2009**, *224*, 679–687. [CrossRef]
16. Wloka, J. *Partial Differential Equations*; Cambridge University Press: Cambridge, UK, 1987.
17. Ladyženskaja, O.A.; Solonnikov, V.A.; Ural'ceva, N.N. *Linear and Quasilinear Equations of Parabolic Type*; American Mathematical Society: Providence, RI, USA, 1967; Volume 23 of Translations of Mathematical Monographs.

5

Evaluation of Fractional Integrals and Derivatives of Elementary Functions

Roberto Garrappa [1,2,*], Eva Kaslik [3] and Marina Popolizio [2,4]

1 Department of Mathematics, University of Bari, Via E. Orabona 4, 70126 Bari, Italy
2 Member of the INdAM Research Group GNCS, Istituto Nazionale di Alta Matematica "Francesco Severi", Piazzale Aldo Moro 5, 00185 Rome, Italy; marina.popolizio@poliba.it
3 Department of Mathematics and Computer Science, West University of Timisoara, Bd. V. Parvan 4, 300223 Timisoara, Romania; eva.kaslik@e-uvt.ro
4 Department of Electrical and Information Engineering, Polytechnic University of Bari, Via E. Orabona 4, 70126 Bari, Italy
* Correspondence: roberto.garrappa@uniba.it

Abstract: Several fractional-order operators are available and an in-depth knowledge of the selected operator is necessary for the evaluation of fractional integrals and derivatives of even simple functions. In this paper, we reviewed some of the most commonly used operators and illustrated two approaches to generalize integer-order derivatives to fractional order; the aim was to provide a tool for a full understanding of the specific features of each fractional derivative and to better highlight their differences. We hence provided a guide to the evaluation of fractional integrals and derivatives of some elementary functions and studied the action of different derivatives on the same function. In particular, we observed how Riemann–Liouville and Caputo's derivatives converge, on long times, to the Grünwald–Letnikov derivative which appears as an ideal generalization of standard integer-order derivatives although not always useful for practical applications.

Keywords: fractional derivative; fractional integral; Mittag–Leffler function; Riemann–Liouville derivative; Caputo derivative; Grünwald–Letnikov derivative

1. Introduction

Fractional calculus, the branch of calculus devoted to the study of integrals and derivatives of non integer order, is nowadays extremely popular due to a large extent of its applications to real-life problems (see, for instance, [1–8]).

Although this subject is as old as the more classic integer-order calculus, its development and diffusion mainly started to take place no more than 20 or 30 years ago. As a consequence, several important results in fractional calculus are still not completely known or understood by non-specialists, and this topic is usually not taught in undergraduate courses.

The presence of more than one type of fractional derivative is sometimes a source of confusion and it is not occasional to find wrong or not completely rigorous results in distinguished journals as well. Even the simple evaluation of a fractional integral or derivative of elementary functions is in some cases not reported in a correct way, which is also due to the difficulty of properly handling the different operators.

For instance, in regards to the exponential, the sine and the cosine functions, the usual and well-known relationships:

$$\frac{\mathrm{d}^n}{\mathrm{d}t^n}\mathrm{e}^{t\Omega} = \Omega^n \mathrm{e}^{t\Omega}, \qquad \frac{\mathrm{d}^n}{\mathrm{d}t^n}\sin t\Omega = \Omega^n \sin\Big(t\Omega + \frac{n\pi}{2}\Big), \qquad \frac{\mathrm{d}^n}{\mathrm{d}t^n}\cos t\Omega = \Omega^n \cos\Big(t\Omega + \frac{n\pi}{2}\Big), \quad (1)$$

which hold for any $n \in \mathbb{N}$ and turn out extremely useful for simplifying a lot of mathematical derivations, are in general no longer true with fractional derivatives, unless a very special definition is used, which presents some not secondary drawbacks.

The main aim of this paper is to provide a tutorial for the evaluation of fractional integrals and derivatives of some elementary functions and to show the main differences resulting from the action of different types of fractional derivatives. At the same time, we present an alternative perspective for the derivation of some of the most commonly used fractional derivatives in order to help the reader to better interpret the results obtained from their application.

In particular, the more widely used definitions of fractional derivatives, namely those known as Grünwald–Letnikov, Riemann–Liouville and Caputo, are introduced according to two approaches: One based on the inversion of the generalization of the integer-order integral and the other based on the more direct generalization of the limit of the difference quotients defining integer-order derivatives. Although they lead to equivalent results, the second and less usual approach allows for a more comprehensive understanding of the nature of the different operators and a better explanation of the effects produced on elementary functions. In particular we will observe when relationships similar to Equation (1) apply to fractional derivatives and the way in which fractional derivatives deviate from Equation (1).

Some of the material presented in this paper is clearly not new (proper references will be given through the paper). Nevertheless, we think that it is important to collect in a single paper a series of results which are scattered among several references or are not clearly exposed, thus to provide a systematic treatment and a guide for researchers approaching fractional calculus for the first time.

The paper is organized as follows: In Section 2 we recall the fractional Riemann–Liouville integral and some definitions of fractional derivatives relying on its inversion. We hence present in Section 3 a different view of the same definitions by showing, in a step-by-step way, how they can be obtained as a generalization of the limit of different quotients defining standard integer-order derivatives after operating a replacement of the function to cope with convergence difficulties. Since the Mittag–Leffler (ML) function plays an important role in fractional calculus, and indeed most of the results on derivatives of elementary functions will be based on this function, Section 4 is devoted to present this function and some of its main properties; in particular, we provide a useful result on the asymptotic behavior of the ML function which allows to investigate the relationships between the action of the different fractional derivatives on the same function. Sections 5–7 are devoted to presenting the evaluation of derivatives of some elementary functions (power, exponential and sine and cosine functions), to study their properties and to highlight the different effects of the various operators. Clearly the results on the few elementary functions considered in this paper may be adopted as a guide to extend the investigation to further and more involved functions. Some concluding remarks are finally presented in Section 8.

2. Fractional Derivatives as Inverses of the Fractional Riemann–Liouville Integral

To simplify the reading of this paper we recall in this Section the most common definitions in fractional calculus and review some of their properties. For a more comprehensive introduction to fractional integrals and fractional derivatives we refer the reader to any of the available textbooks [3,5,9–12] or review papers [13,14]. In particular, we follow here the approach based on the generalization, to any real positive order, of standard integer-order integrals and on the introduction of fractional derivatives as their inverse operators. We therefore start by recalling the well-known definition of the fractional Riemann–Liouville (RL) integral.

Definition 1. *For a function $f \in L^1([t_0, T])$ the RL integral of order $\alpha > 0$ and origin t_0 is defined as:*

$$J_{t_0}^{\alpha} f(t) = \frac{1}{\Gamma(\alpha)} \int_{t_0}^{t} (t-\tau)^{\alpha-1} f(\tau) \mathrm{d}\tau, \quad \forall t \in (t_0, T]. \tag{2}$$

As usual, $L^1([t_0, T])$ denotes the set of Lebesgue integrable functions on $[t_0, T]$ and $\Gamma(x)$ is the Euler gamma function

$$\Gamma(x) = \int_0^{\infty} t^{x-1} \mathrm{e}^{-t} \mathrm{d}t, \tag{3}$$

a function playing an important role in fractional calculus since it generalizes the factorial to real arguments; it is indeed possible to verify that $\Gamma(x+1) = x\Gamma(x)$ and hence, since $\Gamma(1) = 1$, it is:

$$\Gamma(n+1) = n! \text{ for any } n \in \mathbb{N}.$$

It is due to the above fundamental property of the Euler gamma function that the RL integral (2) can be viewed as a straightforward extension of standard n-fold repeated integrals:

$$\int_{t_0}^{t} \int_{t_0}^{\tau_1} \cdots \int_{t_0}^{\tau_{n-1}} f(\tau_n)\, \mathrm{d}\tau_n \cdots \mathrm{d}\tau_2\, \mathrm{d}\tau_1 = \frac{1}{(n-1)!} \int_{t_0}^{t} (t-\tau)^{n-1} f(\tau) \mathrm{d}\tau$$

where it is sufficient to replace the integer n with any real $\alpha > 0$ to obtain RL integral (2).

In the special case of the starting point $t_0 \to -\infty$ the integral on the whole real axis:

$$J_{-\infty}^{\alpha} f(t) = \lim_{t_0 \to -\infty} J_{t_0}^{\alpha} f(t) = \frac{1}{\Gamma(\alpha)} \int_{-\infty}^{t} (t-\tau)^{\alpha-1} f(\tau) \mathrm{d}\tau, \quad t \in \mathbb{R}, \tag{4}$$

is usually referred to as the Liouville (left-sided) fractional integral (see [12] (Chapter 5) or [10] (§2.3)) and satisfies similar properties as the integer-order integral, such as $J_{-\infty}^{\alpha} \mathrm{e}^{\Omega t} = \Omega^{-\alpha} \mathrm{e}^{\Omega t}$.

Once a robust definition for fractional-order integrals is available, as the RL integral (2), fractional derivatives can be introduced as their left-inverses in a similar way as standard integer-order derivatives are the inverse operators of the corresponding integrals.

To this purpose let us denote with $m = \lceil \alpha \rceil$ the smallest integer greater or equal to α and, since $m - \alpha > 0$, consider the RL integral $J_{t_0}^{m-\alpha}$. Thanks to the semigroup property $J_{t_0}^{m-\alpha} J_{t_0}^{\alpha} f(t) = J_{t_0}^{m} f(t)$ [9] (Theorem 2.1) which returns an integer-order integral, it is sufficient to apply the integer-order derivative D^m to obtain the identity:

$$D^m J_{t_0}^{m-\alpha} J_{t_0}^{\alpha} f(t) = D^m J_{t_0}^{m} f(t) = f(t);$$

the concatenation $D^m J_{t_0}^{m-\alpha}$ hence provides the left-inverse of $J_{t_0}^{\alpha}$ and therefore justifies the following definition of the RL fractional derivative.

Definition 2. *Let $\alpha > 0$, $m = \lceil \alpha \rceil$ and $t_0 \in \mathbb{R}$. The RL fractional derivative of order α and starting point t_0 is:*

$$^{\mathrm{RL}}D_{t_0}^{\alpha} f(t) := D^m J_{t_0}^{m-\alpha} f(t) = \frac{1}{\Gamma(m-\alpha)} \frac{\mathrm{d}^m}{\mathrm{d}t^m} \int_{t_0}^{t} (t-\tau)^{m-\alpha-1} f(\tau) \mathrm{d}\tau, \quad t > t_0. \tag{5}$$

The RL derivative (5) is not the only left inverse of $J_{t_0}^{\alpha}$ and in applications, a different operator is usually preferred. One of the major drawbacks of the RL derivative is that it requires to be initialized by means of fractional integrals and fractional derivatives. To fully understand this issue it is useful to consider the following result on the Laplace transform (LT) of the RL derivative [10].

Proposition 1. *Let $\alpha > 0$ and $m = \lceil \alpha \rceil$. The LT of the RL derivative of a function $f(t)$ is:*

$$\mathcal{L}\Big({}^{\mathrm{RL}}D_{t_0}^{\alpha} f(t)\, ;\, s \Big) = s^{\alpha} F(s) - \sum_{j=1}^{m-1} s^{m-1-j}\, {}^{\mathrm{RL}}D_{t_0}^{\alpha-m+j} f(t) \Big|_{t=t_0^+} - s^{m-1} J_{t_0}^{m-\alpha} f(t) \Big|_{t=t_0^+},$$

with $F(s)$ the LT of $f(t)$.

A consequence of this result is that fractional differential equations (FDEs) with the RL derivative need to be initialized with the same kind of values. The uniqueness of the solution $y(t)$ of a FDE requires that initial conditions on $J_{t_0}^{m-\alpha}y(t)\big|_{t=t_0^+}$ and ${}^{\mathrm{RL}}D_{t_0}^{\alpha-m+j}y(t)\big|_{t=t_0^+}$, $j = 1,\ldots,m-1$, are assigned (e.g., see [9] (Theorem 5.1) or [10] (Chapter 3)).

In the majority of applications, however, these values are not available because they do not have a clear physical meaning and therefore the description of the initial state of a system is quite difficult when the RL derivative is involved. This is one of the reasons which motivated the introduction of the alternative fractional Caputo's derivative obtained by simply interchanging differentiation and integration in RL Derivative (5).

Definition 3. *Let $\alpha > 0$, $m = \lceil \alpha \rceil$ and $t_0 \in \mathbb{R}$. For a function $f \in A^m([t_0,T])$, i.e., such that $f^{(m-1)}$ is absolutely continuous, the Caputo's derivative is defined as:*

$$ {}^{\mathrm{C}}D_{t_0}^{\alpha} f(t) := J_{t_0}^{m-\alpha} D^m f(t) = \frac{1}{\Gamma(m-\alpha)} \int_{t_0}^{t} (t-\tau)^{m-\alpha-1} f^{(m)}(\tau) \mathrm{d}\tau, \quad t > t_0, \tag{6} $$

where D^m and $f^{(m)}$ denote integer-order derivatives.

Unlike the RL derivative, the LT of the Caputo's derivative is initialized by standard initial values expressed in terms of integer-order derivatives, as summarized in the following result [9].

Proposition 2. *Let $\alpha > 0$ and $m = \lceil \alpha \rceil$. The LT of the Caputo's derivative of a function $f(t)$ is:*

$$ \mathcal{L}\left({}^{\mathrm{C}}D_{t_0}^{\alpha} f(t)\,;\, s\right) = s^{\alpha} F(s) - \sum_{j=0}^{m-1} s^{\alpha-1-j} f^{(j)}(t_0), $$

with $F(s)$ the LT of $f(t)$.

It is a clear consequence of the above result that FDEs with the Caputo's derivative require, to ensure the uniqueness of the solution $y(t)$, the assignment of initial conditions in the more traditional Cauchy form $y^{(j)}(t_0) = y_{0,j}$, $j = 0,1,\ldots,m-1$, thus allowing a more convenient application to real-life problems.

Although different, the Caputo's derivative shares with the RL derivative the property of being the left inverse of the RL integral since ${}^{\mathrm{C}}D_{t_0}^{\alpha} J_{t_0}^{\alpha} f = f$ [9] (Theorem 3.7). However, ${}^{\mathrm{C}}D_{t_0}^{\alpha}$ is not the right inverse of $J_{t_0}^{\alpha}$ since [9] (Theorem 3.8),

$$ J_{t_0}^{\alpha}\, {}^{\mathrm{C}}D_{t_0}^{\alpha} f(t) = f(t) - T_{m-1}[f;t_0](t), \tag{7} $$

where $T_{m-1}[f;t_0](t)$ is the Taylor polynomial of f centered at t_0,

$$ T_{m-1}[f;t_0](t) = \sum_{k=0}^{m-1} \frac{(t-t_0)^k}{k!} f^{(k)}(t_0). \tag{8} $$

The polynomial $T_{m-1}[f;t_0](t)$ is important for establishing the relationship between fractional derivatives of RL and Caputo type. After differentiating both sides of Formula (7) in the RL sense it is possible to derive:

$$ {}^{\mathrm{C}}D_{t_0}^{\alpha} f(t) = {}^{\mathrm{RL}}D_{t_0}^{\alpha} \big(f(t) - T_{m-1}[f;t_0](t) \big). \tag{9} $$

Although several other definitions of fractional integrals and derivatives have been introduced in the last years, we confine our treatment to the above operators which are the most popular; the utility and the nature of some of the operators recently proposed is indeed still under scientific debate and we refer, for instance, to [15–19] for a critical analysis of the properties which a fractional derivative should (or should not) satisfy.

3. Fractional Derivatives as Limits of Difference Quotients

To better focus on their main characteristic features, we take a look at the fractional derivatives introduced in the previous section from an alternative perspective. We start from recalling the usual definition of the integer-order derivative based on the limit of the difference quotient:

$$f'(t) = \lim_{h \to 0} \frac{f(t) - f(t-h)}{h},$$

where obviously we assume that the above limit exists. By recursion this definition can be generalized to higher orders and, indeed, it is simple to evaluate:

$$f''(t) = \lim_{h \to 0} \frac{f'(t) - f'(t-h)}{h} = \lim_{h \to 0} \frac{f(t) - 2f(t-h) + f(t-2h)}{h^2}$$
$$f'''(t) = \lim_{h \to 0} \frac{f''(t-h) - f''(t)}{h} = \lim_{h \to 0} \frac{f(t) - 3f(t-h) + 3f(t-2h) - f(t-3h)}{h^3}$$

and, more generally, to prove the following result whose proof is straightforward and hence omitted.

Proposition 3. *Let $t \in \mathbb{R}$, $n \in \mathbb{N}$ and assume the function f to be n-times differentiable. Then,*

$$f^{(n)}(t) = \lim_{h \to 0} \frac{1}{h^n} \sum_{j=0}^{n} (-1)^j \binom{n}{j} f(t - jh), \tag{10}$$

where the binomial coefficients are defined as:

$$\binom{n}{j} = \frac{n(n-1)\cdots(n-j+1)}{j!} = \begin{cases} \dfrac{n!}{j!(n-j)!} & j = 0, 1, \ldots, n, \\ 0 & j > n. \end{cases} \tag{11}$$

Formula (10) is of interest since a possible generalization to fractional-order can be proposed by replacing the integer n with any real $\alpha > 0$. While this replacement in the power h^n of Formula (10) is straightforward, some difficulties arise in the other two instances of the integer-order n in Formula (10): the upper limit of the summation cannot be replaced by a real number and the binomial coefficients must be properly defined for real parameters.

The first difficulty can be easily overcome since binomial coefficients vanish when $j > n$. Thus, since no contribution in the summation is given from the presence of terms with $j > n$, the upper limit in Formula (10) can be raised to any value greater than n and, hence, the finite summation in Formula (10) can be replaced with the infinite series:

$$f^{(n)}(t) = \lim_{h \to 0} \frac{1}{h^n} \sum_{j=0}^{\infty} (-1)^j \binom{n}{j} f(t - jh). \tag{12}$$

To extend binomial coefficients and cope with real parameters we use again the Euler gamma function in place of factorials in Formula (11); generalized binomial coefficients are hence defined as:

$$\binom{\alpha}{j} = \frac{\alpha(\alpha-1)\cdots(\alpha-j+1)}{j!} = \frac{\Gamma(\alpha+1)}{j!\Gamma(\alpha-j+1)}, \quad j = 0, 1, \ldots. \tag{13}$$

Note that the above binomial coefficients are the coefficients in the binomial series:

$$(1-x)^\alpha = \sum_{j=0}^{\infty} (-1)^j \binom{\alpha}{j} x^j \tag{14}$$

which for real $\alpha > 0$ converges when $|x| \leq 1$. However, they do not vanish anymore for $j > \alpha$ when $\alpha \notin \mathbb{N}$.

Combining Equation (12) with Equation (13) provides the main justification for the following extension of the integer-order derivative (10) to any real order $\alpha > 0$ which was proposed independently, and almost simultaneously, by Grünwald [20] and Letnikov [21].

Definition 4. *Let $\alpha > 0$. The Grünwald–Letnikov (GL) fractional derivative of order α is:*

$$^{\mathrm{GL}}D^{\alpha} f(t) = \lim_{h \to 0} \frac{1}{h^{\alpha}} \sum_{j=0}^{\infty} (-1)^j \binom{\alpha}{j} f(t - jh), \quad t \in \mathbb{R}. \tag{15}$$

Referring to Equation (15) as the Grünwald–Letnikov fractional derivative is quite common in the literature (e.g., see [10] (§2.8) or [12] (§20)). Moreover, once a starting point t_0 has been assigned, for practical reasons the following (truncated) Grünwald–Letnikov fractional derivative [9,22] is often preferred since it can be applied to functions not defined (or simply not known) in $(-\infty, t_0)$.

Definition 5. *Let $\alpha > 0$ and $t_0 \in \mathbb{R}$. The (truncated) GL fractional derivative of order α is:*

$$^{\mathrm{GL}}D^{\alpha}_{t_0} f(t) = \lim_{h \to 0} \frac{1}{h^{\alpha}} \sum_{j=0}^{N} (-1)^j \binom{\alpha}{j} f(t - jh), \quad N = \left\lceil \frac{t - t_0}{h} \right\rceil, \quad t > t_0. \tag{16}$$

Although they are both named as Grünwald–Letnikov derivatives, $^{\mathrm{GL}}D^{\alpha}$ and $^{\mathrm{GL}}D^{\alpha}_{t_0}$ are different operators. We note however, that $^{\mathrm{GL}}D^{\alpha}$ corresponds to $^{\mathrm{GL}}D^{\alpha}_{t_0}$ when $t_0 \to -\infty$, namely $^{\mathrm{GL}}D^{\alpha} = {}^{\mathrm{GL}}D^{\alpha}_{-\infty}$.

There is a close relationship between the RL derivative and Equation (16). Indeed, it is possible to see that whenever $f \in C^m[t_0, T]$, with $m = \lceil \alpha \rceil$, then [9] (Theorem 2.25),

$$^{\mathrm{GL}}D^{\alpha}_{t_0} f(t) = {}^{\mathrm{RL}}D^{\alpha}_{t_0} f(t), \quad t \in (t_0, T]. \tag{17}$$

The GL derivative (15) possesses similar properties to integer-order derivatives, such as $^{\mathrm{GL}}D^{\alpha} t^k = 0$, for $k < \alpha$, and generalizes in a straightforward way the relationships of Equation (1) since, for instance $^{\mathrm{GL}}D^{\alpha} \mathrm{e}^{\Omega t} = \Omega^{\alpha} \mathrm{e}^{\Omega t}$ when $\mathrm{Re}\,\Omega \geq 0$ (we will better investigate these properties later on). Since this last relationship was the starting point of Liouville for the construction of the fractional calculus, the derivative (15) is sometimes recognized as the Liouville derivative (we refer to some papers on this operator and its application, for instance, in signal theory [23,24]).

It is also worthwhile to remark that the GL derivative (15) is closely related to the Marchaud derivative as discussed, for instance, in [12] (Chapter 20) and [25].

Another interesting feature is the correspondence between the standard Cauchy's integral formula:

$$f^{(n)}(z) = \frac{n!}{2\pi \mathrm{i}} \int_C \frac{f(u)}{(u - z)^{n+1}} \mathrm{d}u, \quad z \in \mathbb{C}, \quad n \in \mathbb{N},$$

and its analogous generalized Cauchy fractional derivative which, as proved in [23], once C is chosen as a complex U-shaped contour encircling the selected branch cut, it is equivalent to $^{\mathrm{GL}}D^{\alpha}$, namely:

$$^{\mathrm{GL}}D^{\alpha} f(z) = \frac{\Gamma(\alpha + 1)}{2\pi \mathrm{i}} \int_C \frac{f(u)}{(u - z)^{\alpha+1}} \mathrm{d}u, \quad z \in \mathbb{C}, \quad \alpha > 0.$$

In view of all these attractive properties, the GL derivative $^{\mathrm{GL}}D^{\alpha}$ may appear as the ideal generalization, to any positive real order, of the integer-order derivative. Unfortunately, there are instead serious issues discouraging the use of the GL derivative in most applications. We observe that:

- The evaluation of $^{\mathrm{GL}}D^{\alpha} f(t)$ at any point t requires the knowledge of the function $f(t)$ over the whole interval $(-\infty, t]$;

- The series (15) converges only for a restricted range of functions, as for instance for bounded functions in $(-\infty, t]$ or functions which do not increase too fast for $t \to -\infty$ (we refer to [12] (§4.20) for a discussion about the convergence of ${}^{\mathrm{GL}}D^{\alpha}$).

To face the above difficulties, the function $f(t)$ can be replaced with some related functions. Two main options are commonly used to perform this replacement and, as we will see, they actually lead to the RL and Caputo's fractional derivatives introduced in the previous Section.

3.1. Replacement with a Discontinuous Function: The RL Derivative

Once a starting point t_0 has been selected, the function $f(t)$ can be replaced, as illustrated in Figure 1, by a function $f^{\mathrm{R}}(t)$ which is equal to f for $t \geq t_0$ and equal to 0 otherwise:

$$f^{\mathrm{R}}(t) = \begin{cases} 0 & t \in (-\infty, t_0) \\ f(t) & t \geq t_0 \end{cases} ;$$

namely all the past history of the function f is assumed to be equal to 0 before t_0.

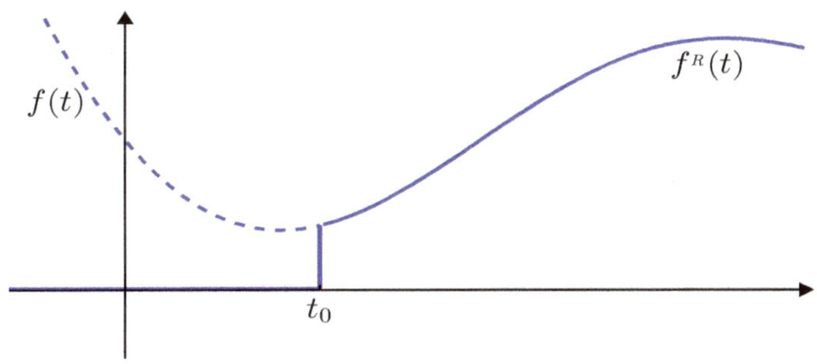

Figure 1. Replacement of $f(t)$ (dotted and solid lines) by $f^{\mathrm{R}}(t)$ (solid line) for a given point t_0.

It is quite intuitive to observe the following relationship between the GL derivative of $f^{\mathrm{R}}(t)$ and the truncated GL derivative (16) of the original function $f(t)$ and, in the end, of its RL derivative.

Proposition 4. *Let $\alpha > 0$, $m = \lceil \alpha \rceil$ and $f \in C^m[t_0, T]$. Then, for any $t \in (t_0, T]$ it is:*

$${}^{\mathrm{GL}}D^{\alpha} f^{\mathrm{R}}(t) = {}^{\mathrm{GL}}D^{\alpha}_{t_0} f(t) = {}^{\mathrm{RL}}D^{\alpha}_{t_0} f(t).$$

Proof. The application of the GL fractional derivative ${}^{\mathrm{GL}}D^{\alpha}$ to $f^{\mathrm{R}}(t)$ leads to:

$${}^{\mathrm{GL}}D^{\alpha} f^{\mathrm{R}}(t) = \lim_{h \to 0} \frac{1}{h^{\alpha}} \sum_{j=0}^{\infty} (-1)^j \binom{\alpha}{j} f^{\mathrm{R}}(t - jh) = \lim_{h \to 0} \frac{1}{h^{\alpha}} \sum_{j=0}^{N} (-1)^j \binom{\alpha}{j} f(t - jh)$$

where $N = \lceil (t - t_0)/h \rceil$ is the smallest integer such that $f^{\mathrm{R}}(t - jh) \equiv 0$ for $j = N+1, N+2, \ldots$ and hence ${}^{\mathrm{GL}}D^{\alpha} f^{\mathrm{R}}(t) = {}^{\mathrm{GL}}D^{\alpha}_{t_0} f(t)$. The second equality comes from (17). □

Unless $f(t_0) = 0$, the replacement of $f(t)$ with $f^{\mathrm{R}}(t)$ introduces a discontinuity at t_0 and, even when $f(t_0) = 0$, the function $f^{\mathrm{R}}(t)$ may suffer from a lack of regularity at t_0 due to the discontinuity of its higher-order derivatives. As we will see, this discontinuity seriously affects the RL derivative of several functions which, indeed, are often unbounded at t_0. Therefore, to provide a regularization and reduce the lack of smoothness introduced by $f^{\mathrm{R}}(t)$, a different replacement is proposed.

3.2. Replacement with a More Regular Function: The Caputo's Derivative

An alternative approach is based on the replacement, as depicted in Figure 2, of $f(t)$ with a function having a more regular behavior at t_0, and whose regularity depends on α. The proposed function is a continuation of $f(t)$ before t_0 in terms of its Taylor polynomial at t_0, namely:

$$f^{\mathrm{C}}(t) = \begin{cases} T_{m-1}[f;t_0](t) & t \in (-\infty, t_0) \\ f(t) & t \geq t_0 \end{cases}$$

where $T_{m-1}[f;t_0](t)$ is the same Taylor polynomial of f centered at t_0 introduced in (8) and $m = \lceil \alpha \rceil$. It is clear that, unlike $f^{\mathrm{R}}(t)$, the function $f^{\mathrm{C}}(t)$ preserves a possible smoothness of $f(t)$ at t_0 since:

$$f^{\mathrm{C}}(t)\Big|_{t \to t_0^-} = f(t_0), \quad \frac{\mathrm{d}}{\mathrm{d}t} f^{\mathrm{C}}(t)\Big|_{t \to t_0^-} = f'(t_0), \quad \ldots, \quad \frac{\mathrm{d}^{m-1}}{\mathrm{d}t^{m-1}} f^{\mathrm{C}}(t)\Big|_{t \to t_0^-} = f^{(m-1)}(t_0).$$

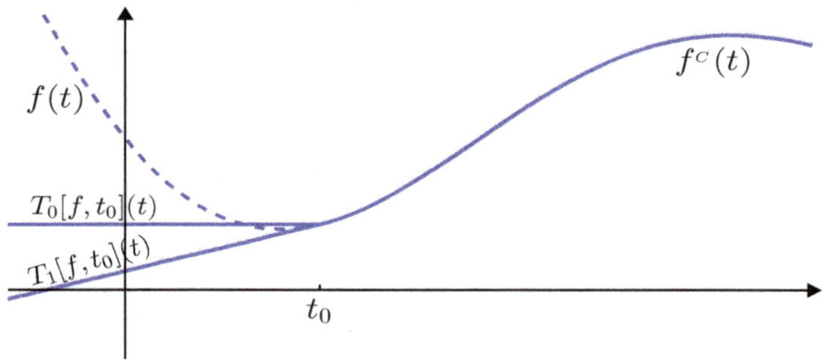

Figure 2. Replacement of $f(t)$ (dotted and solid lines) by $f^{\mathrm{C}}(t)$ (solid line) for a given point t_0 and for $m = 0$ (branch labeled $T_0[f, t_0](t)$) and $m = 1$ (branch labeled $T_1[f, t_0](t)$).

Before showing the effects of the replacement of $f(t)$ by $f^{\mathrm{C}}(t)$ we first have to consider the following preliminary result.

Lemma 1. *Let $\alpha > 0$ and $m = \lceil \alpha \rceil$. For any integer $k = 0, 1, \ldots, m-1$ it is:*

$$\sum_{j=0}^{\infty} (-1) \binom{\alpha}{j} j^k = 0.$$

Proof. When $\alpha \in \mathbb{N}$ we refer to [26] (Proposition 2.1). Assume now $\alpha \notin \mathbb{N}$ and, after using the alternative formulation of the binomial coefficients, one obtains:

$$\sum_{j=0}^{\infty} (-1) \binom{\alpha}{j} j^k = \frac{1}{\Gamma(-\alpha)} \sum_{j=0}^{\infty} \frac{\Gamma(j-\alpha)}{\Gamma(j+1)} j^k.$$

From [26] (Theorem 3.2) we know that the following asymptotic expansion holds:

$$\sum_{j=0}^{n} \frac{\Gamma(j-\alpha)}{\Gamma(j+1)} j^k = \sum_{\ell=0}^{\infty} F_\ell n^{k-\alpha-\ell}, \quad n \to \infty,$$

with coefficients F_ℓ depending on α and k but not on n. The proof hence immediately follows since $k - \alpha < 0$. □

We are now able to study the relationship between the GL derivative of $f^{\mathrm{C}}(t)$ and the truncated GL derivative (16) and the Caputo's derivative (6) of $f(t)$.

Proposition 5. *Let $\alpha > 0$, $m = \lceil \alpha \rceil$ and $f \in C^m[t_0, T]$. Then, for any $t \in (t_0, T]$ it is:*

$${}^{\mathrm{GL}}D^{\alpha} f^{\mathrm{C}}(t) = {}^{\mathrm{GL}}D^{\alpha}_{t_0} \Big(f(t) - T_{m-1}[f;t_0](t) \Big) = {}^{\mathrm{C}}D^{\alpha}_{t_0} f(t).$$

Proof. The application of the GL fractional derivative ${}^{\mathrm{GL}}D^{\alpha}$ to $f^{\mathrm{C}}(t)$ leads to:

$$
\begin{aligned}
{}^{\mathrm{GL}}D^{\alpha} f^{\mathrm{C}}(t) &= \lim_{h\to 0} \frac{1}{h^{\alpha}} \sum_{j=0}^{\infty} (-1)^j \binom{\alpha}{j} f^{\mathrm{C}}(t-jh) \\
&= \lim_{h\to 0} \frac{1}{h^{\alpha}} \left[\sum_{j=0}^{N} (-1)^j \binom{\alpha}{j} f(t-jh) + \sum_{j=N+1}^{\infty} (-1)^j \binom{\alpha}{j} T_{m-1}[f;t_0](t-jh) \right]
\end{aligned}
$$

Observe now that:

$$
T_{m-1}[f;t_0](t-jh) = \sum_{k=0}^{m-1} \frac{(t-jh-t_0)^k}{k!} f^{(k)}(t_0) = \sum_{k=0}^{m-1} \frac{1}{k!} f^{(k)}(t_0) \sum_{\ell=0}^{k} \binom{k}{\ell} (t-t_0)^{\ell} (-jh)^{k-\ell}
$$

and hence,

$$
\sum_{j=N+1}^{\infty} (-1)^j \binom{\alpha}{j} T_{m-1}[f;t_0](t-jh) = \sum_{k=0}^{m-1} \frac{1}{k!} f^{(k)}(t_0) \sum_{\ell=0}^{k} \binom{k}{\ell} (-h)^{k-\ell} (t-t_0)^{\ell} \sum_{j=N+1}^{\infty} (-1)^j \binom{\alpha}{j} j^{k-\ell}.
$$

Since from Lemma 1 it is:

$$
\sum_{j=N+1}^{\infty} (-1)^j \binom{\alpha}{j} j^k = -\sum_{j=0}^{N} (-1)^j \binom{\alpha}{j} j^k
$$

we obtain,

$$
\begin{aligned}
\sum_{j=N+1}^{\infty} (-1)^j \binom{\alpha}{j} T_{m-1}[f;t_0](t-jh) &= -\sum_{k=0}^{m-1} \frac{1}{k!} f^{(k)}(t_0) \sum_{\ell=0}^{k} \binom{k}{\ell} (-h)^{k-\ell} (t-t_0)^{\ell} \sum_{j=0}^{N} (-1)^j \binom{\alpha}{j} j^{k-\ell} \\
&= -\sum_{j=0}^{N} (-1)^j \binom{\alpha}{j} \sum_{k=0}^{m-1} \frac{1}{k!} f^{(k)}(t_0) \sum_{\ell=0}^{k} \binom{k}{\ell} (-hj)^{k-\ell} (t-t_0)^{\ell} \\
&= -\sum_{j=0}^{N} (-1)^j \binom{\alpha}{j} T_{m-1}[f;t_0](t-jh)
\end{aligned}
$$

from which the first equality follows. The second equality is consequence of Proposition 4 together with Equation (9). □

4. The Mittag–Leffler Function

The Mittag–Leffler (ML) function plays a special role in fractional calculus and in the representation of fractional derivatives of elementary functions and will be better investigated later on. It is therefore mandatory to recall some of the main properties of this function.

The definition of the two-parameter ML function is given by:

$$
E_{\alpha,\beta}(z) = \sum_{k=0}^{\infty} \frac{z^k}{\Gamma(\alpha k+\beta)}, \quad z \in \mathbb{C}, \tag{18}
$$

where α and β are two (possibly complex, but with $\mathrm{Re}\,\alpha > 0$) parameters.

The importance of the ML function in fractional calculus is particularly related to the fact that it is the eigenfunction of RL and Caputo's fractional derivatives. It is indeed possible to show that for any $t > t_0$ and $j = 0, 1, \ldots, m-1$, with $m = \lceil \alpha \rceil$, it is:

$$
{}^{\mathrm{RL}}D^{\alpha}_{t_0} y(t) = \Omega y(t), \quad y(t) = (t-t_0)^{\alpha-j-1} E_{\alpha,\alpha-j}\big(\Omega(t-t_0)^{\alpha}\big),
$$
$$
{}^{\mathrm{C}}D^{\alpha}_{t_0} y(t) = \Omega y(t), \quad y(t) = (t-t_0)^{j} E_{\alpha,j+1}\big(\Omega(t-t_0)^{\alpha}\big),
$$

and therefore the ML function has in fractional calculus the same importance as the exponential in the integer-order calculus (indeed, the ML function generalizes the exponential since $E_{1,1}(z) = \mathrm{e}^{z}$).

It is useful to introduce the Laplace transform (LT) of the ML function which, for any real $t > 0$ and $z \in \mathbb{C}$, is given by:

$$
\mathcal{L}\left(t^{\beta-1} E_{\alpha,\beta}(t^{\alpha} z)\,; s\right) = \frac{s^{\alpha-\beta}}{s^{\alpha} - z}, \quad \mathrm{Re}\, s > 0, \quad |z s^{-\alpha}| < 1, \tag{19}
$$

and, for convenience, we impose a branch cut on the negative real semi-axis in order to make the function s^{α} single valued.

The special instance $E_{1,\beta}(z)$ of the ML function will be encountered in the representation of fractional integrals and derivatives of some elementary functions. $E_{1,\beta}(z)$ is closely related to the exponential function and, as for instance emphasized in [27], it is:

$$
E_{1,\beta}(z) = \mathrm{e}^{z} \cdot \widehat{P}_{\beta}(z), \quad \widehat{P}_{\beta}(z) = \frac{1}{\Gamma(\beta-1)} \sum_{k=0}^{\infty} \frac{(-z)^k}{k!(\beta-1+k)}.
$$

It is however more convenient to express the ML function as a deviation from the exponential function according to the following result which will turn out to be useful in the subsequent sections.

Theorem 1. *Let $\Omega \in \mathbb{C}$ with $|\arg(\Omega)| < \pi$. For any $\beta > -1$ and $t \geq 0$ it is:*

$$
t^{-\beta} E_{1,1-\beta}(\Omega t) = \Omega^{\beta} \mathrm{e}^{t\Omega} + F_{\beta}(t;\Omega), \tag{20}
$$

where,

$$
F_{\beta}(t;\Omega) = \frac{\sin(\beta\pi)}{\pi} \int_0^{\infty} \mathrm{e}^{-rt} \frac{r^{\beta}}{r+\Omega} \mathrm{d}r. \tag{21}
$$

Proof. Thanks to Formula (19) for the LT of the ML we observe that:

$$
\mathcal{L}\left(t^{-\beta} E_{1,1-\beta}(\Omega t)\,; s\right) = \frac{s^{\beta}}{s-\Omega};
$$

thus, the formula for the inversion of the LT allows to write this function as:

$$
t^{-\beta} E_{1,1-\beta}(\Omega t) = \frac{1}{2\pi \mathrm{i}} \int_{\sigma-\mathrm{i}\infty}^{\sigma+\mathrm{i}\infty} \mathrm{e}^{st} \frac{s^{\beta}}{s-\Omega} \mathrm{d}s, \quad \sigma > \max\{0, \mathrm{Re}\,\Omega\}.
$$

The Bromwich line $(\sigma - \mathrm{i}\infty, \sigma + \mathrm{i}\infty)$ can be deformed into an Hankel contour $\mathcal{H}_{\epsilon}$ starting at $-\infty$ below the negative real semi-axis and ending at $-\infty$ above the negative real semi-axis after surrounding the origin along a circular disc $|s| = \epsilon$. Since Ω does not lie on the branch cut, the contour $\mathcal{H}_{\epsilon}$ can be collapsed onto the branch-cut by letting $\epsilon \to 0$. The contour thus passes over the singularity Ω and the residue subtraction leads to:

$$
t^{-\beta} E_{1,1-\beta}(\Omega t) = \mathrm{Res}\left(e^{st} \frac{s^{\beta}}{s-\Omega}, \Omega\right) + F_{\beta}(t;\Omega)
$$

where, for shortness, we denoted:

$$
F_{\beta}(t;\Omega) = \lim_{\epsilon \to 0} \frac{1}{2\pi \mathrm{i}} \int_{\mathcal{H}_{\epsilon}} \mathrm{e}^{st} \frac{s^{\beta}}{s-\Omega} \mathrm{d}s.
$$

The residue can be easily computed as:

$$\text{Res}\left(e^{st}\frac{s^{\beta}}{s-\Omega},\Omega\right)=\Omega^{\beta}\mathrm{e}^{\Omega t},$$

whilst to evaluate $F_{\beta}(t;\Omega)$ we first decompose the Hankel contour into its three main paths:

$$\mathcal{H}_{\epsilon}=\gamma_1+\gamma_2+\gamma_3,\quad \begin{cases}\gamma_1: s=r\mathrm{e}^{-\mathrm{i}\pi}, & \infty>r\geq\epsilon,\\ \gamma_2: s=\epsilon\mathrm{e}^{\mathrm{i}\theta}, & -\pi<\theta<\pi,\\ \gamma_3: s=r\mathrm{e}^{\mathrm{i}\pi}, & \epsilon\leq r<\infty,\end{cases}$$

thanks to which we are able to write:

$$\frac{1}{2\pi\mathrm{i}}\int_{\mathcal{H}_{\epsilon}}\mathrm{e}^{st}\frac{s^{\beta}}{s-\Omega}\mathrm{d}s=I_1+I_2+I_3,\quad I_{\ell}=\frac{1}{2\pi\mathrm{i}}\int_{\gamma_{\ell}}\mathrm{e}^{st}\frac{s^{\beta}}{s-\Omega}\mathrm{d}s,\quad \ell=1,2,3.$$

Since $\mathrm{e}^{\mathrm{i}\pi}=\mathrm{e}^{-\mathrm{i}\pi}=-1$, it is possible to compute:

$$\begin{aligned}I_1&=\frac{1}{2\pi\mathrm{i}}\int_{\infty}^{\epsilon}\mathrm{e}^{-rt}\frac{r^{\beta}\mathrm{e}^{-\mathrm{i}\beta\pi}\mathrm{e}^{-\mathrm{i}\pi}}{-r-\Omega}\mathrm{d}r=-\frac{\mathrm{e}^{-\mathrm{i}\beta\pi}}{2\pi\mathrm{i}}\int_{\epsilon}^{\infty}\mathrm{e}^{-rt}\frac{r^{\beta}}{r+\Omega}\mathrm{d}r\\ I_2&=\frac{\epsilon^{\beta+1}}{2\pi}\int_{-\pi}^{\pi}\frac{\mathrm{e}^{\epsilon t\cos\theta+\mathrm{i}[(\beta+1)\theta+\epsilon t\sin\theta]}}{\epsilon\cos\theta-\Omega+\mathrm{i}\epsilon\sin\theta}\mathrm{d}\theta\\ I_3&=\frac{1}{2\pi\mathrm{i}}\int_{\epsilon}^{\infty}\mathrm{e}^{-rt}\frac{r^{\beta}\mathrm{e}^{\mathrm{i}\beta\pi}\mathrm{e}^{\mathrm{i}\pi}}{-r-\Omega}\mathrm{d}r=\frac{\mathrm{e}^{\mathrm{i}\beta\pi}}{2\pi\mathrm{i}}\int_{\epsilon}^{\infty}\mathrm{e}^{-rt}\frac{r^{\beta}}{r+\Omega}\mathrm{d}r\end{aligned}$$

and we observe that, due to the presence of the term $\epsilon^{\beta+1}$, where we assumed $\beta>-1$, the integral I_2 vanishes when $\epsilon\to 0$. For the remaining term I_1+I_2 we note that:

$$\frac{\mathrm{e}^{\mathrm{i}\beta\pi}}{2\pi\mathrm{i}}-\frac{\mathrm{e}^{-\mathrm{i}\beta\pi}}{2\pi\mathrm{i}}=\frac{\sin\beta\pi}{\pi},$$

and, hence, the representation (21) of $F_{\beta}(t;\Omega)$ easily follows. □

The relationship between the ML function and the exponential is even more clear in the presence of an integer second parameter.

Proposition 6. *Let $m\in\mathbb{N}$ and $\Omega\in\mathbb{C}$. For any $t\in\mathbb{R}$ it is:*

$$\begin{aligned}t^{m}E_{1,1+m}(\Omega t)&=\frac{1}{\Omega^{m}}\left(\mathrm{e}^{\Omega t}-\sum_{j=0}^{m-1}\frac{\Omega^{j}t^{j}}{j!}\right),\\ t^{-m}E_{1,1-m}(\Omega t)&=\Omega^{m}\mathrm{e}^{\Omega t}.\end{aligned}$$

Proof. The first equality directly follows from the definition (18) of the ML function since it is:

$$t^{m}E_{1,1+m}(\Omega t)=\frac{1}{\Omega^{m}}\sum_{k=0}^{\infty}\frac{\Omega^{k+m}t^{k+m}}{\Gamma(k+1+m)}=\frac{1}{\Omega^{m}}\sum_{j=m}^{\infty}\frac{\Omega^{j}t^{j}}{\Gamma(j+1)}=\frac{1}{\Omega^{m}}\left(\sum_{j=0}^{\infty}\frac{\Omega^{j}t^{j}}{j!}-\sum_{j=0}^{m-1}\frac{\Omega^{j}t^{j}}{j!}\right)$$

whilst the second equality is a special case of Theorem 1. □

The following result will prove its particular utility when studying the asymptotic behavior of the fractional derivatives of some functions which will be expressed, in the next sections, in terms of special instances of the ML function.

Proposition 7. *Let $\beta>-1$, $\Omega\in\mathbb{C}$, $|\arg\Omega|<\pi$, and $t\geq 0$. Then,*

$$F_{\beta}(t;\Omega)=C_{\beta}\Omega^{-1}t^{-\beta-1}\left(1+\mathcal{O}(t^{-1})\right),\quad t\to\infty,$$

with C_β independent of t and Ω.

Proof. By a change of the integration variable we can write:

$$F_\beta(t;\Omega) = \Omega^\beta \frac{\sin \beta\pi}{\pi} \int_0^\infty \mathrm{e}^{-r\Omega t} \frac{r^\beta}{r+1} \mathrm{d}r = \Omega^\beta \frac{\sin \beta\pi}{\pi} \Gamma(\beta+1) U(\beta+1,\beta+1,\Omega t)$$

where $U(a,b,z)$ is the Tricomi function (often known as the confluent hypergeometric function of the second kind) defined for $\operatorname{Re} a > 0$ and $\operatorname{Re} z > 0$ and by analytic continuation elsewhere [28] (Chapter 48). After putting $C_\beta = \Gamma(\beta+1)\sin(\beta\pi)/\pi$, it is therefore [29] (Chapter 7, § 10.1),

$$F_\beta(t;\Omega) \sim t^{-\beta-1} C_\beta \Omega^{-1} \sum_{j=0}^{\infty} (-1)^j \frac{\Gamma(\beta+1+j)}{\Gamma(\beta+1)} t^{-j} \Omega^{-j}, \quad t \to \infty, \quad |\arg \Omega| \leq \frac{3}{2}\pi - \delta$$

for arbitrary small $\delta > 0$. Hence the proof follows since the selection of the branch cut on the negative real semi-axis. □

5. Fractional Integral and Derivatives of the Power Function

Basic results on fractional integral and derivatives of the power function $(t-t_0)^\beta$, for $\beta > -1$, are available in the literature; see, for instance [9] for the RL integral:

$$J^\alpha_{t_0} (t-t_0)^\beta = \frac{\Gamma(\beta+1)}{\Gamma(\beta+\alpha+1)} (t-t_0)^{\beta+\alpha}, \tag{22}$$

for the RL derivative (as usual, $m = \lceil \alpha \rceil$):

$${}^{\mathrm{RL}}D^\alpha_{t_0} (t-t_0)^\beta = \begin{cases} 0 & \beta \in \{\alpha-m, \alpha-m+1, \ldots, \alpha-1\} \\ \dfrac{\Gamma(\beta+1)}{\Gamma(\beta-\alpha+1)} (t-t_0)^{\beta-\alpha} & \text{otherwise} \end{cases} \tag{23}$$

and for the Caputo's derivative:

$${}^{\mathrm{C}}D^\alpha_{t_0} (t-t_0)^\beta = \begin{cases} 0 & \beta \in \{0,1,\ldots,m-1\} \\ \dfrac{\Gamma(\beta+1)}{\Gamma(\beta-\alpha+1)} (t-t_0)^{\beta-\alpha} & \beta > m-1 \\ \text{non existing} & \text{otherwise} \end{cases} \tag{24}$$

The absence of the Caputo's derivative of $(t-t_0)^\beta$ for real $\beta < m-1$ with $\beta \notin \{0,1,\ldots,m-1\}$ is related to the fact that once the m-th order derivative of $(t-t_0)^\beta$ is evaluated the integrand in Equation (6) is no longer integrable.

For general power functions independent from the starting point, i.e., for t^k instead of $(t-t_0)^k$, we can provide the following results.

Proposition 8. *Let $\alpha > 0$ and $m = \lceil \alpha \rceil$. Then for any $k \in \mathbb{N}$:*

1. ${}^{\mathrm{GL}}D^\alpha t^k = 0$ *for $k < \alpha$;*
2. ${}^{\mathrm{RL}}D^\alpha_{t_0} t^k = \displaystyle\sum_{\ell=0}^{k} \frac{k!}{(k-\ell)!\Gamma(\ell+1-\alpha)} (t-t_0)^{\ell-\alpha} t_0^{k-\ell}$;
3. ${}^{\mathrm{C}}D^\alpha_{t_0} t^k = \begin{cases} 0 & \text{if } k < \alpha \\ \displaystyle\sum_{\ell=m}^{k} \frac{k!}{(k-\ell)!\Gamma(\ell+1-\alpha)} (t-t_0)^{\ell-\alpha} t_0^{k-\ell} & \text{otherwhise} \end{cases}$

Proof. For ${}^{\mathrm{GL}}D^{\alpha} t^k$ we first write:

$$(t - jh)^k = \sum_{\ell=0}^{k} \binom{k}{\ell} t^{\ell} (-h)^{k-\ell} j^{k-\ell}$$

and hence by using the Definition 4 it is possible to see that:

$${}^{\mathrm{GL}}D^{\alpha} t^k = \lim_{h \to 0} \frac{1}{h^{\alpha}} \sum_{j=0}^{\infty} (-1)^j \binom{\alpha}{j} (t - jh)^k = \lim_{h \to 0} \frac{1}{h^{\alpha}} \sum_{\ell=0}^{k} \binom{k}{\ell} t^{\ell} (-h)^{k-\ell} \sum_{j=0}^{\infty} (-1)^j \binom{\alpha}{j} j^{k-\ell} = 0$$

where we applied Lemma 1. For ${}^{\mathrm{RL}}D^{\alpha}_{t_0} t^k$ we expand:

$$t^k = \sum_{\ell=0}^{k} \binom{k}{\ell} (t - t_0)^{\ell} t_0^{k-\ell}$$

and hence the proof immediately follows from Equation (23). Similarly for ${}^{\mathrm{C}}D^{\alpha}_{t_0} t^k$ by using Equation (24). □

Note that ${}^{\mathrm{GL}}D^{\alpha} t^k$ diverges when $k > \alpha$. A representation of ${}^{\mathrm{C}}D^{\alpha}_{t_0} t^{\beta}$, for general real but not integer β, is provided in terms of the hypergeometric ${}_2F_1$ function in [9] (Appendix B).

6. Fractional Integral and Derivatives of the Exponential Function

The exponential function is of great importance in mathematics and in several applications, also to approximate other functions. We therefore study here fractional integral and derivatives of the exponential function.

Proposition 9. *Let $\alpha > 0$, $m = \lceil \alpha \rceil$ and $t_0 \in \mathbb{R}$. For any $\Omega \in \mathbb{C}$ and $t > t_0$ the exponential function $\mathrm{e}^{\Omega(t-t_0)}$ has the following fractional integral and derivatives:*

$$J^{\alpha}_{t_0} \mathrm{e}^{\Omega(t-t_0)} = (t - t_0)^{\alpha} E_{1,1+\alpha}(\Omega(t - t_0)),$$
$${}^{\mathrm{RL}}D^{\alpha}_{t_0} \mathrm{e}^{\Omega(t-t_0)} = (t - t_0)^{-\alpha} E_{1,1-\alpha}(\Omega(t - t_0)),$$
$${}^{\mathrm{C}}D^{\alpha}_{t_0} \mathrm{e}^{\Omega(t-t_0)} = \Omega^m (t - t_0)^{m-\alpha} E_{1,m-\alpha+1}(\Omega(t - t_0)),$$

and, moreover, for any $t \in \mathbb{R}$ and $\mathrm{Re}(\Omega) \geq 0$ it is:

$${}^{\mathrm{GL}}D^{\alpha} \mathrm{e}^{\Omega t} = \Omega^{\alpha} \mathrm{e}^{\Omega t}.$$

Proof. By applying a term-by-term integration to the series expansion of the exponential function:

$$\mathrm{e}^{\Omega t} = \sum_{k=0}^{\infty} \frac{\Omega^k t^k}{k!}, \tag{25}$$

and thanks to Equation (22) and to Definition (18) of the ML function, we obtain:

$$J^{\alpha}_{t_0} \mathrm{e}^{\Omega(t-t_0)} = \sum_{k=0}^{\infty} \frac{\Omega^k}{k!} J^{\alpha}_{t_0} (t - t_0)^k = \sum_{k=0}^{\infty} \frac{\Omega^k}{\Gamma(k + \alpha + 1)} (t - t_0)^{k+\alpha} = (t - t_0)^{\alpha} E_{1,1+\alpha}(\Omega(t - t_0)).$$

For the evaluation of the RL derivative ${}^{\mathrm{RL}}D^{\alpha}_{0} \mathrm{e}^{\Omega(t-t_0)}$ we again consider the series expansion Equation (25) and, by differentiating term by term thanks to Equation (23), it is:

$$ {}^{\mathrm{RL}}D^{\alpha}_{t_0} e^{\Omega(t-t_0)} = {}^{\mathrm{RL}}D^{\alpha}_{t_0} \sum_{k=0}^{\infty} \frac{\Omega^k (t-t_0)^k}{k!} = \sum_{k=0}^{\infty} \frac{\Omega^k}{k!} {}^{\mathrm{RL}}D^{\alpha}_{t_0} (t-t_0)^k = \sum_{k=0}^{\infty} \frac{\Omega^k (t-t_0)^{k-\alpha}}{\Gamma(k-\alpha+1)} $$

from which the proof follows thanks again to Definition (18) of the ML function. We proceed in a similar way for ${}^{\mathrm{C}}D^{\alpha}_{t_0} e^{\Omega(t-t_0)}$ for which it is:

$$ {}^{\mathrm{C}}D^{\alpha}_{t_0} e^{\Omega(t-t_0)} = {}^{\mathrm{C}}D^{\alpha}_{t_0} \left(\sum_{k=0}^{m-1} \frac{\Omega^k (t-t_0)^k}{k!} + \sum_{k=m}^{\infty} \frac{\Omega^k (t-t_0)^k}{k!} \right) = \sum_{k=m}^{\infty} \frac{\Omega^k}{k!} {}^{\mathrm{C}}D^{\alpha}_{t_0} (t-t_0)^k = \sum_{k=m}^{\infty} \frac{\Omega^k (t-t_0)^{k-\alpha}}{\Gamma(k-\alpha+1)} $$

and, after a change $j = k - m$ in the summation index and rearranging some terms we obtain:

$$ {}^{\mathrm{C}}D^{\alpha}_{t_0} e^{\Omega(t-t_0)} = \sum_{j=0}^{\infty} \frac{\Omega^{j+m} (t-t_0)^{j+m-\alpha}}{\Gamma(j+m-\alpha+1)} = \Omega^m (t-t_0)^{m-\alpha} \sum_{j=0}^{\infty} \frac{\Omega^j (t-t_0)^j}{\Gamma(j+m-\alpha+1)} $$

from which, again, the proof follows from Definition (18) of the ML function. To finally evaluate the GL derivative ${}^{\mathrm{GL}}D^{\alpha} e^{\Omega t}$ we first apply its definition from Equation (15)

$$ {}^{\mathrm{GL}}D^{\alpha} e^{\Omega t} = \lim_{h \to 0} \frac{1}{h^{\alpha}} \sum_{j=0}^{\infty} (-1)^j \binom{\alpha}{j} e^{\Omega(t-jh)} = e^{\Omega t} \lim_{h \to 0} \frac{1}{h^{\alpha}} \sum_{j=0}^{\infty} (-1)^j \binom{\alpha}{j} e^{-jh\Omega} $$

and, since we are assuming $\mathrm{Re}(\Omega) \geq 0$, it is $|e^{-h\Omega}| \leq 1$ and hence the binomial series converges:

$$ \sum_{j=0}^{\infty} (-1)^j \binom{\alpha}{j} e^{-jh\Omega} = (1 - e^{-h\Omega})^{\alpha} \tag{26} $$

thanks to which we can easily evaluate:

$$ {}^{\mathrm{GL}}D^{\alpha} e^{\Omega t} = e^{\Omega t} \lim_{h \to 0} \frac{(1-e^{-h\Omega})^{\alpha}}{h^{\alpha}} = e^{\Omega t} \lim_{h \to 0} \left(\frac{1-e^{-h\Omega}}{h} \right)^{\alpha} = \Omega^{\alpha} e^{\Omega t} $$

to conclude the proof: □

Whenever $\alpha \in \mathbb{N}$, and hence $m = \alpha$, the standard integer-order results,

$$ J^{m}_{t_0} e^{\Omega(t-t_0)} = \frac{1}{\Omega^m} \left(e^{\Omega(t-t_0)} - \sum_{j=0}^{m-1} \frac{(t-t_0)^j}{j!} \right), \qquad D^m e^{\Omega(t-t_0)} = \Omega^m e^{\Omega(t-t_0)} $$

are recovered. This is a direct consequence of Proposition 6 for $J^{\alpha}_{t_0}$ and ${}^{\mathrm{RL}}D^{\alpha}_{t_0}$ whilst it comes from the equivalence $e^z = E_{1,1}(z)$ for ${}^{\mathrm{C}}D^{\alpha}_{t_0}$. It is moreover obvious for ${}^{\mathrm{GL}}D^{\alpha}$, for which we just observed that the restriction $\mathrm{Re}(\Omega) \geq 0$ is no longer necessary when $\alpha \in \mathbb{N}$ since the binomial series (26) has just a finite number of nonzero terms and hence converges for any $\Omega \in \mathbb{C}$.

The correspondence ${}^{\mathrm{GL}}D^{\alpha} e^{\Omega t} = \Omega^{\alpha} e^{\Omega t}$ appears as the most natural generalization of the integer-order derivatives but it holds only when $\mathrm{Re}(\Omega) \geq 0$. By combining Proposition 9 and Theorem 1 it is immediately seen that ${}^{\mathrm{RL}}D^{\alpha}_{t_0} e^{\Omega(t-t_0)}$ and ${}^{\mathrm{C}}D^{\alpha}_{t_0} e^{\Omega(t-t_0)}$ can be represented as a deviation from $\Omega^{\alpha} e^{\Omega t}$ as stated in the following result.

Proposition 10. *Let $\alpha > 0$, $m = \lceil \alpha \rceil$ and $t_0 \in \mathbb{R}$. For any $\Omega \in \mathbb{C}$, $|\arg \Omega| < \pi$, and $t > t_0$ it is:*

$$ \begin{aligned} {}^{\mathrm{RL}}D^{\alpha}_{t_0} e^{\Omega(t-t_0)} &= \Omega^{\alpha} e^{\Omega(t-t_0)} + F_{\alpha}(t-t_0;\Omega), \\ {}^{\mathrm{C}}D^{\alpha}_{t_0} e^{\Omega(t-t_0)} &= \Omega^{\alpha} e^{\Omega(t-t_0)} + \Omega^m F_{\alpha-m}(t-t_0;\Omega). \end{aligned} $$

The terms $F_\alpha(t-t_0,\Omega)$ and $\Omega^m F_{\alpha-m}(t-t_0,\Omega)$ describe the deviation of ${}^{\mathrm{RL}}D^{\alpha}_{t_0}\mathrm{e}^{\Omega(t-t_0)}$ and ${}^{\mathrm{C}}D^{\alpha}_{t_0}\mathrm{e}^{\Omega(t-t_0)}$ from the ideal value $\Omega^\alpha \mathrm{e}^{\Omega(t-t_0)}$. From Proposition 7 we know that these deviations decrease in magnitude, until they vanish, as $t \to \infty$. Consequently, ${}^{\mathrm{RL}}D^{\alpha}_{t_0}\mathrm{e}^{\Omega(t-t_0)}$ and ${}^{\mathrm{C}}D^{\alpha}_{t_0}\mathrm{e}^{\Omega(t-t_0)}$ asymptotically tend to $\Omega^\alpha \mathrm{e}^{\Omega(t-t_0)}$ (and hence to ${}^{\mathrm{GL}}D^{\alpha}\mathrm{e}^{\Omega(t-t_0)}$ when $\mathrm{Re}\,\Omega \geq 0$), namely:

$$ {}^{\mathrm{RL}}D^{\alpha}_{t_0}\mathrm{e}^{\Omega(t-t_0)} \sim {}^{\mathrm{C}}D^{\alpha}_{t_0}\mathrm{e}^{\Omega(t-t_0)} \sim \Omega^\alpha \mathrm{e}^{\Omega(t-t_0)}, \quad t \to \infty, \quad |\arg \Omega| < \pi. $$

This asymptotic behavior can be explained by recalling that the above derivatives differ from the way in which the function is assumed before the starting point t_0 and the influence of the function on $(-\infty, t_0)$ clearly becomes of less and less importance as t goes away from t_0, namely as $t \to \infty$.

We observe from Figure 3, where the values $\alpha = 0.7$ and $\Omega = -0.5 + 2\mathrm{i}$ have been considered, that actually both ${}^{\mathrm{RL}}D^{\alpha}_{t_0}\mathrm{e}^{\Omega t}$ and ${}^{\mathrm{C}}D^{\alpha}_{t_0}\mathrm{e}^{\Omega t}$ converge towards $\Omega^\alpha \mathrm{e}^{\Omega t}$, in quite a fast way, as t increases. In all the experiments we used, for ease of presentation, $t_0 = 0$ and the ML function was evaluated by means of the Matlab code described in [30] and based on some ideas previously developed in [31].

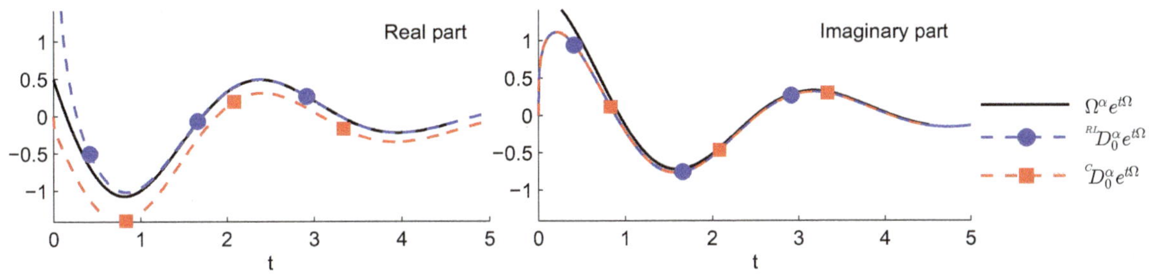

Figure 3. Comparison of ${}^{\mathrm{RL}}D^{\alpha}_{0}\mathrm{e}^{\Omega t}$ and ${}^{\mathrm{C}}D^{\alpha}_{0}\mathrm{e}^{\Omega t}$ with $\Omega^\alpha \mathrm{e}^{\Omega t}$ for $\alpha = 0.7$ and $\Omega = -0.5 + 2\mathrm{i}$.

The unbounded nature of the real part of ${}^{\mathrm{RL}}D^{\alpha}_{t_0}\mathrm{e}^{\Omega t}$ at the origin is due to the presence of the factor $(t-t_0)^{-\alpha}$ (see Proposition 9) but it can also be interpreted as a consequence of the replacement of $f(t)$ by $f^{\mathrm{R}}(t)$ in the RL derivative, as discussed in Section 3.1, which introduces a discontinuity at the starting point; the same phenomena will be observed for the cosine function but, obviously, not for the sine function for which the value at 0 is the same forced in $(-\infty, 0)$.

It is not surprising that ${}^{\mathrm{RL}}D^{\alpha}_{t_0}\mathrm{e}^{\Omega t}$ and ${}^{\mathrm{C}}D^{\alpha}_{t_0}\mathrm{e}^{\Omega t}$ have the same imaginary part (which indeed overlap in the second plot of Figure 3) when $0 < \alpha < 1$. The imaginary part of the exponential is indeed zero at the origin and hence RL and Caputo's derivatives coincide since relation in Equation (9) for $0 < \alpha < 1$ simply reads as ${}^{\mathrm{C}}D^{\alpha}_{t_0} f(t) = {}^{\mathrm{RL}}D^{\alpha}_{t_0}(f(t) - f(t_0))$.

From Figure 3 we observe that the RL derivative converges faster to $\Omega^\alpha \mathrm{e}^{\Omega t}$ than the Caputo derivative. This behavior can be easily explained by observing from Proposition 10 that as $t \to \infty$:

$$ \begin{aligned} {}^{\mathrm{RL}}D^{\alpha}_{t_0}\mathrm{e}^{\Omega(t-t_0)} - \Omega^\alpha \mathrm{e}^{\Omega(t-t_0)} &= F_\alpha(t-t_0;\Omega) \sim (t-t_0)^{-\alpha-1} \\ {}^{\mathrm{C}}D^{\alpha}_{t_0}\mathrm{e}^{\Omega(t-t_0)} - \Omega^\alpha \mathrm{e}^{\Omega(t-t_0)} &= \Omega^m F_{\alpha-m}(t-t_0;\Omega) \sim (t-t_0)^{m-\alpha-1} \end{aligned} $$

which tell us that while ${}^{\mathrm{C}}D^{\alpha}_{t_0}\mathrm{e}^{\Omega(t-t_0)}$ converges towards $\Omega^\alpha \mathrm{e}^{\Omega(t-t_0)}$ according to a power law with exponent $-1 < m - \alpha - 1 < 0$, the RL derivative ${}^{\mathrm{RL}}D^{\alpha}_{t_0}\mathrm{e}^{\Omega(t-t_0)}$ converges according to a power law with exponent $-\alpha - 1 < -1$.

Similar behaviors, showing the convergence for $t \to \infty$ of the different derivatives, are obtained also for $\alpha = 1.7$ and $\Omega = -0.5 + 2\mathrm{i}$ as we can observe from Figure 4. In this case, however, the imaginary parts of ${}^{\mathrm{RL}}D^{\alpha}_{t_0}\mathrm{e}^{\Omega t}$ and ${}^{\mathrm{C}}D^{\alpha}_{t_0}\mathrm{e}^{\Omega t}$ are no longer the same since when $1 < \alpha < 2$ the relationship between the two derivatives is given by ${}^{\mathrm{C}}D^{\alpha}_{t_0} f(t) = {}^{\mathrm{RL}}D^{\alpha}_{t_0}(f(t) - f(t_0) - (t-t_0)f'(t_0))$ and the imaginary part of $f'(t_0)$ is not equal to 0 as the imaginary part of $f(t_0)$.

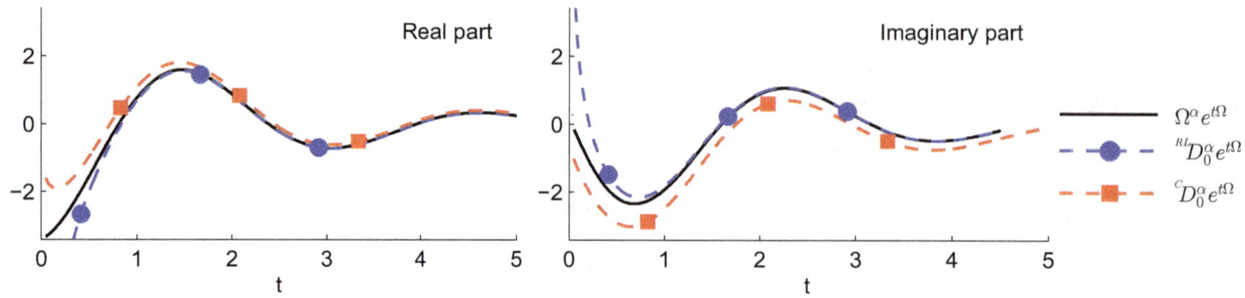

Figure 4. Comparison of ${}^{RL}D_0^{\alpha}e^{\Omega t}$ and ${}^{C}D_0^{\alpha}e^{\Omega t}$ with $\Omega^{\alpha}e^{\Omega t}$ for $\alpha = 1.7$ and $\Omega = -0.5 + 2i$.

7. Fractional Integral and Derivatives of Sine and Cosine Functions

Once fractional derivatives of the exponential are available, the fractional derivatives of the basic trigonometric functions can be easily evaluated by means of the well-known De Moivre formulas:

$$\sin \Omega t = \frac{e^{i\Omega t} - e^{-i\Omega t}}{2i}, \quad \cos \Omega t = \frac{e^{i\Omega t} + e^{-i\Omega t}}{2},$$

which allow to state the following results.

Proposition 11. *Let $\alpha > 0$, $m = \lceil \alpha \rceil$ and $\Omega \in \mathbb{R}$. For $t \geq t_0$ the function $\sin(\Omega(t - t_0))$ has the following fractional integral and derivatives:*

$$\begin{aligned}
J_{t_0}^{\alpha} \sin(\Omega(t - t_0)) &= \frac{(t - t_0)^{\alpha}}{2i} \Big(E_{1,1+\alpha}(+i\Omega(t - t_0)) - E_{1,1+\alpha}(-i\Omega(t - t_0)) \Big) \\
{}^{RL}D_{t_0}^{\alpha} \sin(\Omega(t - t_0)) &= \frac{(t - t_0)^{-\alpha}}{2i} \Big(E_{1,1-\alpha}(+i\Omega(t - t_0)) - E_{1,1-\alpha}(-i\Omega(t - t_0)) \Big) \\
{}^{C}D_{t_0}^{\alpha} \sin(\Omega(t - t_0)) &= i^m \Omega^m \frac{(t - t_0)^{m-\alpha}}{2i} \Big(E_{1,m-\alpha+1}(+i\Omega(t - t_0)) - (-1)^m E_{1,m-\alpha+1}(-i\Omega(t - t_0)) \Big)
\end{aligned}$$

and, moreover, for any $t \in \mathbb{R}$ it is:

$${}^{GL}D^{\alpha} \sin(\Omega t) = \Omega^{\alpha} \sin(\Omega t + \alpha \frac{\pi}{2}).$$

Proof. The proof for $J_{t_0}^{\alpha} \sin(\Omega(t - t_0))$, ${}^{RL}D_{t_0}^{\alpha} \sin(\Omega(t - t_0))$ and ${}^{C}D_{t_0}^{\alpha} \sin(\Omega(t - t_0))$ is a straightforward consequence of Proposition 9. For ${}^{GL}D^{\alpha} \sin(\Omega t)$ we observe that the direct application of Proposition 9 leads to:

$${}^{GL}D^{\alpha} \sin(\Omega t) = \frac{(+i)^{\alpha}\Omega^{\alpha}e^{+i\Omega t} - (-i)^{\alpha}\Omega^{\alpha}e^{-i\Omega t}}{2i} \tag{27}$$

and since $e^{\pm i\Omega t} = \cos \Omega t \pm i \sin \Omega t$ and $(\pm i)^{\alpha} = e^{\pm i\alpha\pi/2}$, the proof follows from the application of basic trigonometric rules. □

Note that the assumption $\operatorname{Re}\Omega \geq 0$ is no longer necessary for ${}^{GL}D^{\alpha} \sin(\Omega t)$ since the arguments of the exponential functions in Equation (27) are always on the imaginary axis. Similar results can also be stated for the cosine function and the proofs are omitted since they are similar to the previous ones.

Proposition 12. *Let $\alpha > 0$, $m = \lceil \alpha \rceil$ and $\Omega \in \mathbb{R}$. For $t \geq t_0$ the function $\cos(\Omega(t-t_0))$ has the following fractional integral and derivatives:*

$$J^{\alpha}_{t_0} \cos(\Omega(t-t_0)) = \frac{(t-t_0)^{\alpha}}{2}\Big(E_{1,1+\alpha}(+\mathrm{i}\Omega(t-t_0)) + E_{1,1+\alpha}(-\mathrm{i}\Omega(t-t_0))\Big)$$
$${}^{\mathrm{RL}}D^{\alpha}_{t_0} \cos(\Omega(t-t_0)) = \frac{(t-t_0)^{-\alpha}}{2}\Big(E_{1,1-\alpha}(+\mathrm{i}\Omega(t-t_0)) + E_{1,1-\alpha}(-\mathrm{i}\Omega(t-t_0))\Big)$$
$${}^{\mathrm{C}}D^{\alpha}_{t_0} \cos(\Omega(t-t_0)) = \mathrm{i}^m \Omega^m \frac{(t-t_0)^{m-\alpha}}{2}\Big(E_{1,m-\alpha+1}(+\mathrm{i}\Omega(t-t_0)) + (-1)^m E_{1,m-\alpha+1}(-\mathrm{i}\Omega(t-t_0))\Big)$$

and, moreover, for any $t \in \mathbb{R}$ it is:

$${}^{\mathrm{GL}}D^{\alpha} \cos(\Omega t) = \Omega^{\alpha} \cos(\Omega t + \alpha \frac{\pi}{2}).$$

As for the exponential function, we observe that with the basic trigonometric functions, the GL derivative ${}^{\mathrm{GL}}D^{\alpha}$ generalizes the known results holding for integer-order derivatives.

Furthermore, in this case, with the help of Proposition 1, it is possible to see that the RL and Caputo's derivatives of $\sin(\Omega(t-t_0))$ and $\cos(\Omega(t-t_0))$ can be expressed as deviations from $\Omega^{\alpha}\sin(\Omega t + \alpha\frac{\pi}{2})$ and $\Omega^{\alpha}\cos(\Omega t + \alpha\frac{\pi}{2})$ respectively. The following results (whose proof is omitted since it is obvious) can indeed be provided.

Proposition 13. *Let $\alpha > 0$, $m = \lceil \alpha \rceil$ and $\Omega \in \mathbb{R}$. Then for any $t \geq t_0$ it is:*

$${}^{\mathrm{RL}}D^{\alpha}_{t_0} \sin(\Omega(t-t_0)) = \Omega^{\alpha} \sin(\Omega(t-t_0) + \alpha\frac{\pi}{2}) + \frac{F_{\alpha}(t-t_0;\mathrm{i}\Omega) - F_{\alpha}(t-t_0;-\mathrm{i}\Omega)}{2\mathrm{i}}$$
$${}^{\mathrm{C}}D^{\alpha}_{t_0} \sin(\Omega(t-t_0)) = \Omega^{\alpha} \sin(\Omega(t-t_0) + \alpha\frac{\pi}{2}) + \mathrm{i}^m\Omega^m \frac{F_{\alpha-m}(t-t_0;\mathrm{i}\Omega) - (-1)^m F_{\alpha-m}(t-t_0;-\mathrm{i}\Omega)}{2\mathrm{i}}$$
$${}^{\mathrm{RL}}D^{\alpha}_{t_0} \cos(\Omega(t-t_0)) = \Omega^{\alpha} \cos(\Omega(t-t_0) + \alpha\frac{\pi}{2}) + \frac{F_{\alpha}(t-t_0;\mathrm{i}\Omega) + F_{\alpha}(t-t_0;-\mathrm{i}\Omega)}{2}$$
$${}^{\mathrm{C}}D^{\alpha}_{t_0} \cos(\Omega(t-t_0)) = \Omega^{\alpha} \cos(\Omega(t-t_0) + \alpha\frac{\pi}{2}) + \mathrm{i}^m\Omega^m \frac{F_{\alpha-m}(t-t_0;\mathrm{i}\Omega) + (-1)^m F_{\alpha-m}(t-t_0;-\mathrm{i}\Omega)}{2}$$

Since the function $F_{\beta}(t;\pm\mathrm{i}\Omega)$ asymptotically vanishes when $t \to \infty$, we can argue that:

$$\begin{aligned} {}^{\mathrm{RL}}D^{\alpha}_{t_0} \sin(\Omega(t-t_0)) &\sim {}^{\mathrm{C}}D^{\alpha}_{t_0} \sin(\Omega(t-t_0)) \sim \Omega^{\alpha}\sin(\Omega t + \alpha\frac{\pi}{2}), \quad t \to \infty \\ {}^{\mathrm{RL}}D^{\alpha}_{t_0} \cos(\Omega(t-t_0)) &\sim {}^{\mathrm{C}}D^{\alpha}_{t_0} \cos(\Omega(t-t_0)) \sim \Omega^{\alpha}\cos(\Omega t + \alpha\frac{\pi}{2}), \quad t \to \infty \end{aligned} \tag{28}$$

as we can clearly observe from Figure 5 and 6.

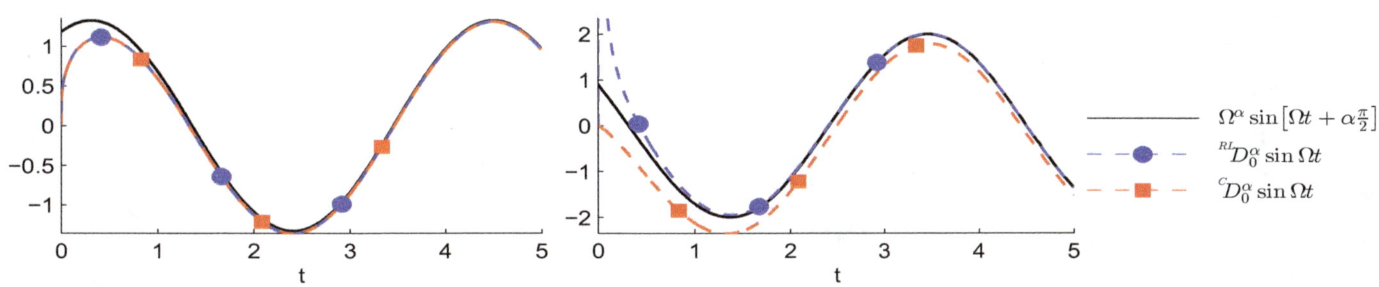

Figure 5. Comparison of ${}^{\mathrm{RL}}D^{\alpha}_{0}\sin\Omega t$ and ${}^{\mathrm{C}}D^{\alpha}_{0}\sin\Omega t$ with $\Omega^{\alpha}\sin(\Omega t + \alpha\frac{\pi}{2})$ for $\Omega = 1.5$, $\alpha = 0.7$ (**left plot**) and $\alpha = 1.7$ (**right plot**).

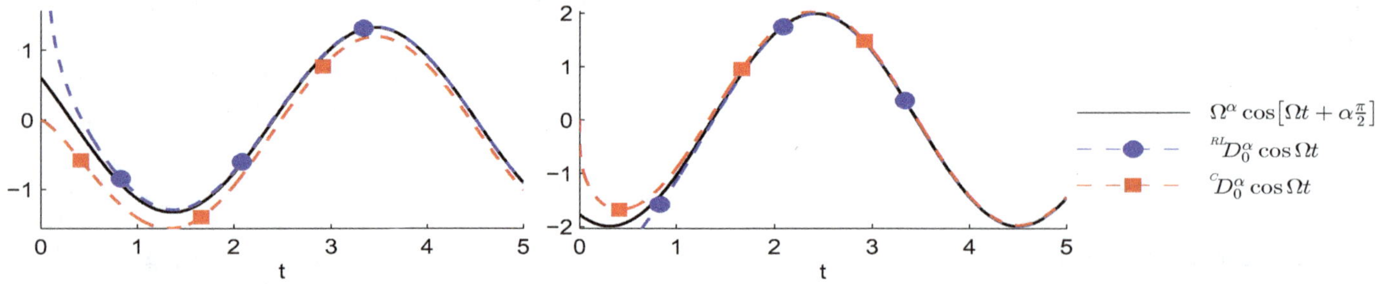

Figure 6. Comparison of ${}^{RL}D_0^\alpha \cos\Omega t$ and ${}^{C}D_0^\alpha \cos\Omega t$ with $\Omega^\alpha \cos(\Omega t + \alpha\frac{\pi}{2})$ for $\Omega = 1.5$, $\alpha = 0.7$ (**left plot**) and $\alpha = 1.7$ (**right plot**).

The above results are mainly useful for studying the asymptotic behavior of different operators applied to the sine and cosine functions. The representation of integrals and derivatives can be simplified thanks to the following results.

Proposition 14. *Let $\alpha > 0$, $m = \lceil \alpha \rceil$ and $\Omega \in \mathbb{R}$. For $t \geq t_0$ the function $\sin(\Omega(t-t_0))$ has the following fractional integral and derivatives:*

$$J_{t_0}^\alpha \sin(\Omega(t-t_0)) = \Omega(t-t_0)^{1+\alpha} E_{2,2+\alpha}(-\Omega^2(t-t_0)^2)$$
$${}^{RL}D_{t_0}^\alpha \sin(\Omega(t-t_0)) = \Omega(t-t_0)^{1-\alpha} E_{2,2-\alpha}(-\Omega^2(t-t_0)^2)$$
$${}^{C}D_{t_0}^\alpha \sin(\Omega(t-t_0)) = \begin{cases} (-1)^{\frac{m-1}{2}} \Omega^m (t-t_0)^{m-\alpha} E_{2,m-\alpha+1}(-\Omega^2(t-t_0)^2) & \text{odd } m \\ (-1)^{\frac{m}{2}} \Omega^{m+1} (t-t_0)^{m-\alpha+1} E_{2,m-\alpha+2}(-\Omega^2(t-t_0)^2) & \text{even } m \end{cases}$$

Proof. By using the series expansion of the ML function in Equation (18), for any $\beta \in \mathbb{C}$ it is:

$$G_{m,\beta}(z) := E_{1,m+\beta}(+\mathrm{i}z) - (-1)^m E_{1,m+\beta}(-\mathrm{i}z) = \sum_{k=0}^{\infty} \frac{\mathrm{i}^k z^k}{\Gamma(k+m+\beta)} (1 - (-1)^{m+k})$$

and since,

$$1 - (-1)^j = \begin{cases} 0 & \text{even } j \\ 2 & \text{odd } j \end{cases},$$

it is simple to evaluate:

$$G_{m,\beta}(z) = \begin{cases} 2 \sum\limits_{\substack{k=0 \\ \text{even } k}}^{\infty} \frac{\mathrm{i}^k z^k}{\Gamma(k+m+\beta)} = 2 \sum\limits_{k=0}^{\infty} \frac{\mathrm{i}^{2k} z^{2k}}{\Gamma(2k+m+\beta)} = 2E_{2,m+\beta}(-z^2) & \text{odd } m \\ 2 \sum\limits_{\substack{k=0 \\ \text{odd } k}}^{\infty} \frac{\mathrm{i}^k z^k}{\Gamma(k+m+\beta)} = 2 \sum\limits_{k=0}^{\infty} \frac{\mathrm{i}^{2k+1} z^{2k+1}}{\Gamma(2k+1+m+\beta)} = 2\mathrm{i}z E_{2,m+\beta+1}(-z^2) & \text{even } m \end{cases}.$$

Moreover it is sufficient to observe, thanks to Proposition 11, that:

$$J_{t_0}^\alpha \sin(\Omega(t-t_0)) = \frac{(t-t_0)^\alpha}{2\mathrm{i}} G_{0,1+\alpha}(\Omega(t-t_0))$$
$${}^{RL}D_{t_0}^\alpha \sin(\Omega(t-t_0)) = \frac{(t-t_0)^{-\alpha}}{2\mathrm{i}} G_{0,1-\alpha}(\Omega(t-t_0))$$
$${}^{C}D_{t_0}^\alpha \sin(\Omega(t-t_0)) = \mathrm{i}^m \Omega^m \frac{(t-t_0)^{m-\alpha}}{2\mathrm{i}} G_{m,1-\alpha}(\Omega(t-t_0))$$

from which the proof immediately follows. □

Proposition 15. *Let $\alpha > 0$, $m = \lceil \alpha \rceil$ and $\Omega \in \mathbb{R}$. For $t \geq t_0$ the function $\cos(\Omega(t-t_0))$ has the following fractional integral and derivatives:*

$$J_{t_0}^{\alpha} \cos(\Omega(t-t_0)) = (t-t_0)^{\alpha} E_{2,1+\alpha}(-\Omega^2(t-t_0)^2)$$
$$^{RL}D_{t_0}^{\alpha} \cos(\Omega(t-t_0)) = (t-t_0)^{-\alpha} E_{2,1-\alpha}(-\Omega^2(t-t_0)^2)$$
$$^{C}D_{t_0}^{\alpha} \cos(\Omega(t-t_0)) = \begin{cases} (-1)^{\frac{m+1}{2}} \Omega^{m+1} (t-t_0)^{m-\alpha+1} E_{2,m-\alpha+2}(-\Omega^2(t-t_0)^2) & \text{odd } m \\ (-1)^{\frac{m}{2}} \Omega^{m} (t-t_0)^{m-\alpha} E_{2,m-\alpha+1}(-\Omega^2(t-t_0)^2) & \text{even } m \end{cases}$$

Proof. The proof is similar to the proof of Proposition 14 where we consider now the function:

$$H_{m,\beta}(z) := E_{1,m+\beta}(+\mathrm{i}z) + (-1)^m E_{1,m+\beta}(-\mathrm{i}z) = \sum_{k=0}^{\infty} \frac{\mathrm{i}^k \Omega^k (t-t_0)^k}{\Gamma(k+m+\beta)} (1+(-1)^{m+k})$$

with,

$$1 + (-1)^j = \begin{cases} 2 & \text{even } j \\ 0 & \text{odd } j \end{cases}$$

and for which we obtain:

$$H_{m,\beta}(z) = \begin{cases} 2 \displaystyle\sum_{\substack{k=0 \\ \text{odd } k}}^{\infty} \frac{\mathrm{i}^k z^k}{\Gamma(k+m+\beta)} = 2 \sum_{k=0}^{\infty} \frac{\mathrm{i}^{2k+1} z^{2k+1}}{\Gamma(2k+1+m+\beta)} = 2\mathrm{i}z E_{2,m+\beta+1}(-z^2) & \text{odd } m \\ 2 \displaystyle\sum_{\substack{k=0 \\ \text{even } k}}^{\infty} \frac{\mathrm{i}^k z^k}{\Gamma(k+m+\beta)} = 2 \sum_{k=0}^{\infty} \frac{\mathrm{i}^{2k} z^{2k}}{\Gamma(2k+m+\beta)} = 2E_{2,m+\beta}(-z^2) & \text{even } m \end{cases}$$

thanks to which the proof follows by applying Proposition 12. □

The representation of $^{RL}D_{t_0}^{\alpha} \sin(\Omega(t-t_0))$ and $^{RL}D_{t_0}^{\alpha} \cos(\Omega(t-t_0))$, together with other related results, was already provided in [32] ([Remark 3). The above Propositions allow to extend to the RL integral and to the Caputo's derivative the results given in [32] solely for the RL derivative.

8. Concluding Remarks

We have discussed the evaluation of fractional integrals and fractional derivatives of some elementary functions. An alternative way of deriving RL and Caputo's derivatives from the GL has also been presented. We have observed that for several functions, the GL derivative generalizes, in a quite direct way, classic rules for integer-order differentiation. The RL and Caputo's derivatives of exponential, sine and cosine function have also been evaluated and represented in terms of special instances of the ML function. We have also shown that they appear as deviations from the GL derivative. The RL derivative converges, as the independent variable $t \to \infty$, faster than the Caputo's counterpart towards the GL derivative and an analytical explanation based on the asymptotic behavior of the ML function has been provided. Thanks to available codes for the evaluation of the ML function, the accurate computation of fractional derivatives of several elementary functions is possible.

Author Contributions: Investigation, R.G., E.K. and M.P.; Supervision, R.G.; Writing—review & editing, R.G., E.K. and M.P.

Acknowledgments: The authors are grateful to the anonymous reviewers for their constructive remarks which helped to improve the quality of the work and, in particular, to provide some simplified representations of fractional integrals and fractional derivatives.

Abbreviations

The following abbreviations are used in this manuscript:

FDE	Fractional differential equation
GL	Grünwald–Letnikov
LT	Laplace transform
ML	Mittag–Leffler
RL	Riemann–Liouville

References

1. Caponetto, R.; Dongola, G.D.; Fortuna, L.; Petráš, I. *Fractional Order Systems: Modeling and Control Applications*; Series on Nonlinear Science, Series A; World Scientific Publishing Co.: Singapore, 2010; Volume 72, p. xxi+178.
2. Herrmann, R. *Fractional Calculus: An Introduction for Physicists*; World Scientific Publishing Co.: Hackensack, NJ, USA, 2018; p. xxiv+610.
3. Mainardi, F. *Fractional Calculus and Waves in Linear Viscoelasticity*; Imperial College Press: London, UK, 2010; p. xx+347.
4. Ostalczyk, P. *Discrete Fractional Calculus: Applications in Control and Image Processing*; Series in Computer Vision; World Scientific Publishing Co.: Hackensack, NJ, USA, 2016; Volume 4, p. xxxi+361.
5. Podlubny, I. Fractional differential equations. In *Mathematics in Science and Engineering*; Academic Press Inc.: San Diego, CA, USA, 1999; Volume 198, p. xxiv+340.
6. Povstenko, Y. *Linear Fractional Diffusion-Wave Equation for Scientists and Engineers*; Springer: Cham, Switzerland, 2015; p. xiv+460.
7. Tarasov, V.E. *Fractional Dynamics*; Nonlinear Physical Science; Springer: Heidelberg, Germany; Higher Education Press: Beijing, China, 2010; p. xvi+504.
8. West, B.J. *Nature's Patterns and the Fractional Calculus*; Fractional Calculus in Applied Sciences and Engineering; De Gruyter: Berlin, Germany, 2017; Volume 2, p. xiii+199.
9. Diethelm, K. *The Analysis of Fractional Differential Equations*; Lecture Notes in Mathematics; Springer: Berlin, Germany, 2010; Volume 2004, p. viii+247.
10. Kilbas, A.A.; Srivastava, H.M.; Trujillo, J.J. *Theory and Applications of Fractional Differential Equations*; North-Holland Mathematics Studies; Elsevier Science B.V.: Amsterdam, The Netherlands, 2006; Volume 204, p. xvi+523.
11. Miller, K.S.; Ross, B. *An Introduction to the Fractional Calculus and Fractional Differential Equations*; John Wiley & Sons, Inc.: New York, NY, USA, 1993; p. xvi+366.
12. Samko, S.G.; Kilbas, A.A.; Marichev, O.I. *Fractional Integrals and Derivatives*; Gordon and Breach Science Publishers: Yverdon, Switzerland, 1993; p. xxxvi+976.
13. Gorenflo, R.; Mainardi, F. Fractional calculus: integral and differential equations of fractional order. In *Fractals and Fractional Calculus in Continuum Mechanics (Udine, 1996)*; CISM Courses and Lect.; Springer: Vienna, Austria, 1997; Volume 378, pp. 223–276.
14. Mainardi, F.; Gorenflo, R. Time-fractional derivatives in relaxation processes: A tutorial survey. *Fract. Calc. Appl. Anal.* **2007**, *10*, 269–308.
15. Hilfer, R.; Luchko, Y. Desiderata for Fractional Derivatives and Integrals. *Mathematics* **2019**, *7*, 149. [CrossRef]
16. Giusti, A. A comment on some new definitions of fractional derivative. *Nonlinear Dyn.* **2018**, *93*, 1757–1763. [CrossRef]
17. Ortigueira, M.D.; Tenreiro Machado, J.A. What is a fractional derivative? *J. Comput. Phys.* **2015**, *293*, 4–13. [CrossRef]
18. Tarasov, V.E. No violation of the Leibniz rule. No fractional derivative. *Commun. Nonlinear Sci. Numer. Simul.* **2013**, *18*, 2945–2948. [CrossRef]
19. Tarasov, V.E. No nonlocality. No fractional derivative. *Commun. Nonlinear Sci. Numer. Simul.* **2018**, *62*, 157–163. [CrossRef]
20. Grünwald, A. Uber "begrenzte" Derivationen und deren Anwendung. *Z. Angew. Math. Phys.* **1867**, *12*, 441–480.

21. Letnikov, A. Theory of differentiation with an arbitrary index. *Mat. Sb.* **1868**, *3*, 1–68. (In Russian)
22. Oldham, K.B.; Spanier, J. *The Fractional Calculus*; Academic Press: New York, NY, USA; London, UK, 1974; p. xiii+234.
23. Ortigueira, M.D.; Coito, F. From differences to derivatives. *Fract. Calc. Appl. Anal.* **2004**, *7*, 459–471.
24. Ortigueira, M.D. Comments on "Modeling fractional stochastic systems as non-random fractional dynamics driven Brownian motions". *Appl. Math. Model.* **2009**, *33*, 2534–2537. [CrossRef]
25. Ferrari, F. Weyl and Marchaud Derivatives: A Forgotten History. *Mathematics* **2018**, *6*, 6. [CrossRef]
26. Garrappa, R. Some formulas for sums of binomial coefficients and gamma functions. *Int. Math. Forum* **2007**, *2*, 725–733. [CrossRef]
27. Paris, R.B. Asymptotics of the special functions of fractional calculus. In *Handbook of Fractional Calculus with Applications Volume 1: Basic Theory*; Kochubei, A., Luchko, Y., Eds.; De Gruyter GmbH: Berlin, Germany, 2019; pp. 297–325.
28. Oldham, K.; Myland, J.; Spanier, J. *An Atlas of Functions*, 2nd ed.; Springer: New York, NY, USA, 2009; p. xii+748.
29. Olver, F.W.J. *Asymptotics and Special Functions*; AKP Classics; A K Peters, Ltd.: Wellesley, MA, USA, 1997; p. xviii+572.
30. Garrappa, R. Numerical evaluation of two and three parameter Mittag-Leffler functions. *SIAM J. Numer. Anal.* **2015**, *53*, 1350–1369. [CrossRef]
31. Garrappa, R.; Popolizio, M. Evaluation of generalized Mittag-Leffler functions on the real line. *Adv. Comput. Math.* **2013**, *39*, 205–225. [CrossRef]
32. Ciesielski, M.; Blaszczyk, T. An exact solution of the second-order differential equation with the fractional/generalised boundary conditions. *Adv. Math. Phys.* **2018**, *2018*, 7283518. [CrossRef]

6

Calculating Nodal Voltages using the Admittance Matrix Spectrum of an Electrical Network

Ioannis Dassios *, Andrew Keane and Paul Cuffe

School of Electrical and Electronic Engineering, University College Dublin, Dublin 4, Ireland; Andrew.Keane@ucd.ie (A.K.); paul.cuffe@ucd.ie (P.C.)
* Correspondence: ioannis.dassios@ucd.ie

Abstract: Calculating nodal voltages and branch current flows in a meshed network is fundamental to electrical engineering. This work demonstrates how such calculations can be performed using the eigenvalues and eigenvectors of the Laplacian matrix which describes the connectivity of the electrical network. These insights should permit the functioning of electrical networks to be understood in the context of spectral analysis.

Keywords: Laplacian matrix; power flow; admittance matrix; voltage profile

---

1. Introduction

Electrical power system calculations rely heavily on the bus admittance matrix, Y_{bus}, which is a Laplacian matrix weighted by the complex-valued admittance of each branch in the network. It is well established that the eigenvalues and eigenvectors (deemed the *spectrum*) of a Laplacian matrix encode meaningful information about a network's structure [1]. Recent work in [2,3] indicates that, in electrical networks, this spectrum can be directly related to nodal voltages and branch current flows. The purpose of the present paper is to *clarify* the derivations provided in [3]. The scope of the present work is narrowly theoretical: Linear algebra is used to articulate the correct relationship between the variables treated in [3].

Notwithstanding these modest ambitions, a key motivation for the present work is to begin to link power flow analysis with the mature literature [4] on spectral graph theory. Extant efforts to apply spectral graph theory to electrical networks are scarce, but include [5,6]. The use of graph theory more generally in this role is reviewed in [7,8]. Notably, simplistic topological approaches do not properly account for the physical realities of electrical power flow, and can thereby fail to identify the critical components in an electrical network [9–11]. The present work seeks to articulate one particular linkage between spectral graph theory and circuit theory, which may offer new ways to understand how power flows in meshed electrical networks.

The rest of this paper is organized as follows: In Section 2 we establish the necessary preliminaries, including electrical flow basics and notation. The main results are presented in Section 3.

2. Preliminaries

2.1. Electrical Flow Basics and Notation

Ohm's law linearly relates the current flowing through an edge in a circuit with the voltage difference between the nodes that the edge connects. Specifically, $I_{kj} = \frac{\Delta V_{kj}}{Z_{kj}}$, and $\sum_{j=1}^{N} I_{kj} = F_k$, $k, j = 1, 2, \ldots, N$, where I_{kj} is the current passing from the k-th node to the j-th in a (typically sparsely connected) network of N nodes, $\Delta V_{kj} = V_k - V_j$ is the voltage difference between the k-th node and the j-th, $Z_{kj} = Z_{jk}$ is branch impedance and F_k are complex-valued net current injections or withdrawals.

From the above notation we arrive easily at $\sum_{j=1}^{N} \frac{\Delta V_{kj}}{Z_{kj}} = F_k, \forall k = 1,2,\ldots,N$, i.e., $\sum_{j=1}^{N} \frac{(V_k - V_j)}{Z_{kj}} = F_k$, $\forall k = 1,2,\ldots,N$, or, equivalently,

$$V_k \sum_{j=1}^{N} \frac{1}{Z_{kj}} - \sum_{j=1}^{N} \frac{V_j}{Z_{kj}} = F_k, \quad \forall k = 1,2,\ldots,N. \tag{1}$$

In the article we will denote with δ_{ij} for the Kronecker delta, i.e., $\delta_{ii} = 1$ and $\delta_{ij} = 0$ for $i \neq j$. With $\bar{u}$ we will denote the complex conjugate of u, and with $^T, ^*$ the conjugate transpose, and conjugate transpose tensor respectively.

2.2. An Exemplary Electrical Network

To provide some context, a nation-spanning electrical power system is shown in Figure 1. This diagram of the `nesta_case2224_edin` test system [12] was created using the techniques described in [13], which uses electrical distances measures, rather than physical geography, to positions nodes. Note the relative spareseness of its connective struture, and how lower nominal voltage levels ($<$ 143 kV) correspond to more tree-like structures. This network of 2224 nodes supplies a total load of up to 60 GW, supplied from 378 different generating sites.

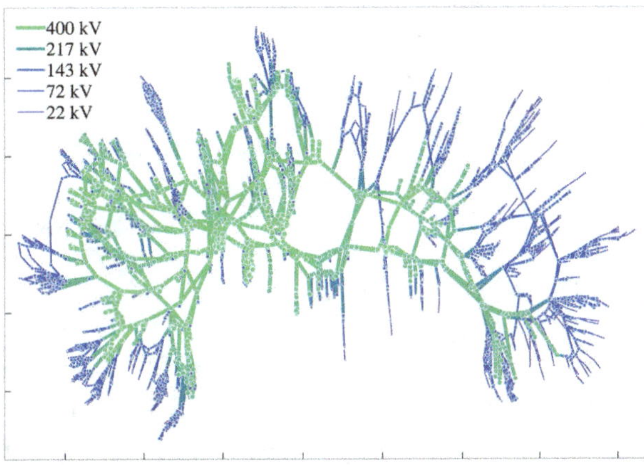

Figure 1. This diagram shows the `nesta_case2224_edin` test power system

3. Derivations

In this section, first we rewrite (1) in matrix form and define the relevant Laplacian matrix. Then we provide a formula which explicitly relates the voltage differences to the eigenvalues and eigenvectors of the Laplacian matrix for meshed electrical networks. We can state now the following theorem.

Theorem 1. *Consider an electrical network with branch currents I_{kj}, $\forall\, k, j = 1,2,\ldots,N$ passing from node k to node j, a complex impedance describing each branch $Z_{kj} = Z_{jk}$, and F_k being the complex-valued net current flow at each bus with $\sum_{k=1}^{N} F_k = 0$. Then the voltage difference ΔV_{mn} between two arbitrary nodes m and n is given by:*

$$\Delta V_{mn} = \sum_{j=2}^{N} \left[\frac{u_{mj} - u_{nj}}{\lambda_j} (\sum_{k=1}^{N} \bar{u}_{kj} F_k) \right] \tag{2}$$

where λ_k, $k = 2,3,\ldots,N$ are the non-zero eigenvalues of the G matrix (equivalent to the Y_{bus} matrix in the power systems context) which describes the connectivity of the electrical network:

$$G = [\delta_{kr}(\sum_{j=1}^{N} \frac{1}{Z_{kj}}) + (\delta_{kr} - 1)\frac{1}{Z_{kr}}]_{k=1,2,\ldots,N}^{r=1,2,\ldots,N} \tag{3}$$

and $\begin{bmatrix} u_{1k} & u_{2k} & \dots & u_{Nk} \end{bmatrix}^T$ *is an eigenvector of the eigenvalue* λ_k.

Proof. For $k = 1, 2, \dots, N$, Equation (1) can be written as

$$\begin{bmatrix} V_1 \sum_{j=1}^{N} \frac{1}{Z_{1j}} - \sum_{j=1}^{N} \frac{V_j}{Z_{1j}} \\ V_2 \sum_{j=1}^{N} \frac{1}{Z_{2j}} - \sum_{j=1}^{N} \frac{V_j}{Z_{2j}} \\ \vdots \\ V_N \sum_{j=1}^{N} \frac{1}{Z_{Nj}} - \sum_{j=1}^{N} \frac{V_j}{Z_{Nj}} \end{bmatrix} = \begin{bmatrix} F_1 \\ F_2 \\ \vdots \\ F_N \end{bmatrix},$$

or, equivalently,

$$\begin{bmatrix} V_1 \sum_{j=1}^{N} \frac{1}{Z_{1j}} - \frac{V_1}{Z_{11}} - \dots - \frac{V_N}{Z_{1N}} \\ V_2 \sum_{j=1}^{N} \frac{1}{Z_{2j}} - \frac{V_1}{Z_{21}} - \dots - \frac{V_N}{Z_{2N}} \\ \vdots \\ V_N \sum_{j=1}^{N} \frac{1}{Z_{Nj}} - \frac{V_1}{Z_{N1}} - \dots - \frac{V_N}{Z_{NN}} \end{bmatrix} = \begin{bmatrix} F_1 \\ F_2 \\ \vdots \\ F_N \end{bmatrix},$$

or, equivalently, by setting $V = \begin{bmatrix} V_i \end{bmatrix}_{i=1,2,\dots,N}$, $F = \begin{bmatrix} F_i \end{bmatrix}_{i=1,2,\dots,N}$, and

$$G = \begin{bmatrix} \sum_{j=1}^{N} \frac{1}{Z_{1j}} - \frac{1}{Z_{11}} & \dots & -\frac{1}{Z_{1N}} \\ -\frac{1}{Z_{21}} & \dots & -\frac{1}{Z_{2N}} \\ \vdots & \ddots & \vdots \\ -\frac{1}{Z_{N1}} & \dots & \sum_{j=1}^{N} \frac{1}{Z_{Nj}} - \frac{1}{Z_{NN}}, \end{bmatrix},$$

We arrive at $GV = F$. We observe that if G_{kj}, $k, j = 1, 2, \dots, N$ is an element of G, then for $k = j$, $G_{kk} = \sum_{j=1}^{N} \frac{1}{Z_{kj}} - \frac{1}{Z_{kk}}$ and for $k \neq j$, $G_{kj} = -\frac{1}{Z_{kj}}$. Hence G is given by (3). We will refer to G as the Laplacian matrix. Note that the rows of G sum to zero, i.e., the matrix has the zero eigenvalue (see [3,14]). The algebraic multiplicity of the zero eigenvalue in the Laplacian is the number of connected components in the network. In the power systems case we deal with only one network which means the algebraic multiplicity of the zero eigenvalue is one. Since the matrix G is symmetric it can be written in the following form:

$$G = PDP^*,$$

where $P = [u_{kj}]_{k=1,2,\dots,N}^{j=1,2,\dots,N}$, P^* is the conjugate transpose of P such that PP^* is the $N \times N$ identity matrix and D is the diagonal matrix $D = diag\{0, \lambda_2, \lambda_3, \dots, \lambda_N\}$. By applying the above expression into the system we get:

$$PDP^*V = F,$$

and since P^* is the inverse of P we have:

$$DP^*V = P^*F,$$

or, equivalently,

$$\begin{bmatrix} 0 \\ \lambda_2 \sum_{k=1}^{N} \bar{u}_{k2} V_k \\ \lambda_3 \sum_{k=1}^{N} \bar{u}_{k3} V_k \\ \vdots \\ \lambda_N \sum_{k=1}^{N} \bar{u}_{kN} V_k \end{bmatrix} = \begin{bmatrix} \sum_{k=1}^{N} \bar{u}_{k1} F_k \\ \sum_{k=1}^{N} \bar{u}_{k2} F_k \\ \sum_{k=1}^{N} \bar{u}_{k3} F_k \\ \vdots \\ \sum_{k=1}^{N} \bar{u}_{kN} F_k \end{bmatrix}.$$

Let $\mathbf{1}_N$ be a column vector that contains exactly N 1's. From the fact that every row of G sums to zero we have the eigenspace of the zero eigenvalue. Indeed $G \cdot \mathbf{1}_N = 0 \cdot \mathbf{1}_N$ which means that $< \mathbf{1}_N >$ is the eigenspace of the zero eigenvalue. Hence there exist $c \in \mathbb{C}$ such that

$$\left[\begin{array}{c} u_{i1} \end{array} \right]_{i=1,2,\ldots,N} = c \cdot \mathbf{1}_N \tag{4}$$

From (4) $u_{k1} = c, \quad \forall k = 1,2,\ldots,N$, or, equivalently, $\bar{u}_{k1} = \bar{c}, \forall k = 1,2,\ldots,N$. In addition, $\sum_{k=0}^{N} F_k = 0$. Hence, $\sum_{k=1}^{N} \bar{u}_{k1} F_k = \bar{c} \sum_{k=1}^{N} F_k = 0$. By ignoring the first row of each column of the above expression we get:

$$\begin{bmatrix} \lambda_2 \sum_{k=1}^{N} \bar{u}_{k2} V_k \\ \lambda_3 \sum_{k=1}^{N} \bar{u}_{k3} V_k \\ \vdots \\ \lambda_N \sum_{k=1}^{N} \bar{u}_{kN} V_k \end{bmatrix} = \begin{bmatrix} \sum_{k=1}^{N} \bar{u}_{k2} F_k \\ \sum_{k=1}^{N} \bar{u}_{k3} F_k \\ \vdots \\ \sum_{k=1}^{N} \bar{u}_{kN} F_k \end{bmatrix}.$$

Which can be rewritten in the following form:

$$\begin{bmatrix} \sum_{k=1}^{N} \bar{c} V_k \\ \lambda_2 \sum_{k=1}^{N} \bar{u}_{k2} V_k \\ \lambda_3 \sum_{k=1}^{N} \bar{u}_{k3} V_k \\ \vdots \\ \lambda_N \sum_{k=1}^{N} \bar{u}_{kN} V_k \end{bmatrix} = \begin{bmatrix} \sum_{k=1}^{N} \bar{c} V_k \\ \sum_{k=1}^{N} \bar{u}_{k2} F_k \\ \sum_{k=1}^{N} \bar{u}_{k3} F_k \\ \vdots \\ \sum_{k=1}^{N} \bar{u}_{kN} F_k \end{bmatrix}.$$

If we set $\Lambda = diag\{\lambda_i\}_{2 \le i \le N}$, $U = \left[\begin{array}{c} \bar{u}_{ij} \end{array} \right]_{i=1,2,\ldots,N}^{j=1,2,\ldots,N}$, we have

$$\begin{bmatrix} 1 & 0_{N-1}^T \\ 0_{N-1} & \Lambda \end{bmatrix} \begin{bmatrix} \bar{c} \cdot \mathbf{1}_N^T \\ \bar{U} \end{bmatrix} V = \begin{bmatrix} \sum_{k=1}^{N} \bar{c} V_k \\ \sum_{k=1}^{N} \bar{u}_{k2} F_k \\ \sum_{k=1}^{N} \bar{u}_{k3} F_k \\ \vdots \\ \sum_{k=1}^{N} \bar{u}_{kN} F_k \end{bmatrix},$$

or, equivalently,

$$\begin{bmatrix} \bar{c} \cdot \mathbf{1}_N^T \\ \bar{U} \end{bmatrix} V = \begin{bmatrix} 1 & 0_{N-1}^T \\ 0_{N-1} & \Lambda^{-1} \end{bmatrix} \begin{bmatrix} \sum_{k=1}^{N} \bar{c} V_k \\ \sum_{k=1}^{N} \bar{u}_{k2} F_k \\ \sum_{k=1}^{N} \bar{u}_{k3} F_k \\ \vdots \\ \sum_{k=1}^{N} \bar{u}_{kN} F_k \end{bmatrix},$$

or, equivalently,

$$V = \left[\begin{array}{cc} c \cdot \mathbf{1}_N & U \end{array} \right] \begin{bmatrix} \sum_{k=1}^{N} \bar{c} V_k \\ \frac{1}{\lambda_2} \sum_{k=1}^{N} \bar{u}_{k2} F_k \\ \frac{1}{\lambda_3} \sum_{k=1}^{N} \bar{u}_{k3} F_k \\ \vdots \\ \frac{1}{\lambda_N} \sum_{k=1}^{N} \bar{u}_{kN} F_k \end{bmatrix}$$

or, equivalently,

$$V = \begin{bmatrix} c\bar{c} \sum_{k=1}^{N} V_k + \sum_{j=2}^{N} \frac{u_{1j}}{\lambda_j} \sum_{k=1}^{N} \bar{u}_{kj} F_k \\ c\bar{c} \sum_{k=1}^{N} V_k + \sum_{j=2}^{N} \frac{u_{2j}}{\lambda_j} \sum_{k=1}^{N} \bar{u}_{kj} F_k \\ \vdots \\ c\bar{c} \sum_{k=1}^{N} V_k + \sum_{j=2}^{N} \frac{u_{Nj}}{\lambda_j} \sum_{k=1}^{N} \bar{u}_{kj} F_k \end{bmatrix}$$

or, equivalently, for $b_j = \sum_{k=1}^{N} \bar{u}_{kj} F_k$:

$$V = \begin{bmatrix} \sum_{k=1}^{N} \bar{u}_{k1} V_k + \sum_{j=2}^{N} \frac{u_{1j}}{\lambda_j} b_j \\ \sum_{k=1}^{N} \bar{u}_{k1} V_k + \sum_{j=2}^{N} \frac{u_{2j}}{\lambda_j} b_j \\ \vdots \\ \sum_{k=1}^{N} \bar{u}_{k1} V_k + \sum_{j=2}^{N} \frac{u_{Nj}}{\lambda_j} b_j \end{bmatrix}.$$

Let V_m, V_n be two arbitrary nodal voltages, i.e.,

$$V_m = c\bar{c} \sum_{k=1}^{N} V_k + \sum_{j=2}^{N} \frac{u_{mj}}{\lambda_j} b_j,$$
$$V_n = c\bar{c} \sum_{k=1}^{N} V_k + \sum_{j=2}^{N} \frac{u_{nj}}{\lambda_j} b_j.$$

Then, the difference between them is given by

$$\Delta V_{mn} = c\bar{c} \sum_{k=1}^{N} V_k + \sum_{j=2}^{N} \frac{u_{mj}}{\lambda_j} b_j - c\bar{c} \sum_{k=1}^{N} V_k - \sum_{j=2}^{N} \frac{u_{nj}}{\lambda_j} b_j,$$

or, equivalently,

$$\Delta V_{mn} = \sum_{j=2}^{N} \frac{u_{mj} - u_{nj}}{\lambda_j} b_j,$$

or, equivalently,

$$\Delta V_{mn} = \sum_{j=2}^{N} \left[\frac{u_{mj} - u_{nj}}{\lambda_j} \left(\sum_{k=1}^{N} \bar{u}_{kj} F_k \right) \right].$$

□

4. Conclusions

This work has clarified the relationship between the admittance matrix spectrum, the current inflows & withdrawals prevailing in an electrical network and the resulting nodal voltage profile. Applying these spectral relationships to practical electrical engineering problems is left to future work.

Author Contributions: Methodology, I.D.; Formal Analysis, I.D., P.C. and A.K.; Writing—Original Draft Preparation, I.D.; Writing—Review and Editing, I.D. and P.C.; Visualization, P.C.; Supervision, P.C. and A.K.

References

1. Mohar, B.; Alavi, Y.; Chartrand, G.; Oellermann, O. The Laplacian spectrum of graphs. *Graph Theory Comb. Appl.* **1991**, *2*, 12.
2. Rubido, N.; Grebogi, C.; Baptista, M.S. Structure and function in flow networks. *EPL (Europhys. Lett.)* **2013**, *101*, 68001. [CrossRef]
3. Rubido, N.; Grebogi, C.; Baptista, M.S. General analytical solutions for DC/AC circuit-network analysis. *Eur. Phys. J. Spec. Top.* **2017**, *226*, 1829–1844. [CrossRef]
4. Chung, F.R.; Graham, F.C. *Spectral Graph Theory*; Number 92; American Mathematical Society: Providence, RI, USA, 1997.
5. Edström, F. On eigenvalues to the Y-bus matrix. *Int. J. Electr. Power Energy Syst.* **2014**, *56*, 147–150. [CrossRef]

6. Edström, F.; Söder, L. On spectral graph theory in power system restoration. In Proceedings of the 2011 2nd IEEE PES International Conference and Exhibition on Innovative Smart Grid Technologies (ISGT Europe), Manchester, UK, 5–7 December 2011; pp. 1–8.
7. Pagani, G.A.; Aiello, M. The power grid as a complex network: A survey. *Phys. A Stat. Mech. Its Appl.* **2013**, *392*, 2688–2700. [CrossRef]
8. Sun, K. Complex networks theory: A new method of research in power grid. In Proceedings of the 2005 IEEE/PES Transmission and Distribution Conference and Exhibition: Asia and Pacific, Dalian, China, 18 August 2005; pp. 1–6.
9. Hines, P.; Cotilla-Sanchez, E.; Blumsack, S. Do topological models provide good information about electricity infrastructure vulnerability? *Chaos Interdiscip. J. Nonlinear Sci.* **2010**, *20*, 033122. [CrossRef] [PubMed]
10. Verma, T.; Ellens, W.; Kooij, R.E. Context-independent centrality measures underestimate the vulnerability of power grids. *arXiv* 2013, arXiv:1304.5402.
11. Cuffe, P. A comparison of malicious interdiction strategies against electrical networks. *IEEE J. Emerg. Sel. Top. Circuits Syst.* **2017**, *7*, 205–217. [CrossRef]
12. Coffrin, D.; Gordon, D.; Scott, P. Nesta the nicta energy system test case archive. *arXiv* 2014, arXiv:1411.0359.
13. Cuffe, P.; Keane, A. Visualizing the electrical structure of power systems. *IEEE Syst. J.* **2017**, *11*, 1810–1821. [CrossRef]
14. Dassios, I.K.; Cuffe, P.; Keane, A. Visualizing voltage relationships using the unity row summation and real valued properties of the F_{LG} matrix. *Electr. Power Syst. Res.* **2016**, *140*, 611–618. [CrossRef]

7

Discrete Two-Dimensional Fourier Transform in Polar Coordinates: Theory and Operational Rules

Natalie Baddour

Department of Mechanical Engineering, University of Ottawa, 161 Louis Pasteur, Ottawa, ON K1N 6N5, Canada; nbaddour@uottawa.ca.

Abstract: The theory of the continuous two-dimensional (2D) Fourier transform in polar coordinates has been recently developed but no discrete counterpart exists to date. In this paper, we propose and evaluate the theory of the 2D discrete Fourier transform (DFT) in polar coordinates. This discrete theory is shown to arise from discretization schemes that have been previously employed with the 1D DFT and the discrete Hankel transform (DHT). The proposed transform possesses orthogonality properties, which leads to invertibility of the transform. In the first part of this two-part paper, the theory of the actual manipulated quantities is shown, including the standard set of shift, modulation, multiplication, and convolution rules. Parseval and modified Parseval relationships are shown, depending on which choice of kernel is used. Similar to its continuous counterpart, the 2D DFT in polar coordinates is shown to consist of a 1D DFT, DHT and 1D inverse DFT.

Keywords: Fourier Theory; DFT in polar coordinates; polar coordinates; multidimensional DFT; discrete Hankel Transform; discrete Fourier Transform; Orthogonality

---

1. Introduction

The Fourier transform (FT) in continuous and discrete forms has seen much application in various disciplines [1]. It easily expands to multiple dimensions, with all the same rules of the one-dimensional (1D) case carrying into the multiple dimensions. Recent work has developed the complete toolkit for working with the continuous multidimensional Fourier transform in two-dimensional (2D) polar and three-dimensional (3D) spherical polar coordinates [2–4]. However, to date no discrete version of the 2D Fourier transform exists in polar coordinates. Hence, the aim of this paper is to develop the discrete version of the 2D Fourier transform in polar coordinates.

For the discrete version of the transform, the values of the transform will be available only at discrete points. To quote Bracewell [5], "we often think of this as though an underlying function of a continuous variable really exists and we are approximating it. From an operational viewpoint, however, it is irrelevant to talk about the existence of values other than those given and those computed (the input and output). Therefore, it is desirable to have a mathematical theory of the actual quantities manipulated". This paper thus aims to develop the mathematical theory of the discrete two-dimensional Fourier transform in polar coordinates. Standard 'operational rules' associated with any Fourier transform (shift, modulation, multiplication, and convolution) will be developed. Parseval and modified Parseval relationships will also be shown, depending on the choice of kernel used.

To the best of the author's knowledge, there is no discrete version of the 2D Fourier transform in polar coordinates. It was shown in [2,4] that the 2D continuous Fourier transform in polar coordinates is actually a combination of a single dimensional Fourier transform, a Hankel transform, followed by an inverse Fourier transform. Of course, the discrete version of the 1D standard Fourier transform is very well known and the literature on this subject alone is vast. Recently, a discrete version of the Hankel transform has been proposed [6,7], yet this discrete transform is still in one dimension. We will show further on that the 2D Fourier transform in polar coordinates requires this transform.

Other researchers have defined the idea of a polar Fourier transform (polar FT), in which the original function in the spatial domain is in Cartesian coordinates but its FT is computed in polar coordinates, meaning discrete polar Fourier data and Cartesian spatial data [8–10]. Fast Fourier transforms (FFT) have also been developed for non-equispaced data, referred to as a unequally spaced FFT (USFFT) or non-uniform FFT (NUFFT) [11–15]. Using this approach, frequencies in a polar frequency domain can be considered to be unequally spaced and hence the problem of evaluating a polar FT can be considered as a special case of the USFFT. Averbuch et al. [8] compared the accuracy results of their proposed approach which used a pseudo-polar grid to those obtained by an USFFT approach and demonstrated that their approach show marked advantage over the USFFT. Fenn et al. [10] examined computing the FT on a polar, modified and pseudo-polar grid using the NUFFT, for both forward and backwards transforms. They demonstrated that the NUFFT was effective at this computation. Although the above demonstrate that the computation of a discrete 2D FT on a polar grid has previously been considered in the literature, there is, to date, no discrete 2D Fourier transform in polar coordinates that exists as a transform in its own right, with its own set of rules of the actual manipulated quantities.

The outline of the paper is as follows. Section 2 presents some of the necessary background material. Section 3 introduces an intuitive 'motivation' for the definition of the 2D Discrete Fourier Transform (DFT) in polar coordinates that will be introduced by considering space and band-limited functions. This leads to an intuitive discretization scheme and an intuitive kernel for the proposed 2D DFT, which is introduced in Section 4. Section 5 introduces the proposed transform while Section 6 derives the transform properties including modulation, shift, multiplication and convolution rules. Section 7 discusses Parseval relations while Section 8 demonstrates that the proposed transform can indeed be decomposed a sequence of DFT, Discrete Hankel Transform (DHT) and inverse DFT (IDFT), in keeping with the approach of the continuous version of the transform. Finally, Section 8 concludes the paper.

2. Background: Continuous 2D Fourier Transforms in Polar Coordinates

The 2D Fourier transform of a function $f(\vec{r}) = f(x, y)$ expressed in 2D Cartesian coordinates is defined as [4]:

$$F(\vec{\omega}) = F(\omega_x, \omega_y) = \int_{-\infty}^{\infty}\int_{-\infty}^{\infty} f(x, y)\, e^{-i\vec{\omega}\cdot\vec{r}}\, dx\, dy \tag{1}$$

The inverse Fourier transform is given by:

$$f(\vec{r}) = f(x, y) = \frac{1}{(2\pi)^2}\int_{-\infty}^{\infty}\int_{-\infty}^{\infty} F(\omega_x, \omega_y) e^{i\vec{\omega}\cdot\vec{r}}\, d\omega_x\, d\omega_y \tag{2}$$

where the shorthand notation of $\vec{\omega} = (\omega_x, \omega_y)$, $\vec{r} = (x, y)$ has been used. For functions with cylindrical or circular symmetry, it is often more convenient to express both the original function $f(\vec{r})$ and its 2D Fourier transform $F(\vec{\omega})$ in polar coordinates. If so, polar coordinates can be introduced as $x = r\cos\theta$, $y = r\sin\theta$ and similarly in the spatial frequency domain as $\omega_x = \rho\cos\psi$ and $\omega_y = \rho\sin\psi$, otherwise written as, $r^2 = x^2 + y^2$, $\theta = \arctan(y/x)$ and $\rho^2 = \omega_x^2 + \omega_y^2$, $\psi = \arctan(\omega_y/\omega_x)$.

Given a function in polar coordinates $f(r, \theta)$, where θ is the angular variable and r is the radial variable, the function can be expanded into a Fourier series as:

$$f(\vec{r}) = f(r, \theta) = \sum_{n=-\infty}^{\infty} f_n(r)\, e^{in\theta} \tag{3}$$

where the Fourier coefficients are given by:

$$f_n(r) = \frac{1}{2\pi}\int_{-\pi}^{\pi} f(r,\theta)\, e^{-in\theta} d\theta \tag{4}$$

Similarly, the 2D Fourier transform of $f(r,\theta)$ is given by $F(\rho,\psi)$. The function $F(\rho,\psi)$, where ψ is the angular frequency variable and ρ is the radial frequency variable, can also be expanded into a Fourier series as:

$$F(\vec{\omega}) = F(\rho,\psi) = \sum_{n=-\infty}^{\infty} F_n(\rho)\, e^{in\psi} \tag{5}$$

where:

$$F_n(\rho) = \frac{1}{2\pi}\int_{-\pi}^{\pi} F(\rho,\psi)e^{-in\psi} d\psi \tag{6}$$

We note that $F_n(\rho)$ is NOT the Fourier transform of $f_n(r)$. The development details can be found in [4], where it is demonstrated that the relationship is given by:

$$\begin{aligned} F_n(\rho) &= 2\pi\, i^{-n}\int_0^{\infty} f_n(r)J_n(\rho r)\, r dr \\ &= 2\pi\, i^{-n}\mathbb{H}_n\{f_n(r)\}, \end{aligned} \tag{7}$$

where $\mathbb{H}_n\{\cdot\}$ denotes an nth order Hankel transform [3], see Appendix A.1 [3]. The inverse relationship is given by:

$$\begin{aligned} f_n(r) &= \tfrac{i^n}{2\pi}\int_0^{\infty} F_n(\rho)J_n(\rho r)\, \rho d\rho \\ &= \tfrac{i^n}{2\pi}\,\mathbb{H}_n\{F_n(\rho)\}. \end{aligned} \tag{8}$$

Thus, the nth term in the Fourier series of the original function will Hankel transform into the nth term of the Fourier series of the Fourier transform function via an nth order Hankel transform for the nth term. Therefore, the steps for finding the 2D Fourier transform $F(\rho,\psi)$ of a function $f(r,\theta)$ are (i) finding its Fourier series coefficients in the angular variable $f_n(r)$, Equation (4), (ii) finding the Fourier series coefficient of the Fourier transform, $F_n(\rho)$ via $F_n(\rho) = 2\pi\, i^{-n}\mathbb{H}_n\{f_n(r)\}$, then (iii) taking the inverse Fourier series transform (summing the series) with respect to the frequency angular variable, Equation (5).

The discrete equivalent to the relationships given by Equations (3) to (8) have not been developed and it is the goal of this paper to develop the discrete counterparts of these equations.

3. Motivation for the Discrete 2D Fourier Transform in Polar Coordinates

3.1. Space-Limited Functions

To motivate the discrete version of a 2D Fourier transform in polar coordinates, we follow the same path used to derive the classical discrete Fourier transform (DFT) and also the recently-proposed discrete Hankel transform (DHT) [6]. This approach starts with a space (or time for the traditional FT) limited function in one domain and then makes the assumption that the transform of the function is also limited in the corresponding frequency domain. While strictly speaking, functions cannot be limited in both space and spatial frequency domains, in practice, they can be made 'effectively' limited in the domain where they are not exactly limited by suitable truncation of an appropriate series. This is how the DFT and DHT were both motivated. The discrete transforms derived in this manner then have properties that exist in their own right, independent of their ability to approximate their continuous transform counterpart.

The same path is followed here. A function $f(r,\theta)$ in polar coordinates, where θ is the angular variable and r is the radial variable, is expanded into a Fourier series given by Equation (3) where $f_n(r)$ is given by Equation (4). It is now supposed that the function $f(r,\theta)$ is space-limited, meaning that $f(r,\theta)$ and, by virtue of Equation (4), all the Fourier coefficients $f_n(r)$ are zero for $r \geq R$. Then, it follows that each of the Fourier coefficients $f_n(r)$ can be written in terms of a Fourier Bessel series (see [6] and Appendix A.2) as:

$$f_n(r) = \begin{cases} \sum\limits_{k=1}^{\infty} C_{nk}^{f} J_n\left(\frac{j_{nk}r}{R}\right) & r < R \\ 0 & r \geq R \end{cases} \tag{9}$$

where the order, n, of the Bessel function in (9) matches the order f_n of the Fourier coefficient, C_{nk}^{f} denotes the kth coefficient of the Fourier–Bessel expansion of $f_n(r)$ and denotes the kth zero of the nth Bessel function. The C_{nk}^{f} can be found from [16]:

$$C_{nk}^{f} = \frac{2}{R^2 J_{n+1}^2(j_{nk})} \int_0^R f_n(r) J_n\left(\frac{j_{nk}r}{R}\right) r\,dr \tag{10}$$

Equation (7) gives the relationship between the Fourier coefficients of the function itself and its 2D Fourier transform. Using Equation (7) and making use of the space limited nature of $f_n(r)$, Equation (10) can be written as:

$$C_{nk}^{f} = \frac{2}{R^2 J_{n+1}^2(j_{nk})} \int_0^\infty f_n(r) J_n\left(\frac{j_{nk}r}{R}\right) r\,dr = \frac{i^n}{\pi R^2 J_{n+1}^2(j_{nk})} F_n\left(\frac{j_{nk}}{R}\right) \tag{11}$$

Therefore, for $r < R$, Equation (9) becomes:

$$f_n(r) = \frac{i^n}{\pi R^2} \sum_{m=1}^{\infty} F_n\left(\frac{j_{nm}}{R}\right) \frac{1}{J_{n+1}^2(j_{nm})} J_n\left(\frac{j_{nm}r}{R}\right) \tag{12}$$

Equation (12) with its infinite summation is *exact*. Now, evaluating Equation (12) at $r = r_{nk} = \frac{j_{nk}R}{j_{nN_1}}$ for *any* N_1 and where $k < N_1$ gives:

$$f_n\left(\frac{j_{nk}R}{j_{nN_1}}\right) = \frac{i^n}{\pi R^2} \sum_{m=1}^{\infty} F_n\left(\frac{j_{nm}}{R}\right) \frac{1}{J_{n+1}^2(j_{nm})} J_n\left(\frac{j_{nm}j_{nk}}{j_{nN_1}}\right) \qquad k < N_1 \tag{13}$$

For $k < N_1$, then $r_{nk} = \frac{j_{nk}R}{j_{nN_1}} < R$, and Equation (13), summing over infinite m, is still exact. For $k \geq N_1$, then $r_{nk} = \frac{j_{nk}R}{j_{nN_1}} \geq R$ and by the assumption of the space-limited nature of the function, $f(r_{nk}) = 0$ for $k \geq N_1$.

We now assume that the function is also effectively band limited, in addition to being space-limited. Now, a function cannot be finite in both space and spatial frequency (equivalently if using a standard Fourier transform it cannot be finite in both time and frequency). However, if a function is *effectively* band-limited, then there exists an integer N_1 for which $F_n\left(\frac{j_{nm}}{R}\right) \approx 0$ for $m > N_1$. In other words, an interval can be found beyond which the Fourier transform coefficients $F_n(\rho)$ become very small. Since the convergence of the Fourier–Bessel series in (13) is known, then $\lim\limits_{m\to\infty} F_n\left(\frac{j_{nm}}{R}\right) = 0$. In other words, for any arbitrarily small ρ, there exists an integer N_1 for which $F_n\left(\frac{j_{nm}}{R}\right) < \rho$ for $m > N_1$.

Hence, using notion of an effective band-limit as stated in the preceding paragraph, the series in Equation (13) can be terminated at a suitably chosen N_1, thus giving an effective band limit. Termination of the series at $m = N_1$ is equivalent to assuming that $F_n(\rho) \approx 0$ for $\rho > W_\rho = \frac{j_{nN_1}}{R}$. It is noted that at $m = N_1$, the last term in Equation (13) is $J_n\left(\frac{j_{nN_1} j_{nk}}{j_{nN_1}}\right) = J_n(j_{nk}) = 0$, so that termination of the series at N_1 implies that Equation (13) becomes

$$f_n\left(\frac{j_{nk}R}{j_{nN_1}}\right) = \frac{i^n}{\pi R^2} \sum_{m=1}^{N_1-1} \frac{1}{J_{n+1}^2(j_{nm})} J_n\left(\frac{j_{nm} j_{nk}}{j_{nN_1}}\right) F_n\left(\frac{j_{nm}}{R}\right) \quad k = 1..N_1 - 1 \tag{14}$$

Equation (14) is the discrete equivalent of Equation (8) in that it demonstrates that the relationship between discrete samples of $f_n(r)$ and $F_n(r)$ is given by a discrete Hankel transform type of relationship, whereas the continuous relationship involved a continuous Hankel transform. The termination of the series at N_1 is equivalent to assuming an "effective" band-limit on the function. In other words, it states that for $m > N_1$, the values of $F_n\left(\frac{j_{nm}}{R}\right)$, which from Equation (11) are proportional to the Fourier–Bessel coefficients, are negligibly small. Of course, this is never exactly true, however, since the Fourier–Bessel series converges, it is always possible to choose N_1 so that the approximation introduced by truncating the series at N_1 is good [16].

The truncation of the series at N_1 also permits Equation (14) to be easily inverted. Multiplying both sides of (14) by $\frac{4J_n\left(\frac{j_{nk} j_{np}}{j_{nN_1}}\right)}{j_{nN_1}^2 J_{n+1}^2(j_{nk})}$ and summing over k gives:

$$\sum_{k=1}^{N_1-1} f_n\left(\frac{j_{nk}R}{j_{nN_1}}\right) \frac{4J_n\left(\frac{j_{nk} j_{np}}{j_{nN_1}}\right)}{j_{nN_1}^2 J_{n+1}^2(j_{nk})} = \sum_{m=1}^{N_1-1} \frac{i^n}{\pi R^2} \underbrace{\sum_{k=1}^{N_1-1} \frac{4J_n\left(\frac{j_{nm} j_{nk}}{j_{nN_1}}\right) J_n\left(\frac{j_{nk} j_{np}}{j_{nN_1}}\right)}{j_{nN_1}^2 J_{n+1}^2(j_{nm}) J_{n+1}^2(j_{nk})}}_{=\delta_{mp}} F_n\left(\frac{j_{nm}}{R}\right) \tag{15}$$

where we have used the discrete orthogonality of the Bessel functions as given in Appendix A.4. Hence,

$$F_n\left(\frac{j_{np}}{R}\right) = i^{-n} \pi R^2 \sum_{k=1}^{N_1-1} f_n\left(\frac{j_{nk}R}{j_{nN_1}}\right) \frac{4J_n\left(\frac{j_{nk} j_{np}}{j_{nN_1}}\right)}{j_{nN_1}^2 J_{n+1}^2(j_{nk})} \tag{16}$$

Equations (14) and (16) offer the basic structure on which to base the discrete transform formulation. Equation (16) is the basic structure to define the forward transform and Equation (14) offers the basic structure to define the inverse transform.

To proceed further, we need ways to compute $f_n\left(\frac{j_{nk}R}{j_{nN_1}}\right)$ and $F_n\left(\frac{j_{nm}}{R}\right)$. Here, the theory of discrete Fourier transforms can be used. For $n \in [-M, M]$ where $N_2 = 2M + 1$, it is shown in [17] that the Fourier coefficients $f_n(r)$ and $F_n(\rho)$ can be well approximated with expressions given by (see Appendix A.5):

$$\begin{aligned} F_n(\rho) &\approx \frac{1}{N_2} \sum_{p=-M}^{M} F\left(\rho, \frac{2\pi p}{N_2}\right) e^{-i\frac{2\pi n p}{N_2}} \\ f_n(r) &\approx \frac{1}{N_2} \sum_{p=-M}^{M} f\left(r, \frac{2\pi p}{N_2}\right) e^{-i\frac{2\pi n p}{N_2}} \end{aligned} \tag{17}$$

Hence, we will use Equation (17) to write:

$$\begin{aligned} F_n\left(\frac{j_{nm}}{R}\right) &= \frac{1}{N_2} \sum_{p=-M}^{M} F\left(\frac{j_{pm}}{R}, \frac{2\pi p}{N_2}\right) e^{-i\frac{2\pi n p}{N_2}} \\ f_n\left(\frac{j_{nk}R}{j_{nN_1}}\right) &= \frac{1}{N_2} \sum_{p=-M}^{M} f\left(\frac{j_{pk}R}{j_{pN_1}}, \frac{2\pi p}{N_2}\right) e^{-i\frac{2\pi n p}{N_2}} \end{aligned} \tag{18}$$

Equation (18) is a key assumption of the development. Note that in both cases, the function is sampled in the summation over p at the radial variable $\left(\frac{j_{pk,m}}{R}\right)$, that is, it is included in the summation index. However, the function on the left hand side of Equation (18) is sampled at $\left(\frac{j_{nm}}{R}\right)$ We show in Appendix A.6 that this assumption is valid. This assumption is what also permits the invertibility of the discrete transforms, since without this assumption it would not be possible to propose an invertible, orthogonal discrete transform. Equation (18) will be used to derive the forward and inverse discrete transforms.

3.1.1. Forward Transform

For the forward transform, we can start with Equation (16), and use the key relationships given by Equation (18). Under these conditions, Equation (16) becomes:

$$\underbrace{\frac{1}{N_2}\sum_{l=-M}^{M} F\left(\frac{j_{lm}}{R},\frac{2\pi l}{N_2}\right)e^{-i\frac{2\pi nl}{N_2}}}_{F_n\left(\frac{j_{nm}}{R}\right)} = i^{-n}\pi R^2\sum_{k=1}^{N_1-1}\frac{4J_n\left(\frac{j_{nk}j_{nm}}{j_{nN_1}}\right)}{j_{nN_1}^2 J_{n+1}^2(j_{nk})}\underbrace{\left\{\frac{1}{N_2}\sum_{p=-M}^{M} f\left(\frac{j_{pk}R}{j_{pN_1}},\frac{2\pi p}{N_2}\right)e^{-i\frac{2\pi np}{N_2}}\right\}}_{f_n\left(\frac{j_{nk}R}{j_{nN_1}}\right)} \tag{19}$$

Equation (19) is the discrete equivalent of Equation (7). From Equation (19), multiply both sides by $e^{+i\frac{2\pi nq}{N_2}}$ and sum from $n = -M..M$ gives:

$$\sum_{n=-M}^{M}\sum_{l=-M}^{M} F\left(\frac{j_{lm}}{R},\frac{2\pi l}{N_2}\right)e^{-i\frac{2\pi nl}{N_2}}e^{+i\frac{2\pi nq}{N_2}} = \sum_{n=-M}^{M} i^{-n}\pi R^2\sum_{k=1}^{N_1-1}\frac{4J_n\left(\frac{j_{nk}j_{nm}}{j_{nN_1}}\right)}{j_{nN_1}^2 J_{n+1}^2(j_{nk})}\sum_{p=-M}^{M} f\left(\frac{j_{pk}R}{j_{pN_1}},\frac{2\pi p}{N_2}\right)e^{-i\frac{2\pi np}{N_2}}e^{+i\frac{2\pi nq}{N_2}} \tag{20}$$

Interchanging the order of summation on the left hand side of (20) and using the orthogonality relationship of the complex exponential (Appendix A.3) gives:

$$F\left(\frac{j_{qm}}{R},\frac{2\pi q}{N_2}\right) = \frac{2\pi R^2}{N_2}\sum_{n=-M}^{M}\sum_{k=1}^{N_1-1}\sum_{p=-M}^{M} f\left(\frac{j_{pk}R}{j_{pN_1}},\frac{2\pi p}{N_2}\right)\frac{2i^{-n}J_n\left(\frac{j_{nk}j_{nm}}{j_{nN_1}}\right)}{j_{nN_1}^2 J_{n+1}^2(j_{nk})}e^{-i\frac{2\pi np}{N_2}}e^{+i\frac{2\pi nq}{N_2}} \tag{21}$$

3.1.2. Inverse Transform

For the inverse transform, we start with the structure of Equation (14) and then use the key approximations given in Equation (18) to obtain:

$$\underbrace{\frac{1}{N_2}\sum_{p=-M}^{M} f\left(\frac{j_{pk}R}{j_{pN_1}},\frac{2\pi p}{N_2}\right)e^{-i\frac{2\pi np}{N_2}}}_{f_n\left(\frac{j_{nk}R}{j_{nN_1}}\right)} = \frac{i^n}{\pi R^2}\sum_{m=1}^{N_1-1}\frac{J_n\left(\frac{j_{nm}j_{nk}}{j_{nN_1}}\right)}{J_{n+1}^2(j_{nm})}\underbrace{\left\{\frac{1}{N_2}\sum_{q=-M}^{M} F\left(\frac{j_{qm}}{R},\frac{2\pi q}{N_2}\right)e^{-i\frac{2\pi nq}{N_2}}\right\}}_{F_n\left(\frac{j_{nm}}{R}\right)} \tag{22}$$

Multiplying both sides of Equation (22) by $e^{+i\frac{2\pi np}{N_2}}$, summing from $n = -M..M$, interchanging the order of summation on the left hand side and using the orthogonality relationship of the discrete complex exponential gives:

$$f\left(\frac{j_{pk}R}{j_{pN_1}},\frac{2\pi p}{N_2}\right) = \frac{1}{2\pi R^2 N_2}\sum_{n=-M}^{M}\sum_{m=1}^{N_1-1}\sum_{q=-M}^{M} F\left(\frac{j_{qm}}{R},\frac{2\pi q}{N_2}\right)\frac{2i^n J_n\left(\frac{j_{nm}j_{nk}}{j_{nN_1}}\right)}{J_{n+1}^2(j_{nm})}e^{-i\frac{2\pi nq}{N_2}}e^{+i\frac{2\pi np}{N_2}} \tag{23}$$

3.2. Band-Limited Functions

The process in the previous section can be repeated by starting with the assumption that the function is band-limited. That is, we suppose that the 2D Fourier transform $F(\rho, \psi)$ of $f(r, \theta)$ is band-limited, meaning that $F(\rho, \psi)$ itself and therefore by virtue of the equivalent of Equation (9), all of its Fourier coefficients $F_n(\rho)$ are zero for $\rho \geq W_\rho = 2\pi W$. Typically, W would be given in units of Hz (cycles per second) if using temporal units, or cycles per meter if using spatial units. Hence, the definition of W_ρ (with a multiplication by 2π) ensures that the final units are given in s^{-1} or m^{-1}. The details of this development follow the same steps as for the space-limited function but start with the assumption of a band-limited function and then impose a space-limit (i.e., truncation of the series). The results of this are summarized below.

3.3. Summary of Above Relationships

From the above, we summarize the derived relationships. In the case of a space-limited function, it is found that the forward transform is given by:

$$F\left(\frac{j_{qm}}{R}, \frac{2\pi q}{N_2}\right) = \frac{2\pi R^2}{N_2} \sum_{n=-M}^{M} \sum_{k=1}^{N_1-1} \sum_{p=-M}^{M} f\left(\frac{j_{pk}R}{j_{pN_1}}, \frac{2\pi p}{N_2}\right) \frac{2i^{-n} J_n\left(\frac{j_{nk} j_{nm}}{j_{nN_1}}\right)}{j_{nN_1}^2 J_{n+1}^2(j_{nk})} e^{-i\frac{2\pi np}{N_2}} e^{+i\frac{2\pi nq}{N_2}} \tag{24}$$

and the inverse transform is given by:

$$f\left(\frac{j_{pk}R}{j_{pN_1}}, \frac{2\pi p}{N_2}\right) = \frac{1}{2\pi R^2 N_2} \sum_{n=-M}^{M} \sum_{m=1}^{N_1-1} \sum_{q=-M}^{M} F\left(\frac{j_{qm}}{R}, \frac{2\pi q}{N_2}\right) \frac{2i^{n} J_n\left(\frac{j_{nm} j_{nk}}{j_{nN_1}}\right)}{J_{n+1}^2(j_{nm})} e^{-i\frac{2\pi nq}{N_2}} e^{+i\frac{2\pi np}{N_2}} \tag{25}$$

Similarly, starting from the assumption of a bandlimited function, the forward transform is given by:

$$F\left(\frac{j_{qm} W_\rho}{j_{qN_1}}, \frac{2\pi q}{N_2}\right) = \frac{2\pi}{W_\rho^2 N_2} \sum_{n=-M}^{M} \sum_{k=1}^{N_1} \sum_{p=-M}^{M} f\left(\frac{j_{pk}}{W_\rho}, \frac{2\pi p}{N_2}\right) \frac{2\, i^{-n} J_n\left(\frac{j_{nk} j_{nm}}{j_{nN_1}}\right)}{J_{n+1}^2(j_{nk})} e^{-i\frac{2\pi np}{N_2}} e^{+i\frac{2\pi nq}{N_2}} \tag{26}$$

and the inverse transform is given by:

$$f\left(\frac{j_{pk}}{W_\rho}, \frac{2\pi p}{N_2}\right) = \frac{W_\rho^2}{2\pi N_2} \sum_{n=-M}^{M} \sum_{m=1}^{N_1-1} \sum_{q=-M}^{M} F\left(\frac{j_{qm} W_\rho}{j_{qN_1}}, \frac{2\pi q}{N_2}\right) \frac{2i^{n} J_n\left(\frac{j_{nm} j_{nk}}{j_{nN_1}}\right)}{j_{nN_1}^2 J_{n+1}^2(j_{nm})} e^{-i\frac{2\pi nq}{N_2}} e^{+i\frac{2\pi np}{N_2}} \tag{27}$$

It is noted that the forward-inverse transform pair defined by Equations (24) and (25) is similar to the transform pair defined by (26) and (27), with a few differences. First, the sampling points *appear* to be slightly different, depending on whether we started with the assumption of a space-limited function or a bandlimited function. The second observation is that the form of the transform itself might appear to be slightly different, depending on whether a space-limited or a band-limited function was assumed as a starting point. However, it was shown in [6] that for a nth order discrete Hankel transform, the required relationship between the band limit and space limit is given by $W_\rho R = j_{nN_1}$. If the substitution $W_\rho R = j_{nN_1}$ is used in Equations (24) and (25), then it yields the same discrete transform as the transform pair defined by (26) and (27). Also, the relationship $W_\rho R = j_{nN_1}$ arose naturally in the development above when the truncation of the Fourier–Bessel series at N_1 was implemented, meaning that the truncation of the series at N_1 is the same as assuming $W_\rho R = j_{nN_1}$.

Formally using the relationship $W_\rho R = j_{nN_1}$, the expressions in Equations (24) and (25) can also be written using a symmetric forward/inverse transform pair, where the forward transform is given by:

$$F\left(\frac{j_{qm}}{R}, \frac{2\pi q}{N_2}\right) = \frac{2\pi R}{N_2 W_\rho} \sum_{n=-M}^{M} \sum_{k=1}^{N_1-1} \sum_{p=-M}^{M} f\left(\frac{j_{pk}R}{j_{pN_1}}, \frac{2\pi p}{N_2}\right) \frac{2i^{-n} J_n\left(\frac{j_{nk} j_{nm}}{j_{nN_1}}\right)}{j_{nN_1} J_{n+1}^2(j_{nk})} e^{-i\frac{2\pi np}{N_2}} e^{+i\frac{2\pi nq}{N_2}} \tag{28}$$

For the inverse transform, we can similarly write:

$$f\left(\frac{j_{pk}R}{j_{pN_1}}, \frac{2\pi p}{N_2}\right) = \frac{W_\rho}{2\pi R N_2} \sum_{n=-M}^{M} \sum_{m=1}^{N_1-1} \sum_{q=-M}^{M} F\left(\frac{j_{qm}}{R}, \frac{2\pi q}{N_2}\right) \frac{2i^{n} J_n\left(\frac{j_{nm} j_{nk}}{j_{nN_1}}\right)}{j_{nN_1} J_{n+1}^2(j_{nm})} e^{-i\frac{2\pi nq}{N_2}} e^{+i\frac{2\pi np}{N_2}} \tag{29}$$

The advantage of the formulation in Equations (28) and (29) shall be noted in the next section in that it suggests a symmetric form of the kernel for the 2D discrete transform in polar coordinates.

The above demonstrates that a natural, $(N_1 - 1) \times N_2$ dimensional discretization scheme in finite space and finite frequency space is given by:

$$r_{pk} = \frac{j_{pk}R}{j_{pN_1}} \text{ or } r_{pk} = \frac{j_{pk}}{W_\rho}, \quad \text{and} \quad \theta_p = \frac{p2\pi}{N_2} \tag{30}$$

and:

$$\rho_{qm} = \frac{j_{qm}}{R} \quad \text{or} \quad \rho_{qm} = \frac{j_{qm}W_\rho}{j_{qN_1}}, \quad \text{and} \quad \psi_q = \frac{q2\pi}{N_2} \tag{31}$$

where p, k, q, m, n, N_1, and N_2 are integers such that $-M \le n \le M$, where $2M+1 = N_2, 1 \le m, k, \le N_1 - 1$ and $-M \le p, q \le M$. The relationship $W_\rho = \frac{j_{nN_1}}{R}$ can be used to formally switch from a finite frequency domain to a finite space domain. This is a 'formal' approach because in making this substitution, the index of the Bessel function is not fixed whereas W_ρ and R are assumed fixed values. Nevertheless, it demonstrates the approach to switching from a space-limited based discretization scheme to a band-limited discretization scheme.

4. Proposed Kernel for the Discrete Transform

4.1. Proposed Kernel for 2D Polar Discrete Fourier Transform

To work with the polar 2D DFT, a kernel for the transformation is required. Inspired by the formulations shown in Equations (24) and (25), we propose the following kernels:

$$\begin{aligned} E^-_{qm;pk} &= \frac{1}{N_2} \sum_{n=-M}^{M} \frac{J_n\left(\frac{j_{nk} j_{nm}}{j_{nN_1}}\right)}{j_{nN_1}^2 J_{n+1}^2(j_{nk})} 2i^{-n} e^{-in\frac{2\pi p}{N_2}} e^{+in\frac{2\pi q}{N_2}} \\ E^+_{qm;pk} &= \frac{1}{N_2} \sum_{n=-M}^{M} \frac{J_n\left(\frac{j_{nm} j_{nk}}{j_{nN_1}}\right)}{J_{n+1}^2(j_{nm})} 2i^{n} e^{+i\frac{2\pi np}{N_2}} e^{-i\frac{2\pi nq}{N_2}} \end{aligned} \tag{32}$$

where p, k, q, m, n, N_1, and N_2 are integers such that $-M \le n \le M$, where $2M+1 = N_2, 1 \le m, k, \le N_1 - 1$ and $-M \le p, q \le M$. It is noted that the proposed kernels in Equation (32) are *almost* complex conjugates of each other save for a factor of $j_{nN_1}^2$ in the denominator of $E^-_{qm;pk}$. The formulation in Equation (32) is proposed in order to emulate Equations (24) and (25). A symmetric formulation of the kernels, with one j_{nN_1} in the denominator of each of $E^-(qm;pk)$ and $E^+(qm;pk)$ would also be possible and would make $E^\pm_{qm;pk}$ complex conjugates of each other; however, such a kernel would be more of a departure from a discretization of the continuous transform. The integers N_1, and N_2 denote the size of the working spaces, with N_2 giving the size in the angular direction and N_1 giving the size in the radial direction. Since $N_2 = 2M+1$, it follows that N_2 must be an odd integer. The notation for $E^-(qm;pk)$

and $E^{+}(qm;pk)$ are chosen deliberately. The subscript (+ or -) indicate the sign on the $i^{\pm}$ and on the exponent containing the p variable; the q variable exponent then takes the opposite sign.

4.2. Another Choice of Kernel

A second, more symmetric choice of kernel is also possible. We will see that this choice of kernel will allow for a more traditional version of Parseval's theorem. All the following expressions will hold with either form of kernel. Using as inspiration the forms written in Equations (28) and (29), then we suggest for a kernel the following expression:

$$\begin{aligned} E^{(s)-}_{qm;pk} &= \frac{1}{N_2}\sum_{n=-M}^{M}\frac{J_n\left(\frac{j_{nk}j_{nm}}{j_{nN_1}}\right)}{j_{nN_1}J^2_{n+1}(j_{nk})}2i^{-n}e^{-in\frac{2\pi p}{N_2}}e^{+in\frac{2\pi q}{N_2}} \\ E^{(s)+}_{qm;pk} &= \frac{1}{N_2}\sum_{n=-M}^{M}\frac{J_n\left(\frac{j_{nm}j_{nk}}{j_{nN_1}}\right)}{j_{nN_1}J^2_{n+1}(j_{nm})}2i^{n}e^{+i\frac{2\pi np}{N_2}}e^{-i\frac{2\pi nq}{N_2}} \end{aligned} \tag{33}$$

As before, p, k, q, m, n, N_1, and N_2 are integers such that $-M \leq n \leq M$, where $2M + 1 = N_2$ $1 \leq m, k, \leq N_1 - 1$ and $-M \leq p, q \leq M$. In Equation (33), $E^{(s)+}_{qm;pk}$ is now the complex conjugate of $E^{(s)-}_{qm;pk}$, as mentioned above.

4.3. Orthogonality of the Proposed Kernel

In what follows, we assume the ranges of the variables are such that p, k, q, m, n, N_1, and N_2 are integers such that $-M \leq n \leq M$, where $2M + 1 = N_2$, $1 \leq m, k, \leq N_1 - 1$ and $-M \leq p, q \leq M$. We state and prove that the following relationship is true:

$$\sum_{m=1}^{N_1-1}\sum_{q=-M}^{M} E^{-}_{qm;pk}E^{+}_{qm;p'k'} = \delta_{pp'}\delta_{kk'} \tag{34}$$

where $\delta_{pp'}$ is the Kronecker-delta function, defined as $\delta_{pp'} = 1$ if $p = p'$ and $\delta_{pp'} = 0$ otherwise. It is known that the continuous complex exponential expression can be written as:

$$e^{i\vec{\omega}\cdot\vec{r}} = \sum_{n=-\infty}^{\infty} i^n J_n(\rho r)\, e^{in\theta} e^{-in\psi} \tag{35}$$

Hence, the form of the discrete kernel as proposed in (32) or (33) *resembles* discrete samples of the right hand side of Equation (35). It then follows that our proposed kernels in (32) or (33) can be considered to be the (discrete) corresponding form of the complex exponential kernel for the proposed discrete transform. The orthogonality relationship in (34) can then be considered to be the discrete version of:

$$\int_0^{\infty}\int_0^{2\pi} e^{-i\vec{\omega}\cdot\vec{r}} e^{i\vec{\omega}\cdot\vec{r}'}\, d\vec{\omega} = \delta\left(\vec{r} - \vec{r}'\right) \tag{36}$$

where the integration over the frequency vector $\vec{\omega}$ has been replaced with a discrete sum over the frequency vector indices (q,m). The proof of Equation (34) uses the orthogonality of the discrete complex exponential and the discrete Hankel transform and can be found in Appendix A.7.

It can be similarly shown that the following orthogonality relationship is also true:

$$\sum_{k=1}^{N_1-1}\sum_{p=-M}^{M} E^{-}_{qm;pk}E^{+}_{q'm';pk} = \delta_{qq'}\delta_{mm'} \tag{37}$$

which is similarly to be considered to be the discrete version of:

$$\int_0^\infty \int_0^{2\pi} e^{-i\vec{\omega}\cdot\vec{r}} e^{i\vec{\omega}'\cdot\vec{r}} d\vec{r} = \delta\left(\vec{\omega} - \vec{\omega}'\right) \tag{38}$$

Once again, the integration over the vector $\vec{r}$ has been replaced with a discrete sum over the $\vec{r}$ vector indices (p,k). The proof of Equation (37) can also be found in Appendix A.7. The orthogonality expressions in Equations (34) and (37) still hold if $E^{\pm}_{qm;pk}$ is replaced with the symmetric $E^{(s)\pm}_{qm;pk}$ since the only difference between the $E^{\pm}_{qm;pk}$ and $E^{(s)\pm}_{qm;pk}$ is the attribution of a j_{nN_1} term in the denominator and this makes no difference when the two kernels are multiplied.

5. Proposed Transform

In this section, we propose a definition of the 2D discrete Fourier transform (DFT) in polar coordinates which is motivated by the results of the 2D Fourier transform applied to space-limited and band-limited functions and also by the proposed kernel. The 2D DFT in polar coordinates will be a transform that transforms a 2-subscript set of numbers (ie matrix) f_{pk} to another set of values, matrix F_{qm} where $p, k, q, m,$ are integers such that $1 \le m, k, \le N_1 - 1$ and $-M \le p, q \le M$ where $N_2 = 2M + 1$ for integers N_1, and N_2.

Forward and Inverse Transform

The proposed forward transform, $f_{pk} \to F_{qm}$ is given by:

$$\begin{aligned} F_{qm} &= \frac{1}{N_2} \sum_{n=-M}^{M} \sum_{k=1}^{N_1-1} \sum_{p=-M}^{M} 2i^{-n} f_{pk} \frac{J_n\left(\frac{j_{nk} j_{nm}}{j_{nN_1}}\right)}{j^2_{nN_1} J^2_{n+1}(j_{nk})} e^{-i\frac{2\pi np}{N_2}} e^{+i\frac{2\pi nq}{N_2}} \\ &= \sum_{k=1}^{N_1-1} \sum_{p=-M}^{M} f_{pk} E^{-}_{qm;pk} \end{aligned} \tag{39}$$

where $N_2 = 2M + 1$ for some integer M. Similarly, for the inverse transform we propose:

$$\begin{aligned} f_{pk} &= \frac{1}{N_2} \sum_{n=-M}^{M} \sum_{m=1}^{N_1-1} \sum_{q=-M}^{M} 2i^{n} F_{qm} \frac{J_n\left(\frac{j_{nm} j_{nk}}{j_{nN_1}}\right)}{J^2_{n+1}(j_{nm})} e^{-i\frac{2\pi nq}{N_2}} e^{+i\frac{2\pi np}{N_2}} \\ &= \sum_{m=1}^{N_1-1} \sum_{q=-M}^{M} F_{qm} E^{+}_{qm;pk} \end{aligned} \tag{40}$$

In the proposed transform, $E^{(s)\pm}_{qm;pk}$ could easily be used in placed of $E^{\pm}_{qm;pk}$ and all the following expressions will still be valid.

Proof. Substituting Equation (39) into the right-hand side of (40), interchanging the order of summation and using the orthogonality relationships of the kernel given in Equation (34) gives:

$$\sum_{m=1}^{N_1-1} \sum_{q=-M}^{M} \underbrace{\left\{ \sum_{l=1}^{N_1-1} \sum_{s=-M}^{M} f_{sl} E^{-}_{qm;sl} \right\}}_{F_{qm}} E^{+}_{qm;pk} = \sum_{l=1}^{N_1-1} \sum_{s=-M}^{M} f_{sl} \delta_{sp} \delta_{lk} = f_{pk} \tag{41}$$

Similarly, substituting Equation (40) into the right-hand side of (39), interchanging the order of summation and using the orthogonality of the kernel given in Equation (37) gives:

$$\sum_{k=1}^{N_1-1}\sum_{p=-M}^{M}\underbrace{\left\{\sum_{l=1}^{N_1-1}\sum_{s=-M}^{M}F_{sl}E^{+}_{sl;pk}\right\}}_{f_{pk}}E^{-}_{qm;pk}=\sum_{l=1}^{N_1-1}\sum_{s=-M}^{M}F_{sl}\delta_{sq}\delta_{lm}=F_{qm} \tag{42}$$

Hence, Equation (39) and (40) are inverses of each other. These expressions would also hold if $E^{(s)\pm}_{qm;pk}$ were used instead of $E^{\pm}_{qm;pk}$.

6. Properties of the Transform—Transform Rules

6.1. The Complex Exponential

For the discrete case, the functions $E^{-}_{qm;pk}$ and $E^{+}_{qm;pk}$ as introduced above are the complex exponentials for this space, satisfying the required orthogonality condition and functioning as the kernel for the 2D-DFT in polar coordinates. These kernels are *not* $e^{\pm i\vec{\omega}\cdot\vec{r}}$ evaluated at particular points because the evaluation of the discrete radial variables in regular and frequency space varies with the order of the Bessel function. Nevertheless, these functions are the 'effective' complex exponentials for the space under consideration. From the orthogonality condition of the 2D polar DFT kernel, it can be shown that the expected Fourier rule of a complex exponential transforming to a delta function applies. Specifically, the 2D DFT of $f_{pk}=E^{+}_{q_0m_0;pk}$ for some fixed, given values (q_0,m_0) is given by:

$$\begin{aligned}F_{qm}&=\sum_{k=1}^{N_1-1}\sum_{p=-M}^{M}f_{pk}E^{-}_{qm;pk}=\sum_{k=1}^{N_1-1}\sum_{p=-M}^{M}E^{+}_{q_0m_0;pk}E^{-}_{qm;pk}\\&=\delta_{qq_0}\delta_{mm_0}\end{aligned} \tag{43}$$

Hence, $f_{pk}=E^{+}_{q_0m_0;pk}$ transforms to $\delta_{qq_0}\delta_{mm_0}$ or in compact notation, $E^{+}_{q_0m_0;pk}\Leftrightarrow\delta_{qq_0}\delta_{mm_0}$. This is the discrete version of the transform of $\exp\left(\vec{\omega}_0\cdot\vec{r}\right)$.

6.2. The Delta Function

Clearly, the discrete equivalent of the Dirac-delta function is the Kronecker-delta function and in 2D, this needs to be a 2-subscript function. Thus, the discrete function whose 2D DFT is sought is given by $f_{pk}=\delta_{pp_0}\delta_{kk_0}$, which defines a matrix indexed by (p,k) where all the entries are zero except for the index where $p=p_0$ and $k=k_0$. The dimensions of this matrix are in keeping with all the dimensions assumed for the space which are p,k,q,m,n, N_1, and N_2 are integers such that $-M\le n\le M$, $1\le m,k,\le N_1-1$ and $-M\le p,q\le M$ and where $2M+1=N_2$. Finding the 2D DFT of this function gives:

$$\begin{aligned}F_{qm}&=\sum_{k=1}^{N_1-1}\sum_{p=0}^{N_2-1}f_{pk}E^{-}_{qm;pk}=\sum_{k=1}^{N_1-1}\sum_{p=0}^{N_2-1}\delta_{pp_0}\delta_{kk_0}E^{-}_{qm;pk}\\&=E^{-}_{qm;p_0k_0}\end{aligned} \tag{44}$$

Hence, as in the continuous case, the delta function transforms to the complex exponential (with a negative sign in the exponent). Hence we have another the Fourier pair $\delta_{pp_0}\delta_{kk_0}\Leftrightarrow E^{-}_{qm;p_0k_0}$.

6.3. The Generalized Shift Operator

For a one dimensional Fourier transform, the shift rule is one of the known transform rules. This rule says that a shift in time is equivalent to a modulation in frequency. Mathematically, this is stated as:

$$f(t-a) = \mathbb{F}^{-1}\left\{e^{-ia\omega}\hat{f}(\omega)\right\} = \frac{1}{2\pi}\int_{-\infty}^{\infty}\left\{e^{-ia\omega}\hat{f}(\omega)\right\}e^{i\omega t}d\omega \tag{45}$$

Using this result as motivation, a generalized-shift operator is defined by finding the inverse DFT of the DFT of the function multiplied by the DFT kernel (modulation). A generalized shift operator was first proposed by Levitan [18], and our definition is a discretized version of this definition. Levitan suggested the complex conjugate of the Fourier operator as a generalized shift operator, which for Fourier transforms is the inverse transform operator. This approach to a generalized shift operator has previously been used with the Hankel transform itself [6,19]. Thus, we *define* the definition of a generalized-shifted function $f_{pk}^{p_0k_0}$ as the inverse Fourier transform of the function multiplied by the inverse transform operator. That is, it is defined as:

$$f_{pk}^{p_0k_0} := \sum_{q=-M}^{M}\sum_{m=1}^{N_1-1}\left\{F_{qm}E^{-}_{qm;p_0k_0}\right\}E^{+}_{qm;pk} \tag{46}$$

Here, f_{pk} is the original (unshifted) function with 2D DFT F_{qm} such that $f_{pk} \to F_{qm}$. $f_{pk}^{p_0k_0}$ is the shifted function where p_0k_0 denotes the amount of the shift (the equivalent of a in Equation (45)).

The shifted function $f_{pk}^{p_0k_0}$ can also be expressed in terms of the unshifted function f_{pk} by writing F_{qm} in terms of f_{pk} such as:

$$\begin{aligned} f_{pk}^{p_0k_0} &= \sum_{q=-M}^{M}\sum_{m=1}^{N_1-1} F_{qm}E^{-}_{qm;p_0k_0}E^{+}_{qm;pk} \\ &= \sum_{q=-M}^{M}\sum_{m=1}^{N_1-1}\left\{\sum_{p'=-M}^{M}\sum_{k'=1}^{N_1-1} f_{p'k'}E^{-}_{qm;p'k'}\right\}E^{-}_{qm;p_0k_0}E^{+}_{qm;pk} \end{aligned} \tag{47}$$

By interchanging the order of summation, this can be rewritten as:

$$f_{pk}^{p_0k_0} = \sum_{p'=-M}^{M}\sum_{k'=1}^{N_1-1} f_{p'k'}\underbrace{\sum_{q=-M}^{M}\sum_{m=1}^{N_1-1} E^{-}_{qm;p'k'}E^{+}_{qm;pk}E^{-}_{qm;p_0k_0}}_{\substack{\text{shift operator in space domain} \\ = S^{p_0k_0}_{p'k',pk}}} \tag{48}$$

Equation (48) permits the definition of a shift operator so that the shift operator in the spatial domain is defined as:

$$S^{p_0k_0}_{p'k',pk} = \sum_{q=-M}^{M}\sum_{m=1}^{N_1-1} E^{-}_{qm;p'k'}E^{+}_{qm;pk}E^{-}_{qm;p_0k_0} \tag{49}$$

This triple-product shift operator resembles previous definitions of shift operators for multidimensional Fourier transforms [2,3], generalized Hankel convolutions [20–22] and also discrete Hankel transforms [6].

6.4. Forward Transform of the Generalized Shift

We now consider the forward 2D Fourier transform of the generalized shifted function $f_{pk}^{p_0k_0}$. From the definition of the shifted function given in Equation (46), it is obvious that the forward transform of the shifted function is given by:

$$\mathbb{F}^{2D}\left(f_{pk}^{p_0k_0}\right) = F_{qm}E^{-}_{qm;p_0k_0} \tag{50}$$

The above can also be verified directly. The 2D Fourier transform of the shifted function can be found from:

$$\begin{aligned}\mathbb{F}^{2D}\left(f_{pk}^{p_0k_0}\right) &= \sum_{p=-M}^{M}\sum_{k=1}^{N_1-1} f_{pk}^{p_0k_0}E^{-}_{qm;pk} \\ &= \sum_{p=-M}^{M}\sum_{k=1}^{N_1-1}\left\{\sum_{q'=-M}^{M}\sum_{m'=1}^{N_1-1} F_{q'm'}E^{-}_{q'm';p_0k_0}E^{+}_{q'm';pk}\right\}E^{-}_{qm;pk}\end{aligned} \tag{51}$$

where the definition in (46) was used. Interchanging the order of summation and using the orthogonality result in (37) gives:

$$\mathbb{F}^{2D}\left(f_{pk}^{p_0k_0}\right) = \sum_{q'=-M}^{M}\sum_{m'=1}^{N_1-1} F_{q'm'}E^{-}_{q'm';p_0k_0}\underbrace{\sum_{p=-M}^{M}\sum_{k=1}^{N_1-1} E^{+}_{q'm';pk}E^{-}_{qm;pk}}_{=\delta_{qq'}\delta_{mm'}} = F_{qm}E^{-}_{qm;p_0k_0} \tag{52}$$

This gives another transform pair and also defines the shift-modulation rule. This rule is in analogy with the shift-modulation rule for regular Fourier transforms that states that a shift in the spatial/time domain is equivalent to modulation in the frequency domain:

$$f_{pk}^{p_0k_0} \Leftrightarrow F_{qm}E^{-}_{qm;p_0k_0} \tag{53}$$

Equation (53) is equivalent to the standard 1D continuous transform rule of:

$$\mathbb{F}\{f(t-a)\} = e^{-ia\omega}\hat{f}(\omega) \tag{54}$$

6.5. Modulation

We suppose that the forward 2D-DHT of a function g_{pk} is 'modulated' in the space domain so that the function whose transform we seek is $f_{pk} = E^{+}_{q_0m_0;pk}g_{pk}$. This is the discrete equivalent of a function $g(t)$ modulated as $e^{iat}g(t)$ Here, the interpretation of $f_{pk} = E^{+}_{q_0m_0;pk}g_{pk}$ is as follows:

$$\begin{aligned} f_{pk} &= E^{+}_{q_0m_0;pk}g_{pk} \\ f_{pk} &= g_{pk}\frac{2}{N_2}\sum_{n=-M}^{M}\frac{J_n\left(\frac{j_{nm_0}j_{nk}}{j_{nN_1}}\right)}{J_{n+1}(j_{nm_0})}i^n e^{+i\frac{2\pi np}{N_2}}e^{-i\frac{2\pi nq_0}{N_2}}\end{aligned} \tag{55}$$

Again, we implement the definition of the forward transform on the modulated function $f_{pk} = E^{+}_{q_0m_0;pk}g_{pk}$ so that:

$$F_{qm} = \sum_{k=1}^{N_1-1}\sum_{p=-M}^{M} f_{pk}E^{-}_{qm;pk} = \sum_{k=1}^{N_1-1}\sum_{p=-M}^{M} E^{+}_{q_0m_0;pk}g_{pk}E^{-}_{qm;pk} \tag{56}$$

and write g_{pk} in terms of its inverse transform:

$$g_{pk} = \sum_{m=1}^{N_1-1} \sum_{q=-M}^{M} G_{qm} E^{+}_{qm;pk} \tag{57}$$

So that Equation (56) becomes:

$$F_{qm} = \sum_{k=1}^{N_1-1} \sum_{p=-M}^{M} E^{+}_{q_0 m_0;pk} \sum_{m'=1}^{N_1-1} \sum_{q'=-M}^{M} G_{q'm'} E^{+}_{q'm';pk} E^{-}_{qm;pk} \tag{58}$$

Interchanging the order of summation gives:

$$F_{qm} = \sum_{m'=1}^{N_1-1} \sum_{q'=-M}^{M} G_{q'm'} \underbrace{\sum_{k=1}^{N_1-1} \sum_{p=-M}^{M} E^{+}_{q'm';pk} E^{-}_{qm;pk} E^{+}_{q_0 m_0;pk}}_{\text{shift operator in the frequency domain}} = G^{q_0 m_0}_{qm} \tag{59}$$

By comparing Equation (59) with Equation (49), we recognize the shift operator as shown in (59). This follows from a shift over the (q,m) variables and defines a shift operator in the frequency domain as:

$$S^{q_0 m_0}_{q'm',qm} = \sum_{k=1}^{N_1-1} \sum_{p=-M}^{M} E^{+}_{q'm';pk} E^{-}_{qm;pk} E^{+}_{q_0 m_0;pk} \tag{60}$$

Hence, Equation (59) can be written as:

$$F_{qm} = \sum_{m'=1}^{N_1-1} \sum_{q'=-M}^{M} G_{q'm'} S^{q_0 m_0}_{q'm',qm} = G^{q_0 m_0}_{qm} \tag{61}$$

The shift operator in the frequency domain over the (q,m) variables as given by Equation (60) can be compared to the shift operator over the (p,k) variables in the space domain as shown in (49). We note that operations in the spatial domains are operations that involve the (p,k) variables or the second group of variables in $E^{\pm}_{qm;pk}$. Similarly, operations in the frequency domain involve operations over the (q,m) variables or the first set of variables in $E^{\pm}_{qm;pk}$.

Hence, the above development shows the derivation of a modulation-shift rule, where the forward 2D-DHT of a modulated function is equivalent to a generalized shift in the frequency domain. This gives the following transform pair:

$$E^{+}_{q_0 m_0;pk} g_{pk} \Leftrightarrow G^{q_0 m_0}_{qm} \tag{62}$$

Otherwise stated, Equation (62) shows that modulation in the space domain is equivalent to shift in the frequency domain, in keeping with expectations for a (generalized) Fourier transform.

6.6. Convolution–Multiplication

For a 2D convolution/multiplication rule, we consider a 2D convolution in the space domain. The convolution is defined in the traditional manner as the product of a shifted function with another unshifted function, and then the summation over all possible shifts. Specifically, we write it as:

$$f_{pk} = h_{pk} ** g_{pk} = \underbrace{\sum_{p_0=-M}^{M} \sum_{k_0=1}^{N_1-1}}_{\substack{\text{summation over} \\ \text{all possible shifts}}} \underbrace{h_{pk}^{p_0k_0}}_{\substack{\text{shifted} \\ \text{function}}} \underbrace{g_{p_0k_0}}_{\substack{\text{unshifted} \\ \text{function}}} \tag{63}$$

where $h_{pk}^{p_0k_0}$ is the h_{pk} shifted by p_0k_0 given by:

$$h_{pk}^{p_0k_0} = \sum_{q=-M}^{M} \sum_{m=1}^{N_1-1} H_{qm} E_{qm;p_0k_0}^{-} E_{qm;pk}^{+} \tag{64}$$

The summation in Equation (63) is then over all the possible shifts. Taking the forward transform of f_{pk} as defined in (63) gives:

$$\begin{aligned} F_{qm} &= \sum_{k=1}^{N_1-1} \sum_{p=-M}^{M} f_{pk} E_{qm;pk}^{-} = \sum_{k=1}^{N_1-1} \sum_{p=-M}^{M} \left\{ \sum_{p_0=-M}^{M} \sum_{k_0=1}^{N_1-1} h_{pk}^{p_0k_0} g_{p_0k_0} \right\} E_{qm;pk}^{-} \\ &= \sum_{k=1}^{N_1-1} \sum_{p=-M}^{M} \sum_{p_0=-M}^{M} \sum_{k_0=1}^{N_1-1} \underbrace{\sum_{q'=-M}^{M} \sum_{m'=1}^{N_1-1} H_{q'm'} E_{q'm';p_0k_0}^{-} E_{q'm';pk}^{+}}_{h_{pk}^{p_0k_0}} \underbrace{\sum_{q''=-M}^{M} \sum_{m''=1}^{N_1-1} G_{q''m''} E_{q''m'';p_0k_0}^{+} E_{qm;pk}^{-}}_{g_{p_0k_0}} \end{aligned} \tag{65}$$

Interchanging the order of summation so that the summation over p,k is performed first and using the orthogonality of the kernel gives:

$$\begin{aligned} F_{qm} &= \sum_{p_0=-M}^{M} \sum_{k_0=1}^{N_1-1} \sum_{q'=-M}^{M} \sum_{m'=1}^{N_1-1} \sum_{q''=-M}^{M} \sum_{m''=1}^{N_1-1} H_{q'm'} E_{q'm';p_0k_0}^{-} G_{q''m''} E_{q''m'';p_0k_0}^{+} \delta_{qq'} \delta_{mm'} \\ &= \sum_{p_0=-M}^{M} \sum_{k_0=1}^{N_1-1} \sum_{q''=-M}^{M} \sum_{m''=1}^{N_1-1} H_{qm} E_{qm;p_0k_0}^{-} G_{q''m''} E_{q''m'';p_0k_0}^{+} \end{aligned} \tag{66}$$

Now summing over p_0, k_0 and again using the orthogonality of the kernel gives:

$$F_{qm} = \sum_{q''=-M}^{M} \sum_{m''=1}^{N_1-1} H_{qm} G_{q''m''} \delta_{qq''} \delta_{mm''} = H_{qm} G_{qm} \tag{67}$$

In other words, we have the result that:

$$h_{pk} ** g_{pk} \Leftrightarrow H_{qm} G_{qm} \tag{68}$$

Equation (68) is, of course, the expected convolution–multiplication rule where convolution in the space domain is equivalent to multiplication in the frequency domain.

6.7. Multiplication–Convolution Rule

We now consider the forward 2D FT of a term-by-term product in the space domain so that $f_{pk} = h_{pk} g_{pk}$. Then, the forward transform of the term-by-term product is given by:

$$F_{qm} = \sum_{k=1}^{N_1-1} \sum_{p=-M}^{M} f_{pk} E_{qm;pk}^{-} = \sum_{k=1}^{N_1-1} \sum_{p=-M}^{M} \left\{ h_{pk} g_{pk} \right\} E_{qm;pk}^{-} \tag{69}$$

Using the definitions of the inverse 2D FT to write h_{pk} and g_{pk} then:

$$\begin{aligned} F_{qm} &= \sum_{k=1}^{N_1-1}\sum_{p=-M}^{M} f_{pk}E^{-}_{qm;pk} = \sum_{k=1}^{N_1-1}\sum_{p=-M}^{M} \underbrace{\sum_{q'=-M}^{M}\sum_{m'=1}^{N_1-1} H_{q'm'}E^{+}_{q'm';pk}}_{h_{pk}} g_{pk}E^{-}_{qm;pk} \\ &= \sum_{q'=-M}^{M}\sum_{m'=1}^{N_1-1} H_{q'm'} \underbrace{\sum_{k=1}^{N_1-1}\sum_{p=-M}^{M} g_{pk}E^{+}_{q'm';pk}E^{-}_{qm;pk}}_{=G^{q'm'}_{qm}} \end{aligned} \tag{70}$$

In Equation (70), we have used the modulation rule $E^{+}_{q_0m_0;pk}g_{pk} \Leftrightarrow G^{q_0m_0}_{qm}$. In other words, Equation (70) states that:

$$F_{qm} = \sum_{q'=-M}^{M}\sum_{m'=1}^{N_1-1} H_{q'm'}G^{q'm'}_{qm} = H_{qm} * *G_{qm} \tag{71}$$

Hence, $h_{pk}g_{pk} \Leftrightarrow H_{qm} * *G_{qm}$ which is the multiplication-convolution rule where multiplication in the space domain is equivalent to convolution in the frequency domain.

6.8. Rotation

It is generally known that rotating a function in 2D space also rotates its 2D Fourier transform. We demonstrate that this is still true with our definition of the discrete 2D DFT in polar coordinates. To see this, we consider a shift of the function in frequency space, meaning consider $F_{(q-q_0)m}$ where a shift by q_0 in the angular coordinate has been implemented. In this case, since the circular direction is circularly periodic, we interpret $q - q_0$ in the sense of modulo N_2. So consider the inverse discrete 2D DFT of $F_{(q-q_0)m}$, that is from the definition in Equation (40)

$$\mathbb{F}^{-1}_{2D}\{F_{(q-q_0)m}\} = \frac{1}{N_2}\sum_{n=-M}^{M} j_{nN_1} i^n e^{+i\frac{2\pi np}{N_2}} \sum_{m=1}^{N_1-1} \frac{2J_n\left(\frac{j_{nm}j_{nk}}{j_{nN_1}}\right)}{j_{nN_1}J^2_{n+1}(j_{nm})}\left\{\sum_{q=-M}^{M} F_{(q-q_0)m}e^{-i\frac{2\pi nq}{N_2}}\right\} \tag{72}$$

Now suppose that $q' = q - q_0$ so that $q = q' + q_0$ and $q = -M$ implies $q' = -q_0 - M$ and also $q = +M$ implies $q' = -q_0 + M$. Hence, Equation (72) becomes:

$$\frac{1}{N_2}\sum_{n=-M}^{M} j_{nN_1} i^n e^{+i\frac{2\pi np}{N_2}} \sum_{m=1}^{N_1-1} \frac{2J_n\left(\frac{j_{nm}j_{nk}}{j_{nN_1}}\right)}{j_{nN_1}J^2_{n+1}(j_{nm})}\left\{\sum_{q'=-q_0-M}^{q'=-q_0+M} F_{q'm}e^{-i\frac{2\pi n(q'+q_0)}{N_2}}\right\} \tag{73}$$

But because of the circular (N_2) periodicity of the function, then:

$$\sum_{q'=-q_0-M}^{q'=-q_0+M} F_{q'm}e^{-i\frac{2\pi n(q'+q_0)}{N_2}} = e^{-i\frac{2\pi nq_0}{N_2}} \sum_{q'=-q_0-M}^{q'=-q_0+M} F_{q'm}e^{-i\frac{2\pi nq'}{N_2}} = e^{-i\frac{2\pi nq_0}{N_2}} \sum_{q'=-M}^{q'=+M} F_{q'm}e^{-i\frac{2\pi nq'}{N_2}} \tag{74}$$

Hence, Equation (72) becomes:

$$\frac{1}{N_2}\sum_{n=-M}^{M} j_{nN_1} i^n \underbrace{e^{+i\frac{2\pi np}{N_2}} e^{-i\frac{2\pi nq_0}{N_2}}}_{=e^{+i\frac{2\pi n(p-q_0)}{N_2}}} \sum_{m=1}^{N_1-1} \frac{2J_n\left(\frac{j_{nm}j_{nk}}{j_{nN_1}}\right)}{j_{nN_1}J^2_{n+1}(j_{nm})}\left\{\sum_{q=-M}^{M} F_{qm}e^{-i\frac{2\pi nq}{N_2}}\right\} = f_{(p-q_0)k} \tag{75}$$

As above, $f_{(p-q_0)k}$ is to be interpreted in the sense of modulo N_2. However, what this clearly demonstrates is that rotating the Fourier transform by q_0 is equivalent to rotating the original function by q_0, as is expected of a 2D Fourier transform.

7. Generalized Parseval Theorem

Under the proposed transform, inner products are preserved and, therefore, energies are preserved with the symmetric version of the transform. With the non-symmetric version of the transform, a modified version of Parseval's theorem is possible. This will be demonstrated in the following subsections.

7.1. Parseval's Theorem with the Symmetric Kernel

Consider the total energy of the term-by-term product (Hadamard product) of two matrices in the spatial domain $f_{pk} = h_{pk}\overline{g_{pk}}$. We use the overbar notation to denote the complex conjugate, so that $\overline{g_{pk}}$ denotes the complex conjugate of g_{pk}. We recall that in the case of the symmetric kernel, the complex conjugate of $E^{(s)+}_{qm;pk}$ is $E^{(s)-}_{qm;pk}$, which is what will enable the Parseval relationship to exist in its expected form, as will be shown. More specifically, it is noted that:

$$\begin{aligned}\sum_{k=1}^{N_1-1}\sum_{p=-M}^{M} h_{pk}\overline{g_{pk}} &= \sum_{k=1}^{N_1-1}\sum_{p=-M}^{M}\left\{\sum_{m'=1}^{N_1-1}\sum_{q'=-M}^{M} H_{q'm'}E^{(s)+}_{q'm';pk}\right\}\left\{\sum_{m''=1}^{N_1-1}\sum_{q''=-M}^{M}\overline{G_{q''m''}E^{(s)-}_{q''m'';pk}}\right\} \\ &= \sum_{m'=1}^{N_1-1}\sum_{q'=-M}^{M} H_{q'm'}\sum_{m''=1}^{N_1-1}\sum_{q''=-M}^{M}\overline{G_{q''m''}}\sum_{k=1}^{N_1-1}\sum_{p=-M}^{M} E^{(s)+}_{q'm';pk}E^{(s)-}_{q''q'';pk}\end{aligned} \tag{76}$$

However,

$$\sum_{k=1}^{N_1-1}\sum_{p=-M}^{M} E^{(s)+}_{q'm';pk}E^{(s)-}_{q''m'';pk} = \delta_{q'q''}\delta_{m'm''} \tag{77}$$

Hence, Equation (76) becomes:

$$\begin{aligned}\sum_{k=1}^{N_1-1}\sum_{p=-M}^{M} h_{pk}\overline{g_{pk}} &= \sum_{m'=1}^{N_1-1}\sum_{q'=-M}^{M} H_{q'm'}\sum_{m''=1}^{N_1-1}\sum_{q''=-M}^{M}\overline{G_{q''m''}}\delta_{q'q''}\delta_{m'm''} \\ &= \sum_{m'=1}^{N_1-1}\sum_{q'=-M}^{M} H_{q'm'}\overline{G_{q'm'}}\end{aligned} \tag{78}$$

For the special case that $g_{pk} = h_{pk}$ then Equation (78) yields:

$$\sum_{k=1}^{N_1-1}\sum_{p=-M}^{M}|h_{pk}|^2 = \sum_{m=1}^{N_1-1}\sum_{q=-M}^{M}|H_{qm}|^2 \tag{79}$$

Equations (78) and (79) are the expected for of the Parseval relationship, which essentially states that the energy computed in one domain is equivalent to the energy computed in the other domain. The reader is reminded that the symmetric kernel was used for the derivation in (79).

7.2. Parseval's Theorem with the Non-Symmetric Kernel

For the non-symmetric kernel, some modifications to the above Parseval relationship are necessary. Again, we consider the total energy of a Hadamard product of two matrices in the spatial domain. However, now we need to define a more 'general' version of a complex conjugate expression in order for the Parseval relationship to exist. We denote this more general version as $\overline{g_{pk}}^*$ (over bar *and* star) and define this expression as:

$$\overline{g_{pk}}^* := \sum_{m''=1}^{N_1-1} \sum_{q''=-M}^{M} \overline{G_{q''m''}} E^-_{q''m'';pk} \quad \text{(definition)} \tag{80}$$

We note in Equation (80) that $\overline{g_{pk}}^*$ uses $E^-_{q''m'';pk}$ instead of $\overline{E^+_{q''m'';pk}}$ (where the latter would normally be used for the complex conjugate). The reason for this is that with the non-symmetric kernel, using $\overline{E^+_{q''m'';pk}}$ will not lead to the required orthogonality condition. However, with our 'modified' version of the complex conjugate as denoted by the $\overline{g_{pk}}^*$ and defined in Equation (80), it then follows that:

$$\begin{aligned} \sum_{k=1}^{N_1-1} \sum_{p=-M}^{M} h_{pk} \overline{g_{pk}}^* &= \sum_{k=1}^{N_1-1} \sum_{p=-M}^{M} \left\{ \sum_{m'=1}^{N_1-1} \sum_{q'=-M}^{M} H_{q'm'} E^+_{q'm';pk} \right\} \left\{ \sum_{m''=1}^{N_1-1} \sum_{q''=-M}^{M} \overline{G_{q''m''}} E^-_{q''m'';pk} \right\} \\ &= \sum_{m'=1}^{N_1-1} \sum_{q'=-M}^{M} H_{q'm'} \sum_{m''=1}^{N_1-1} \sum_{q''=-M}^{M} \overline{G_{q''m''}} \underbrace{\sum_{k=1}^{N_1-1} \sum_{p=-M}^{M} E^+_{q'm';pk} E^-_{q''m'';pk}}_{\delta_{q'q''}\delta_{m'm''}} \end{aligned} \tag{81}$$

Using the orthogonality of the kernel, Equation (81) becomes:

$$\begin{aligned} \sum_{k=1}^{N_1-1} \sum_{p=-M}^{M} h_{pk} \overline{g_{pk}}^* &= \sum_{m'=1}^{N_1-1} \sum_{q'=-M}^{M} H_{q'm'} \sum_{m''=1}^{N_1-1} \sum_{q''=-M}^{M} \overline{G_{q''m''}} \delta_{q'q''} \delta_{m'm''} \\ &= \sum_{m'=1}^{N_1-1} \sum_{q'=-M}^{M} H_{q'm'} \overline{G_{q'm'}} \end{aligned} \tag{82}$$

Similarly, we can consider the special product in the frequency domain $F_{qm} = H_{qm}\overline{G_{qm}}^*$ where again the special expression $\overline{G_{qm}}^*$ needs to be defined as follows:

$$\overline{G_{qm}}^* := \sum_{k''=1}^{N_1-1} \sum_{p''=-M}^{M} \overline{g_{p''k''}} E^+_{qm;p''k''} \quad \text{(definition)} \tag{83}$$

Consider:

$$\sum_{m=1}^{N_1-1} \sum_{q=-M}^{M} H_{qm} \overline{G_{qm}}^* = \sum_{m=1}^{N_1-1} \sum_{q=-M}^{M} \left\{ \sum_{k'=1}^{N_1-1} \sum_{p'=-M}^{M} h_{p'k'} E^-_{qm;p'k'} \right\} \cdot \left\{ \sum_{k''=1}^{N_1-1} \sum_{p''=-M}^{M} \overline{g_{p''k''}} E^+_{qm;p''k''} \right\} \tag{84}$$

Interchanging the order of summation and summing over the (q,m) variables first gives:

$$\sum_{m=1}^{N_1-1} \sum_{q=-M}^{M} H_{qm} \overline{G_{qm}}^* = \sum_{k'=1}^{N_1-1} \sum_{p'=-M}^{M} h_{p'k'} \sum_{k''=1}^{N_1-1} \sum_{p''=-M}^{M} \overline{g_{p''k''}} \sum_{m=1}^{N_1-1} \sum_{q=-M}^{M} E^-_{qm;p'k'} E^+_{qm;p''k''} \tag{85}$$

Using the orthogonality of the kernel, the last line can be rewritten as:

$$\begin{aligned} \sum_{m=1}^{N_1-1} \sum_{q=-M}^{M} H_{qm} \overline{G_{qm}}^* &= \sum_{k'=1}^{N_1-1} \sum_{p'=-M}^{M} h_{p'k'} \sum_{k'=1}^{N_1-1} \sum_{p''=-M}^{M} \overline{g_{p''p''}} \delta_{p'p''} \delta_{k'k''} \\ &= \sum_{k'=1}^{N_1-1} \sum_{p'=-M}^{M} h_{p'k'} \overline{g_{p'k'}} \end{aligned} \tag{86}$$

In summary, Equation (82) shows how to interpret $\sum_{m'=1}^{N_1-1}\sum_{q'=-M}^{M} H_{q'm'}\overline{G_{q'm'}}$ and Equation (86) shows how to interpret $\sum_{k'=1}^{N_1-1}\sum_{p'=-M}^{M} h_{p'k'}\overline{g_{p'k'}}$ and also shows that they are not quite equivalent as was the case for the symmetric kernel. In summary,

$$\begin{aligned}\sum_{m=1}^{N_1-1}\sum_{q=-M}^{M} H_{qm}\overline{G_{qm}} &= \sum_{k=1}^{N_1-1}\sum_{p=-M}^{M} h_{pk}\overline{g_{pk}^{*}} \\ \sum_{k=1}^{N_1-1}\sum_{p=-M}^{M} h_{pk}\overline{g_{pk}} &= \sum_{m=1}^{N_1-1}\sum_{q=-M}^{M} H_{qm}\overline{G_{qm}^{*}}\end{aligned} \tag{87}$$

In the special case that $h = g$, then Equation (87) becomes:

$$\begin{aligned}\sum_{m=1}^{N_1-1}\sum_{q=-M}^{M} \left|H_{qm}\right|^2 &= \sum_{k=1}^{N_1-1}\sum_{p=-M}^{M} h_{pk}\overline{h_{pk}^{*}} \\ \sum_{k=1}^{N_1-1}\sum_{p=-M}^{M} \left|h_{pk}\right|^2 &= \sum_{m=1}^{N_1-1}\sum_{q=-M}^{M} H_{qm}\overline{H_{qm}^{*}}\end{aligned} \tag{88}$$

8. Discussion: Interpretation of the Transform

In the previous sections, we demonstrated that the 2D DFT in polar coordinates is most conveniently defined in terms of the kernels $E^{\pm}_{qm;pk}$ or $E^{(s)\pm}_{qm;pk}$, and indeed this definition allows many of the proofs of the DFT properties to assume a straightforward form that exploits the properties of the kernel. In this section, we demonstrate that the proposed forms of the 2D DFT can be interpreted in terms of a sequence of 1D DFT, DHT and IDFT discrete transforms, thereby demonstrating that the proposed transform follows the same path as the continuous 2D transform in that it can be decomposed into a sequence of Fourier, Hankel and inverse Fourier transforms [2].

8.1. Interpretation of the 2D Forward DFT in Polar Coordinates

Let us reconsider the definition of the forward 2D DFT, Equation (39), and rewrite it as:

$$F_{qm} = \frac{1}{N_2}\sum_{n=-M}^{M} e^{+in\frac{2\pi q}{N_2}}\,\frac{i^{-n}}{j_{nN_1}}\sum_{k=1}^{N_1-1}\frac{2J_n\left(\frac{j_{nk}j_{nm}}{j_{nN_1}}\right)}{j_{nN_1}J_{n+1}^2(j_{nk})}\left\{\sum_{p=-M}^{M} f_{pk}e^{-in\frac{2\pi p}{N_2}}\right\} \tag{89}$$

We can consider these as a sequence of 1D discrete Fourier transforms, with a discrete Hankel transform, as explained in the following. The first step is a forward 1D DFT transforming $f_{pk} \to \widetilde{f}_{nk}$ where the p subscript is transformed to the n subscript as:

$$\widetilde{f}_{nk} = \sum_{p=-M}^{M} f_{pk}e^{-in\frac{2\pi p}{N_2}} \qquad \text{for } n = -M..M,\ k = 1..N_1-1 \tag{90}$$

The tilde is used to indicate a standard 1D DFT. In matrix terms, this says that each *column* of f_{pk} is DFT'ed to yield $\widetilde{f}_{nk}$. The second step of Equation (89) is a discrete Hankel transform of order n that transforms $\widetilde{f}_{nk} \to \widetilde{\widetilde{f}}_{nm}$, where the k subscript is Hankel transformed to the m subscript via:

$$\widetilde{\widetilde{f}}_{nm} = \sum_{k=1}^{N_1-1}\frac{2J_n\left(\frac{j_{nk}j_{nm}}{j_{nN_1}}\right)}{j_{nN_1}J_{n+1}^2(j_{nk})}\widetilde{f}_{nk} \qquad \text{for } n = -M..M,\ m = 1..N_1-1 \tag{91}$$

The overhat denotes a DHT, as defined in [6]. Using the transformation matrix notation defined in [6], we define:

$$Y_{m,k}^{nN_1} = \frac{2}{j_{nN_1} J_{n+1}^2(j_{nk})} J_n\left(\frac{j_{nm} j_{nk}}{j_{nN_1}}\right) \quad 1 \le m,k \le N_1 - 1 \tag{92}$$

Hence Equation (91) can be written as

$$\tilde{\hat{f}}_{nm} = \sum_{k=1}^{N_1-1} Y_{m,k}^{nN_1} \tilde{f}_{nk} \quad \text{for } n = -M..M,\ m = 1..N_1 - 1 \tag{93}$$

In matrix terms, this shows that each *row* of $\tilde{f}_{nk}$ is nth-order DHT'ed to yield $\tilde{\hat{f}}_{nm}$. The nth row is nth order DHT'ed (with some loose interpretation of row counters since in this case the index n takes on negative values). A scaling operation then gives the Fourier coefficients of the 2D DFT $\tilde{\hat{f}}_{nm} \to \tilde{F}_{nm}$ such that:

$$\tilde{F}_{nm} = \frac{i^{-n}}{j_{nN_1}} \tilde{\hat{f}}_{nm} = \frac{i^{-n}}{j_{nN_1}} \sum_{k=1}^{N_1-1} Y_{m,k}^{nN_1} \tilde{f}_{nk} \quad \text{for } n = -M..M,\ m = 1..N_1 - 1 \tag{94}$$

It is noted that Equation (94) exactly parallels the equivalent step of the continuous form of the transform where $F_n(\rho) = 2\pi i^{-n} \mathbb{H}_n\{f_n(r)\}$, see [4,6]. If the symmetric form of the kernel is used, that is, Equation (33), then Equation (94) is replaced with $\tilde{F}_{nm} = i^{-n} \tilde{\hat{f}}_{nm}$.

The final step to compute the forward 2D DFT in polar coordinates is then a standard *inverse* 1D DFT. Here, each *column* of $\tilde{F}_{nm} \to F_{qm}$ is transformed so that the n subscript is (inverse) transformed to the q subscript via

$$F_{qm} = \frac{1}{N_2} \sum_{n=-M}^{M} \tilde{F}_{nm} e^{+in\frac{2\pi q}{N_2}} \quad \text{for } q = -M..M,\ m = 1..N_1 - 1 \tag{95}$$

This last step is a 1D IDFT for each *column* of $\tilde{F}_{nm}$ to obtain F_{qm}. It was shown in [2,4] that a continuous 2D Fourier transform in polar coordinates is a sequence of operations consisting of (i) a Fourier series transform (transforming the continuous function to its discrete set of Fourier coefficients), (ii) a Hankel transform for each Fourier coefficient (an nth order transform for the nth coefficient), and (iii) an inverse Fourier series transform (a set of Fourier coefficients is transformed back to a continuous function via the infinite Fourier series summation). Hence, we have shown here that the proposed 2D DFT in polar coordinates consists of the same sequence of transforms: a forward DFT, a forward DHT and then an inverse DFT.

8.2. Interpretation of the 2D Inverse DFT in Polar Coordinates

Similarly, we can decompose the inverse 2D DFT in polar coordinates, from Equation (40) written as:

$$f_{pk} = \frac{1}{N_2} \sum_{n=-M}^{M} j_{nN_1} i^n e^{+i\frac{2\pi np}{N_2}} \sum_{m=1}^{N_1-1} \frac{2 J_n\left(\frac{j_{nm} j_{nk}}{j_{nN_1}}\right)}{j_{nN_1} J_{n+1}^2(j_{nm})} \left\{ \sum_{q=-M}^{M} F_{qm} e^{-i\frac{2\pi nq}{N_2}} \right\} \tag{96}$$

The steps of the inverse 2D DFT are the reverse of those outlined for the forward 2DDFT. First $F_{qm} \to \tilde{F}_{nm}$ via a forward 1D DFT:

$$\tilde{F}_{nm} = \sum_{q=-M}^{M} F_{qm} e^{-i\frac{2\pi nq}{N_2}} \quad n = -M..M,\ m = 1..N_1 - 1 \tag{97}$$

This is followed by a discrete Hankel transform to obtain $\widetilde{F}_{nm} \to \hat{\widetilde{F}}_{nk}$

$$\hat{\widetilde{F}}_{nk} = \sum_{m=1}^{N_1-1} Y_{k,m}^{nN_1} \widetilde{F}_{nm} \qquad \text{for } n = -M..M,\ k = 1..N_1 - 1 \tag{98}$$

The next step is a scaling operation to obtain $\hat{\widetilde{F}}_{nk} \to \widetilde{f}_{nk}$ via:

$$\widetilde{f}_{nk} = j_{nN_1} i^{+n} \hat{\widetilde{F}}_{nk} \qquad \text{for } n = -M..M,\ k = 1..N_1 - 1 \tag{99}$$

The step in Equation (99) follows the pattern of the continuous form transform, where $f_n(r) = \frac{i^n}{2\pi} \mathbb{H}_n\{F_n(\rho)\}$, see [2,4,6]. As before, if the symmetric form of the kernel is used (Equation (33)), then Equation (99) is replaced with $\widetilde{f}_{nk} = i^{+n} \hat{\widetilde{F}}_{nk}$. Finally, an inverse 1D DFT is used to obtain $\widetilde{f}_{nk} \to f_{pk}$ via:

$$f_{pk} = \frac{1}{N_2} \sum_{n=-M}^{M} \widetilde{f}_{nk} e^{+i\frac{2\pi np}{N_2}} \qquad \text{for } p = -M..M,\ k = 1..N_1 - 1 \tag{100}$$

As previously mentioned, this parallels the steps taken for the continuous case, with each continuous operation (Fourier series, Hankel transform) replaced by its discrete counterpart (DFT, DHT).

For both forward and inverse 2D DFT, the same sequence of steps are followed. The operations are a 1D DFT of each column of the given matrix, then a DHT of each row, then a term-by-term scaling, and finally an IDFT of each column.

9. Conclusions

In this paper, a discrete 2D Fourier transform in polar coordinates was motivated and proposed by applying a discretization and truncation approach to the continuous 2D Fourier transform in polar coordinates. This new transform stands in its own right and, unlike previous approaches to a polar FT, is not an evaluation of the Cartesian form of the transform on a polar grid. This approach yields two possible kernels for the discrete 2D transform in polar coordinates. One of these two kernels is closer to the continuous version of the transform and the second kernel is symmetric, in that the kernel for the forward transform is the complex conjugate of the kernel for the inverse transform. Both versions of the kernel yield a 2D transform that transform a 2-subscripted entity (matrix) to another one. The standard set of shift, modulation, multiplication and convolution rules were derived for both kernels and are the same for either form of the kernel. However, only the symmetric kernel yields the expected Parseval relationship. It was also shown that the 2D discrete transform can be interpreted as a 1D discrete Fourier transform (DFT), followed by a 1D discrete Hankel transform (DHT), followed by a 1D inverse DFT. This DFT-DHT-IDFT pattern mimics the manner in which the continuous 2D Fourier transform in polar coordinates is evaluated. In conclusion, part I of the paper proposes the form of the 2D DFT in polar coordinates, and demonstrates the expected operational rules for this transform. Part II of the paper will examine how the proposed 2D DFT in polar coordinates can be used to approximate the continuous FT at certain discrete points.

Appendix A

The following appendices give important definitions for Hankel transforms (Appendix A.1), Fourier Bessel series (Appendix A.2), finite Fourier transforms (Appendix A.5) and also contain statements of the orthogonality of the discrete complex exponential (Appendix A.3) and the discrete Bessel functions (Appendix A.4). Section Appendix A.6 contains a discussion on the sampling points and how they affect the proposed evaluation of the discrete Fourier coefficients at the chosen sampling points. Proofs of the orthogonality of the proposed kernel can be found in Appendix A.7.

Appendix A.1. Hankel Transform

The nth order Hankel transform is defined by the integral [3]:

$$\overset{\frown}{f}^{n}(\rho) = \int_0^\infty f(r) J_n(\rho r) r dr \tag{A1}$$

where $J_n(z)$ is the nth order Bessel function with the overhat indicating a Hankel transform as shown in Equation (A1). Here, n may be an arbitrary real or complex number. The Hankel transform is self-reciprocating and the inversion formula is given by:

$$f(r) = \int_0^\infty \overset{\frown}{f}^{n}(\rho) J_n(\rho r) \rho d\rho \tag{A2}$$

The Hankel transform exists only if the Dirichlet condition is satisfied, i.e., $\int_0^\infty \left| r^{1/2} f(r) \right| dr$ exists and is particularly useful for problems involving cylindrical symmetry.

Appendix A.2. Fourier–Bessel Series

Functions defined on a finite portion of the real line $[0, R]$ in the radial coordinate can be expanded in terms of a Fourier–Bessel series [16] given by:

$$f(r) = \begin{cases} \sum_{k=1}^{\infty} f_k J_n\left(\frac{j_{nk} r}{R}\right) & r \leq R \\ 0 & r > R \end{cases} \tag{A3}$$

where $J_n(z)$ is the nth order Bessel function, the order of the Bessel function in (A3) is arbitrary and j_{nk} denotes the kth root of the nth Bessel function. The kth order Fourier–Bessel coefficients f_k of the function $f(r)$ can be found from:

$$f_k = \frac{2}{R^2 J_{n+1}^2(j_{nk})} \int_0^R f(r) J_n\left(\frac{j_{nk} r}{R}\right) r \, dr \tag{A4}$$

Equations (A3) and (A4) can be considered to be a transform pair where the continuous function $f(r)$ is forward-transformed to the discrete vector f_k given by the finite integral in (A4). The summation in Equation (A3) is then taken as the inverse transformation which returns $f(r)$ when starting with f_k. The Fourier–Bessel series is the cylindrical coordinate counterpart of the Fourier series. Just as the Fourier series is defined for a finite interval and has a counterpart, the continuous Fourier transform over an infinite interval, so the Fourier–Bessel series has a counterpart over an infinite interval, namely the Hankel transform.

Appendix A.3. Orthogonality of the Discrete Complex Exponential

The success of the discrete Fourier transform (DFT) is based on the exploitation of known discrete orthogonality relationships for the complex exponential evaluated at a finite number of certain special points [23]. This relationship is given by:

$$\sum_{p=0}^{N-1} e^{-\frac{ip2\pi n}{N}} e^{+\frac{ip2\pi m}{N}} = N\delta_{mn} \tag{A5}$$

where m, n, p, N are integers. In Equation (A5), δ_{mn} is the Kronecker delta function, defined as:

$$\delta_{mn} = \begin{cases} 1 & \text{if } m = n \\ 0 & \text{otherwise} \end{cases} \tag{A6}$$

It can easily be shown by a simple change of variables that the following orthogonality relationship is true:

$$\sum_{p'=-M}^{M} e^{-\frac{ip'2\pi n}{N}} e^{+\frac{ip'2\pi m}{N}} = N\delta_{mn} \tag{A7}$$

Appendix A.4. Discrete Orthogonality of the Bessel Functions

It is shown in [24] that the following discrete orthogonality relationship is true:

$$\sum_{k=1}^{N-1} \frac{J_n\left(\frac{j_{nm} j_{nk}}{j_{nN}}\right) J_n\left(\frac{j_{ni} j_{nk}}{j_{nN}}\right)}{J_{n+1}^2(j_{nk})} = \frac{j_{nN}^2}{4} J_{n+1}^2(j_{nm})\delta_{mi} \tag{A8}$$

where j_{nm} represents the mth zero of $J_n(x)$.

It is noted that Equation (A8) is the discrete version of the Bessel orthogonality relationship on a finite interval given by:

$$\begin{aligned} \int_0^1 J_n(rj_{nm}) J_n(rj_{ni}) r dr &= \frac{J_{n+1}^2(j_{nm})}{2}\delta_{mi} \\ \int_0^b J_n\left(\frac{r' j_{nm}}{b}\right) J_n\left(\frac{r' j_{ni}}{b}\right) r' dr' &= \frac{b^2 J_{n+1}^2(j_{nm})}{2}\delta_{mi} \end{aligned} \tag{A9}$$

From Watson in [25], the following expressions are also valid:

$$\begin{aligned} \int_0^{W_\rho} J_n\left(\frac{j_{nk}\rho}{W_\rho}\right) J_n(r\rho)\rho d\rho &= \frac{j_{nk}}{\frac{j_{nk}^2}{W_\rho^2} - r^2} J_{n+1}(j_{nk}) J_n\left(rW_\rho\right) \\ \int_0^R J_n\left(\frac{j_{nk} r}{R}\right) J_n(\rho r) r dr &= \frac{j_{nk}}{\frac{j_{nk}^2}{R^2} - \rho^2} J_{n+1}(j_{nk}) J_n(\rho R) \end{aligned} \tag{A10}$$

Appendix A.5. Fourier Series and Finite Fourier Transform

A function of angular position $f(\theta)$, where $-\pi \le \theta \le \pi$ can be expanded into a Fourier series as:

$$f(\theta) = \sum_{n=-\infty}^{\infty} f_n e^{in\theta} \tag{A11}$$

where the Fourier coefficients are given by:

$$f_n = \frac{1}{2\pi}\int_{-\pi}^{\pi} f(\theta)\, e^{-in\theta} d\theta \tag{A12}$$

A principal application of the finite Fourier transform (FFT) is to approximately compute samples of the Fourier transform of a function. We define the FFT partial sum of the samples $f\left(\frac{2\pi p}{N_2}\right)$ of the continuous function $f(\theta)$ as:

$$\hat{f}_n = \frac{1}{N_2}\sum_{p=-M}^{M} f\left(\frac{2\pi p}{N_2}\right) e^{-i\frac{2\pi np}{N_2}} \qquad n \in [-M, M] \tag{A13}$$

where N_2 is an integer such that $N_2 = 2M + 1$ for some other integer M. The over square-hat notation $\hat{f}$ indicates the taking of a (finite) Fourier transform. Clearly, Equation (A13) is a Riemann sum for the integral in (A12). It is generally asserted in the signal processing literature that $\hat{f}_n \approx f_n$, and it is specifically shown in [17] that $\hat{f}_n$ provides a uniformly good estimate for f_n for $n \in [-M, M]$.

It is also shown in [17] that the finite Fourier transform partial sum given by:

$$\hat{f}(\theta) = \sum_{n=-M}^{M} \hat{f}_n e^{in\theta} \tag{A14}$$

is almost as good an approximation to $f(\theta)$ as the usual partial sum:

$$f^{N_2}(\theta) = \sum_{n=-M}^{M} f_n\, e^{in\theta} \tag{A15}$$

Appendix A.6. Sampling Points

In this section, the difference between including the radial sampling points in the index of summation for the discrete Fourier transform is discussed. We noted above in Equation (18) that the radial sampling point is included in the index of summation of the discrete Fourier transform. In other words, we wrote for $n \in [-M, M]$ that:

$$F_n\left(\frac{j_{nl}}{R}\right) = \frac{1}{N_2}\sum_{p=-M}^{M} F\left(\frac{j_{pl}}{R}, \frac{2\pi p}{N_2}\right) e^{-i\frac{2\pi np}{N_2}} \qquad (i) \tag{A16}$$

However, strictly speaking, the radial sampling points should be fixed to the value of the radial sampling point on the left hand side, that is the *expected* discrete definition of $F_n\left(\frac{j_{nl}}{R}\right)$ should be given by:

$$F_n\left(\frac{j_{nl}}{R}\right) = \frac{1}{N_2}\sum_{p=-M}^{M} F\left(\frac{j_{nl}}{R}, \frac{2\pi p}{N_2}\right) e^{-i\frac{2\pi np}{N_2}} \qquad (ii) \tag{A17}$$

Note that in both Equations (A16) and (A17), the index of summation is p, and the radial sampling point is j_{pl} in (A16) but j_{nl} in (A17).

Which of the definitions for $F_n\left(\frac{j_{nl}}{R}\right)$ is correct? Definition (i), as given in Equation (A16), or definition (ii) as given in Equation (A17)? Traditionally, (ii) of Equation (A17) would be expected but taking this form does not allow the 2D discrete transform that ensues to be invertible. We showed above in the main text of the manuscript, that version (i) with Equation (A16) leads to an invertible, discrete 2D transform. We show in this section that if we confine ourselves to the chosen sampling points, then both versions are equivalent.

Considering the reconstruction formula based on (A14) which says:

$$F\left(\frac{j_{ml}}{R},\theta\right)=\sum_{n=-M}^{M}F_n\left(\frac{j_{ml}}{R}\right)e^{in\theta} \tag{A18}$$

Then, sampling at $\theta=\frac{2\pi m}{N_2}$ gives:

$$F\left(\frac{j_{ml}}{R},\frac{2\pi m}{N_2}\right)=\sum_{n=-M}^{M}F_n\left(\frac{j_{ml}}{R}\right)e^{i\frac{2\pi nm}{N_2}} \tag{A19}$$

So now consider the right hand side of Equation (A19) under the two different sampling assumptions implied by (i) or (ii). That is:

$$\left.\begin{array}{lr}\sum_{n=-M}^{M}\left\{\underbrace{\frac{1}{N_2}\sum_{p=-M}^{M}F\left(\frac{j_{pl}}{R},\frac{2\pi p}{N_2}\right)e^{-i\frac{2\pi np}{N_2}}}_{F_n\left(\frac{j_{ml}}{R}\right)\text{ using }(i)}\right\}e^{i\frac{2\pi nm}{N_2}} & (i)\\ \sum_{n=-M}^{M}\left\{\underbrace{\frac{1}{N_2}\sum_{p=-M}^{M}F\left(\frac{j_{ml}}{R},\frac{2\pi p}{N_2}\right)e^{-i\frac{2\pi np}{N_2}}}_{F_n\left(\frac{j_{ml}}{R}\right)\text{ using }(ii)}\right\}e^{i\frac{2\pi nm}{N_2}} & (ii)\end{array}\right. \tag{A20}$$

Equation (A20) (i) gives:

$$\begin{aligned}&\sum_{p=-M}^{M}F\left(\tfrac{j_{ml}}{R},\tfrac{2\pi p}{N_2}\right)\underbrace{\frac{1}{N_2}\sum_{n=-M}^{M}e^{-i\frac{2\pi np}{N_2}}e^{i\frac{2\pi nm}{N_2}}}_{N_2\delta_{pm}}\\&=\sum_{p=-M}^{M}F\left(\tfrac{j_{ml}}{R},\tfrac{2\pi p}{N_2}\right)\delta_{pm}=F\left(\tfrac{j_{ml}}{R},\tfrac{2\pi m}{N_2}\right)\end{aligned} \tag{A21}$$

Therefore, the (i) version works the way it is expected to work. Now considering the (ii) version:

$$\begin{aligned}&\sum_{p=-M}^{M}F\left(\tfrac{j_{ml}}{R},\tfrac{2\pi p}{N_2}\right)\underbrace{\frac{1}{N_2}\sum_{n=-M}^{M}e^{-i\frac{2\pi np}{N_2}}e^{i\frac{2\pi nm}{N_2}}}_{N_2\delta_{pm}}\\&=\sum_{p=-M}^{M}F\left(\tfrac{j_{ml}}{R},\tfrac{2\pi p}{N_2}\right)\delta_{pm}=F\left(\tfrac{j_{ml}}{R},\tfrac{2\pi m}{N_2}\right)\end{aligned} \tag{A22}$$

Therefore, the (ii) version also works the way it is expected to work. Therefore, both (i) and (ii) work properly.

However, if we try evaluating at different values of angular position, say $\theta = \frac{2\pi r}{N_2}$ where now the sampling index on the angle and the Bessel function do not match, in other words:

$$F\left(\frac{j_{ml}}{R}, \frac{2\pi r}{N_2}\right) = \sum_{n=-M}^{M} F_n\left(\frac{j_{ml}}{R}\right)e^{i\frac{2\pi nr}{N_2}} \tag{A23}$$

Then,

$$\begin{aligned} F\left(\tfrac{j_{ml}}{R}, \tfrac{2\pi r}{N_2}\right) &= \sum_{n=-M}^{M}\left\{\frac{1}{N_2}\sum_{p=-M}^{M} F\left(\tfrac{j_{pl}}{R}, \tfrac{2\pi p}{N_2}\right)e^{-i\frac{2\pi np}{N_2}}\right\}e^{i\frac{2\pi nr}{N_2}} \\ &= \sum_{p=-M}^{M} F\left(\tfrac{j_{pl}}{R}, \tfrac{2\pi p}{N_2}\right)\delta_{pr} = F\left(\tfrac{j_{rl}}{R}, \tfrac{2\pi r}{N_2}\right) \qquad (i) \\ F\left(\tfrac{j_{ml}}{R}, \tfrac{2\pi r}{N_2}\right) &= \sum_{n=-M}^{M}\left\{\frac{1}{N_2}\sum_{p=-M}^{M} F\left(\tfrac{j_{ml}}{R}, \tfrac{2\pi p}{N_2}\right)e^{-i\frac{2\pi np}{N_2}}\right\}e^{i\frac{2\pi nr}{N_2}} \\ &= \sum_{p=-M}^{M} F\left(\tfrac{j_{ml}}{R}, \tfrac{2\pi p}{N_2}\right)\delta_{pr} = F\left(\tfrac{j_{ml}}{R}, \tfrac{2\pi r}{N_2}\right) \qquad (ii) \end{aligned} \tag{A24}$$

In this case, (ii) does not yield the expected result, but (i) does. So the question of (i) vs (ii) becomes a question of where on the theta (angular position) the total function needs to be evaluated—not only a question of evaluating on a discrete radial position. However, if the fixed set of sampling points that have been proposed for the discrete 2D transform are used, where the indices on radial and angular position match, then the results are as expected.

Appendix A.7. Proofs of Orthogonality of the Proposed Kernel

In what follows, we assume the ranges of the variables are such that p, k, q, m, n, N_1, and N_2 are integers such that $-M \le n \le M$, where $2M + 1 = N_2$ $1 \le m, k, \le N_1 - 1$ and $0 \le p, q \le N_2 - 1$.

Appendix A.7.1. Proof of Orthogonality of the Kernel over the Frequency Indices

We state and prove that the following relationship is true:

$$\sum_{m=1}^{N_1-1}\sum_{q=-M}^{M} E^{-}_{qm;pk}E^{+}_{qm;p'k'} = \delta_{pp'}\delta_{kk'} \tag{A25}$$

The proof is as follows. We start by substituting the definition of the kernel into the expression:

$$\begin{aligned} &\sum_{m=1}^{N_1-1}\sum_{q=-M}^{M} E^{-}_{qm;pk}E^{+}_{qm;p'k'} = \\ &\sum_{m=1}^{N_1-1}\sum_{q=-M}^{M}\frac{4}{N_2^2}\sum_{n=-M}^{M}\frac{J_n\left(\frac{j_{nk}j_{nm}}{j_{nN_1}}\right)}{j_{nN_1}^2 J_{n+1}^2(j_{nk})} i^{-n}e^{-in\frac{2\pi p}{N_2}}e^{+in\frac{2\pi q}{N_2}}\sum_{n'=-M}^{M}\frac{J_{n'}\left(\frac{j_{n'm}j_{n'k'}}{j_{n'N_1}}\right)}{J_{n'+1}^2(j_{n'm})} i^{n'}e^{+i\frac{2\pi n'p'}{N_2}}e^{-i\frac{2\pi n'q}{N_2}} \end{aligned} \tag{A26}$$

Summing over the index q and using the orthogonality of the discrete complex exponential (Appendix A.3) returns a $N_2\delta_{nn'}$ so that $n' = n$ and Equation (A26) becomes:

$$\frac{4}{N_2}\sum_{m=1}^{N_1-1}\sum_{n=-M}^{M}\frac{J_n\left(\frac{j_{nk}j_{nm}}{j_{nN_1}}\right)}{j_{nN_1}^2 J_{n+1}^2(j_{nk})}e^{-in\frac{2\pi p}{N_2}}\frac{J_n\left(\frac{j_{nm}j_{nk'}}{j_{nN_1}}\right)}{J_{n+1}^2(j_{nm})}e^{+i\frac{2\pi np'}{N_2}} \tag{A27}$$

This can be rewritten as:

$$\frac{1}{N_2}\sum_{n=-M}^{M} e^{-i\frac{2\pi np}{N_2}} e^{+i\frac{2\pi np'}{N_2}} \underbrace{\sum_{m=1}^{N_1-1} \frac{4J_n\left(\frac{j_{nk}j_{nm}}{j_{nN_1}}\right)J_n\left(\frac{j_{nm}j_{nk'}}{j_{nN_1}}\right)}{j_{nN_1}^2 J_{n+1}^2(j_{nk})J_{n+1}^2(j_{nm})}}_{=\delta_{kk'}} \tag{A28}$$

Now, summing over the index and using the discrete orthogonality relationship of the Bessel functions (Appendix A.4) gives:

$$\frac{1}{N_2}\sum_{n=-M}^{M} e^{-i\frac{2\pi np}{N_2}} e^{+i\frac{2\pi np'}{N_2}} \delta_{kk'} = \delta_{pp'}\delta_{kk'} \tag{A29}$$

where the orthogonality relationship of the discrete complex exponential has been used again.

Appendix A.7.2. Proof of Orthogonality of the Kernel over the Spatial Indices

It can be similarly shown that the following orthogonality relationship is also true:

$$\sum_{k=1}^{N_1-1}\sum_{p=-M}^{M} E^{-}_{qm;pk}E^{+}_{q'm';pk} = \delta_{qq'}\delta_{mm'} \tag{A30}$$

The proof is as follows. We start by substituting the definition of the kernel into the expression:

$$\begin{aligned}&\sum_{k=1}^{N_1-1}\sum_{p=-M}^{M} E^{-}_{qm;pk}E^{+}_{q'm';pk}\\ &= \sum_{k=1}^{N_1-1}\sum_{p=-M}^{M}\frac{1}{N_2}\sum_{n=-M}^{M}\frac{J_n\left(\frac{j_{nk}j_{nm}}{j_{nN_1}}\right)2i^{-n}e^{-in\frac{2\pi p}{N_2}}e^{+in\frac{2\pi q}{N_2}}}{j_{nN_1}^2 J_{n+1}^2(j_{nk})}\frac{1}{N_2}\sum_{n'=-M}^{M}\frac{J_{n'}\left(\frac{j_{n'm'}j_{n'k}}{j_{n'N_1}}\right)2i^{n'}e^{+i\frac{2\pi n'p}{N_2}}e^{-i\frac{2\pi n'q'}{N_2}}}{J_{n'+1}^2(j_{n'm'})}\end{aligned} \tag{A31}$$

Summation over p gives:

$$\begin{aligned}&\sum_{k=1}^{N_1-1}\sum_{p=-M}^{M} E^{-}_{qm;pk}E^{+}_{q'm';pk}\\ &= \sum_{k=1}^{N_1-1}\frac{1}{N_2}\sum_{n=-M}^{M}\frac{J_n\left(\frac{j_{nk}j_{nm}}{j_{nN_1}}\right)}{j_{nN_1}^2 J_{n+1}^2(j_{nk})}2i^{-n}e^{+in\frac{2\pi q}{N_2}}\frac{1}{N_2}\sum_{n'=-M}^{M}\frac{J_{n'}\left(\frac{j_{n'm'}j_{n'k}}{j_{n'N_1}}\right)}{J_{n'+1}^2(j_{n'm'})}2i^{n'}e^{-i\frac{2\pi n'q'}{N_2}}N_2\delta_{nn'}\\ &= \sum_{k=1}^{N_1-1}\frac{1}{N_2}\sum_{n=-M}^{M}\frac{J_n\left(\frac{j_{nk}j_{nm}}{j_{nN_1}}\right)}{j_{nN_1}^2 J_{n+1}^2(j_{nk})}2e^{+in\frac{2\pi q}{N_2}}\frac{J_n\left(\frac{j_{nm'}j_{nk}}{j_{nN_1}}\right)}{J_{n+1}^2(j_{nm'})}2e^{-i\frac{2\pi nq'}{N_2}}\end{aligned} \tag{A32}$$

Now summation over k gives the right hand side of (A32) and using the discrete orthogonality of the Bessel functions (Appendix A.4) gives for the right hand side:

$$\frac{1}{N_2}\sum_{n=-M}^{M}\delta_{mm'}e^{+in\frac{2\pi q}{N_2}}e^{-i\frac{2\pi nq'}{N_2}} \tag{A33}$$

Then, finally summation over n and using the orthogonality of the discrete complex exponential (Appendix A.3) finally gives:

$$\sum_{k=1}^{N_1-1}\sum_{p=-M}^{M} E^{-}_{qm;pk}E^{+}_{q'm';pk} = \delta_{mm'}\delta_{qq'} \tag{A34}$$

as required.

References

1. Sneddon, N. *The Use of Integral Transforms*, 2nd ed.; McGraw Hill: New York, NY, USA, 1972.
2. Baddour, N. Operational and convolution properties of two-dimensional Fourier transforms in polar coordinates. *J. Opt. Soc. Am. A* **2009**, *26*, 1767–1777. [CrossRef]
3. Baddour, N. Operational and convolution properties of three-dimensional Fourier transforms in spherical polar coordinates. *J. Opt. Soc. Am. A* **2010**, *27*, 2144–2155. [CrossRef] [PubMed]
4. Baddour, N. Two-Dimensional Fourier Transforms in Polar Coordinates. *Adv. Imaging Electron Phys.* **2011**, *165*, 1–45.
5. Bracewell, R. *The Fourier Transform and its Applications*; McGraw Hill: New York, NY, USA, 1999.
6. Baddour, N.; Chouinard, U. Theory and operational rules for the discrete Hankel transform. *J. Opt. Soc. Am. A* **2015**, *32*, 611–622. [CrossRef] [PubMed]
7. Baddour, N.; Chouinard, U. Matlab Code for the Discrete Hankel Transform. *J. Open Res. Softw.* **2017**, *5*, 4.
8. Averbuch, A.; Coifman, R.R.; Donoho, D.L.; Elad, M.; Israeli, M. Fast and accurate Polar Fourier transform. *Appl. Comput. Harmon. Anal.* **2006**, *21*, 145–167. [CrossRef]
9. Abbas, S.A.; Sun, Q.; Foroosh, H. An Exact and Fast Computation of Discrete Fourier Transform for Polar and Spherical Grid. *IEEE Trans. Signal Process.* **2017**, *65*, 2033–2048. [CrossRef]
10. Fenn, M.; Kunis, S.; Potts, D. On the computation of the polar FFT. *Appl. Comput. Harmon. Anal.* **2007**, *22*, 257–263. [CrossRef]
11. Dutt, A.; Rokhlin, V. Fast Fourier Transforms for Nonequispaced Data. *SIAM J. Sci. Comput.* **1993**, *14*, 1368–1393. [CrossRef]
12. Fourmont, K. Non-Equispaced Fast Fourier Transforms with Applications to Tomography. *J. Fourier Anal. Appl.* **2003**, *9*, 431–450. [CrossRef]
13. Dutt, A.; Rokhlin, V. Fast Fourier Transforms for Nonequispaced Data, II. *Appl. Comput. Harmon. Anal.* **1995**, *2*, 85–100. [CrossRef]
14. Potts, D.; Steidl, G.; Tasche, M. Fast Fourier Transforms for Nonequispaced Data: A Tutorial. In *Modern Sampling Theory: Mathematics and Applications*; Benedetto, J.J., Ferreira, P.J.S.G., Eds.; Birkhäuser Boston: Boston, MA, USA, 2001; pp. 247–270.
15. Fessler, J.A.; Sutton, B.P. Nonuniform fast Fourier transforms using min-max interpolation. *IEEE Trans. Signal Process.* **2003**, *51*, 560–574. [CrossRef]
16. Schroeder, J. Signal Processing via Fourier-Bessel Series Expansion. *Digit. Signal Process.* **1993**, *3*, 112–124. [CrossRef]
17. Epstein, C.L. How well does the finite Fourier transform approximate the Fourier transform? *Commun. Pure Appl. Math.* **2005**, *58*, 1421–1435. [CrossRef]
18. Levitan, B.M. Generalized Displacement Operators. In *Encyclopedia of Mathematics*; Springer: Heidelberg, Germany, 2002.
19. Baddour, N. Application of the generalized shift operator to the Hankel transform. *SpringerPlus* **2014**, *3*, 1–6. [CrossRef] [PubMed]
20. Belhadj, M.; Betancor, J.J. Hankel convolution operators on entire functions and distributions. *J. Math. Anal. Appl.* **2002**, *276*, 40–63. [CrossRef]
21. de Sousa Pinto, J. A Generalised Hankel Convolution. *SIAM J. Math. Anal.* **1985**, *16*, 1335–1346. [CrossRef]
22. Malgonde, S.P.; Gaikawad, G.S. On a generalized Hankel type convolution of generalized functions. *Proc. Indian Acad. Sci. Math. Sci.* **2001**, *111*, 471–487. [CrossRef]
23. Arfken, G.; Weber, H. *Mathematical Methods for Physicists*; Elsevier Academic Press: New York, NY, USA, 2005.
24. Johnson, H.F. An improved method for computing a discrete Hankel transform. *Comput. Phys. Commun.* **1987**, *43*, 181–202. [CrossRef]
25. Watson, G.N. *A Treatise on the Theory of Bessel Functions*; Cambridge University Press: Cambridge, UK, 1995.

8

Chebyshev Spectral Collocation Method for Population Balance Equation in Crystallization

Chunlei Ruan

School of Mathematics& Statistics, Henan University of Science & Technology, Luoyang 471023, China; ruanchunlei@haust.edu.cn

Abstract: The population balance equation (PBE) is the main governing equation for modeling dynamic crystallization behavior. In the view of mathematics, PBE is a convection–reaction equation whose strong hyperbolic property may challenge numerical methods. In order to weaken the hyperbolic property of PBE, a diffusive term was added in this work. Here, the Chebyshev spectral collocation method was introduced to solve the PBE and to achieve accurate crystal size distribution (CSD). Three numerical examples are presented, namely size-independent growth, size-dependent growth in a batch process, and with nucleation, and size-dependent growth in a continuous process. Through comparing the results with the numerical results obtained via the second-order upwind method and the HR-van method, the high accuracy of Chebyshev spectral collocation method was proven. Moreover, the diffusive term is also discussed in three numerical examples. The results show that, in the case of size-independent growth (PBE is a convection equation), the diffusive term should be added, and the coefficient of the diffusive term is recommended as $2G \times 10^{-3}$ to $G \times 10^{-2}$, where G is the crystal growth rate.

Keywords: spectral collocation method; population balance equation; Chebyshev points; crystallization

1. Introduction

Crystallization operations are widely used in chemical and pharmaceutical industries. Crystallization kinetics is the basis for the development of crystallization theory. Simultaneously, studies on crystalline kinetics also guide the design of crystallizers and the quality control of crystalline products [1]. The population balance equation (PBE) is a widely accepted equation to model the dynamic crystallization behavior [2]. Therefore, it is of great significance to accurately solve the PBE and to achieve accurate crystal size distribution (CSD).

Currently, there are a many numerical results on the study of PBE. In terms of numerical methods, they can be roughly divided into the following five categories [1,2]: (1) the method of moments [3]; (2) the method of characteristics [4,5]; (3) the method of weight residuals/orthogonal collocation [6,7]; (4) the Monte Carlo method [8,9]; (5) the finite-difference schemes/discrete population balances [10]. The review by Ramkrishna [2] is helpful in this regard. The method of moments approximates the CSD through its moments. The predictions under certain conditions are closed. However, the moment closure conditions are violated for more complex systems. The method of characteristics tries reducing the partial differential equations (PDEs) to ordinary differential equations (ODEs) using lines of characteristics. The approach is efficient with simple physics, but it is not suitable for complex physics. The method of weight residuals/orthogonal collocation approximates the distribution by the basis function; the main weakness is the dependence on the choice of basis function. The Monte Carlo method tracks the histories of individual particles. It has an incomparable degree of omnipotence. However, its computational cost is relatively expensive. The finite-difference/volume methods are approximated by finite-difference/volume schemes. They can be applied to complex cases. However, they usually require a high number of grid points in order to achieve desirable accuracy.

Recently, a large number of numerical methods with high accuracy were constructed to solve PBE, e.g., the HR-van method with a flux limiter [1,11], the Lattice Boltzmann method [12], the weighted essentially non-oscillatory (WENO) method [13], the spectral method [14,15], etc. Among these, the spectral method has the advantage of being able to accurately solve the PBE using a lower number of grid points. It was also found that the spectral collocation method is more efficient than other spectral methods [14] and finite-volume schemes [15].

Because the PBE is a convection–reaction equation in mathematics, its strong hyperbolic property may challenge the numerical methods. The numerical methods may gain unstable results, which are not applicable for physics. In order to weaken the hyperbolic property of PBE, a diffusive term is added herein. Although the diffusive term makes the numerical solution deviate from the exact solution to some extent, it guarantees the stability of the numerical results.

In this paper, Chebyshev spectral collocation is presented to solve the one-dimensional PBE. The CSD of three types of crystallization in batch crystallizers and continuous crystallizers were calculated. By comparing the analytical results and other numerical results, the validity and high accuracy of the Chebyshev spectral collocation method were verified. Furthermore, the effects of the diffusive term are also discussed using three numerical examples.

2. One-Dimensional PBE and Numerical Schemes

2.1. One-Dimensional PBE

PBE is a differential equation which describes the evolution of a population of particles. In the crystallization process, it can reflect the crystal nucleation, growth, agglomeration, and breakage [2].

One-dimensional PBE [1,2,11] can be written as follows:

$$\frac{\partial f(L,t)}{\partial t} + \frac{\partial \{G(L,t)f(L,t)\}}{\partial L} = h(L,t,f), \tag{1}$$

where $f(L,t)$ is the crystal size distribution function, L is the crystal size variable, t is the time, $G(L,t)$ is the growth rate of crystals, and $h(L,t,f)$ is the source term and may include the aggregation, nucleation, breakage, etc. Here, crystal size variable L is typically the characteristic length, volume, or mass, but it can also represent age, composition, and other characteristics of an entity in a distribution [1]. The growth rate $G(L,t)$ can be a function of size and other variables, such as the temperature and the concentration of chemical species in solution [1]. The source term $h(L,t,f)$ is a creation/depletion rate, and it can be a function of other variables including the distribution, which occurs in nucleation processes resulting from particle–particle interactions and in agglomeration processes [1]. Since many of these processes involve integrals, PBE is considered an integro-differential equation describing the complex crystallization processes [2].

Since Equation (1) is a convection–reaction equation, its strong hyperbolic property may challenge the numerical methods. In order to weaken the hyperbolic property of Equation (1), a diffusive term "$\varepsilon\frac{\partial^2 f(L,t)}{\partial L^2}$" is added to the left in Equation (1), leading to

$$\frac{\partial f(L,t)}{\partial t} + \frac{\partial \{G(L,t)f(L,t)\}}{\partial L} + \varepsilon\frac{\partial^2 f(L,t)}{\partial L^2} = h(L,t,f). \tag{2}$$

Here, ε is a small positive number. Theoretically, the closer ε is to 0, the more accurate the result is. However, in numerical experiments, when ε approaches 0, the effect of the diffusive term becomes smaller, which leads to an unstable numerical solution. Therefore, in this paper, the value of ε was determined by numerical tests.

2.2. Chebyshev Spectral Collocation Method

The Chebyshev spectral collocation method is a kind of collocation method [16,17]. On the one hand, as a spectral method, it has high accuracy. On the other hand, as a collocation method, it transforms the derivative term into a differential matrix, which makes the equations simple and easy to solve. To date, this method was successfully applied to solve differential equations in the fields of fluid mechanics [18,19], quantum mechanics [20,21], etc.

Equation (2) can be written as follows:

$$\frac{\partial f(L,t)}{\partial t} = R(L,t,f), \tag{3}$$

where $R(L,t,f) = h(L,t,f) - \frac{\partial\{G(L,t)f(L,t)\}}{\partial L} - \varepsilon\frac{\partial^2 f(L,t)}{\partial L^2}$. The fourth-order Runge–Kutta method [22] is used to discretize the time, leading to

$$\begin{aligned} f^{n+1} &= f^n + \frac{1}{6}(k_1 + 2k_2 + 2k_3 + k_4) \\ k_1 &= \Delta t \cdot R(t_n, f^n(L)) \\ k_2 &= \Delta t \cdot R(t_n + \frac{\Delta t}{2}, f^n(L) + \frac{1}{2}k_1) \\ k_3 &= \Delta t \cdot R(t_n + \frac{\Delta t}{2}, f^n(L) + \frac{1}{2}k_2) \\ k_4 &= \Delta t \cdot R(t_{n+1}, f^n(L) + k_3) \end{aligned} \quad . \tag{4}$$

Here, Δt is the time step, $t_n = n\Delta t$ is the time on the nth step ($n = 0, 1, \cdots$), f^n is the value of f on t_n and the initial distribution is written as f^0. It is worth mentioning that the variable f^n in Equation (4) is only a function of the spatial variable L. In this way, the space can be discretized using the spectral collocation method.

In the spectral collocation method, the solution of $f^n(L)$ can be written as follows [16,17]:

$$f^n(L) = \sum_{k=0}^{N} P_k(L)u_k, \tag{5}$$

where $P_k(L)$ is the Lagrange interpolation polynomial at the node $L_k(k = 0, 1, \cdots N)$, which is

$$P_k(L) = \frac{(L - L_0)\cdots(L - L_{k-1})(L - L_{k+1})\cdots(L - L_N)}{(L_k - L_0)\cdots(L_k - L_{k-1})(L_k - L_{k+1})\cdots(L_k - L_N)}, \tag{6}$$

and u_k is the function value at interpolation node L_k.

In the Chebyshev spectral collocation method, Chebyshev points are chosen as the interpolation nodes. In the standard calculation domain $[-1,1]$, the Chebyshev points are $\hat{L}_j = \cos(j\pi/N)(j = 0, 1, \cdots N)$. As for the general calculation domain $[a,b]$, the Chebyshev points can be obtained by the coordinate transformation of standard Chebyshev points, namely $L_j = \frac{b-a}{2}\hat{L}_j + \frac{a+b}{2}$.

Here, we take the calculation of $R(t_n, f^n(L))$ in Equation (4) as an example to show the implementation of the Chebyshev spectral collocation method. In fact, the right-hand term $R(t_n, f^n(L))$ is equal to $h(t_n, f^n(L)) - \frac{dG(L)}{dL}f^n(L) - G(L)\frac{df^n(L)}{dL} - \varepsilon\frac{d^2 f^n(L)}{dL^2}$. Here, we need to show the calculation of $\frac{df^n(L)}{dL}$ and $\frac{d^2f^n(L)}{dL^2}$ terms. In the Chebyshev spectral collocation method [16,17], the derivative term $\frac{df^n(L)}{dL}$ can be transformed into a differential matrix as follows:

$$\frac{df^n(L)}{dL} = \boldsymbol{D_N u}. \tag{7}$$

Here, $\boldsymbol{D_N}$ is a $(N+1)\times(N+1)$ Chebyshev derivative matrix, and $\boldsymbol{u} = (u_0, u_1, \cdots, u_N)^T$ is a $N+1$ function value vector at the interpolation nodes. In the standard calculation domain $[-1,1]$, the Chebyshev derivative matrix $\boldsymbol{\hat{D}_N}$ satisfies the following properties [16,17]:

$$\begin{aligned}
&(\boldsymbol{\hat{D}_N})_{00} = \frac{2N^2+1}{6}, (\boldsymbol{\hat{D}_N})_{NN} = -\frac{2N^2+1}{6} \\
&(\boldsymbol{\hat{D}_N})_{jj} = \frac{-x_j}{2(1-x_j^2)}, j = 1, \cdots, N-1 \\
&(\boldsymbol{\hat{D}_N})_{ij} = \frac{c_i(-1)^{i+j}}{c_j(x_i - x_j)}, i \neq j, i, j = 0, \cdots, N
\end{aligned}, \tag{8}$$

where $(\boldsymbol{\hat{D}_N})_{ij}$ represents the element of the $i+1$th row and $j+1$th column in the Chebyshev matrix, and c_i is the constant which is determined by the following equation:

$$c_i = \begin{cases} 2, & i = 0, N \\ 1, & i = 1, \cdots, N-1 \end{cases}. \tag{9}$$

As for the general calculation domain $[a,b]$, the Chebyshev derivative matrix $\boldsymbol{D_N}$ can be calculated using the standard Chebyshev derivative matrix $\boldsymbol{\hat{D}_N}$, which is

$$\boldsymbol{D_N} = \frac{2}{a+b}\boldsymbol{\hat{D}_N}. \tag{10}$$

Similarly, in the Chebyshev spectral collocation method [16,17], the second-order derivative term $\frac{d^2f^n(L)}{dL^2}$ can be written as follows:

$$\frac{d^2f^n(L)}{dL^2} = \boldsymbol{D_N^2 u}. \tag{11}$$

Here, $\boldsymbol{D_N^2}$ is the second-order Chebyshev derivative matrix. The relationship between the second-order Chebyshev derivative matrix $\boldsymbol{D_N^2}$ and the first-order Chebyshev derivative matrix $\boldsymbol{D_N}$ is

$$\boldsymbol{D_N^2} = (\boldsymbol{D_N})^2 = (\frac{2}{a+b}\boldsymbol{\hat{D}_N})^2. \tag{12}$$

2.3. Other Numerical Schemes

We mainly focus our attention on the numerical schemes of spatial discretization. The numerical method for time discretization is the classic Runge–Kutta method which is shown in Equation (4). The diffusive term "$\varepsilon\frac{\partial^2 f(L,t)}{\partial L^2}$" is no longer taken into account in the following two schemes. Thus, the right-hand term in Equation (4) becomes $R(t_n, f^n(L)) = h(t_n, f^n(L)) - \frac{\partial}{\partial L}(G^n(L)f^n(L))$.

2.3.1. Second Upwind Scheme

In the calculation of $\frac{\partial}{\partial L}(G^n(L)f^n(L))$, we use the following second-order upwind scheme [22]:

$$\left.\frac{\partial(G^n(L)f^n(L))}{\partial L}\right|_{L=L_i} = \begin{cases} \dfrac{3(Gf)_i^n - 4(Gf)_{i-1}^n + (Gf)_{i-2}^n}{2\Delta L} & G^n(L_i) \geq 0 \\ \dfrac{3(Gf)_{i+2}^n - 4(Gf)_{i+1}^n + (Gf)_i^n}{2\Delta L} & G^n(L_i) < 0 \end{cases}. \tag{13}$$

Here, f_i^n represents the value of f^n at node L_i, G_i^n represents the value of G^n at node L_i, and ΔL represents the distance between two adjacent nodes. During our simulation, the nodes are taken as uniformly distributed nodes, rather than Chebyshev points in the Chebyshev spectral collocation method.

2.3.2. HR-Van Scheme

In the calculation of $\frac{\partial}{\partial L}(G^n(L)f^n(L))$, the following HR-van scheme is used [1]:

$$\left.\frac{\partial(G^n(L)f^n(L))}{\partial L}\right|_{L=L_i} = \begin{cases} \frac{1}{\Delta L}[G_i^n\left(f_i^n + \frac{1}{2}\phi(r_i)(f_{i+1}^n - f_i^n)\right) \\ -G_{i-1}^n\left(f_{i-1}^n + \frac{1}{2}\phi(r_{i-1})(f_i^n - f_{i-1}^n)\right)] & G^n(L_i) \geq 0 \\ \frac{1}{\Delta L}[G_{i+1}^n\left(f_{i+1}^n + \frac{1}{2}\phi(r_{i+1})(f_{i+2}^n - f_{i+1}^n)\right) \\ -G_i^n\left(f_i^n + \frac{1}{2}\phi(r_i)(f_{i+1}^n - f_i^n)\right)] & G^n(L_i) < 0 \end{cases}, \tag{14}$$

where $r_i = \dfrac{f_i^n - f_{i-1}^n}{f_{i+1}^n - f_i^n}$, $\phi(r) = \dfrac{|r| + r}{1 + |r|}$, with $\phi(r)$ being the flux limiter proposed by Van. Similarly, this scheme is implemented under uniformly distributed nodes rather than Chebyshev points.

3. Numerical Experiments

In order to compare the accuracy of each numerical method, we define the following errors:

$$\begin{aligned} L_1 - error &= \sum_{k=0}^{N} |f_k^e - f_k^n|\Delta L_k / \sum_{k=0}^{N} |f_k^e|\Delta L_k \\ L_2 - error &= \sqrt{\sum_{k=0}^{N} (f_k^e - f_k^n)^2 \Delta L_k / \sum_{k=0}^{N} |f_k^e|\Delta L_k} \end{aligned} \tag{15}$$

Here, f_k^e is the exact solution at node L_k.

3.1. Size-Independent Growth in a Batch Process

The first example is the crystallization in a batch process with the crystal growth rate G independent of the crystal size L. Here, we do not consider the aggregation, nucleation, and breakage, and we set $h(L,t,f) = 0$. The crystal growth rate is set to $G = 1.0\ \mu m/s$ and the initial CSD is given by with the following Gaussian distribution:

$$f(L,0) = \frac{10^{10}}{\sqrt{2\pi}\sigma} e^{-\frac{(x-\mu)^2}{2\sigma^2}}, \tag{16}$$

with $\mu = 20\ \mu m$, $\sigma = 3$. The other parameters in the calculation are set to $N = 200$, $dt = 0.001$, $\varepsilon = 0.005$, $t_{end} = 60$ s. The exact solution of this example is $f(L,t) = f(L - Gt, 0)$.

Figure 1 shows the comparison of exact solution with the results of three numerical scheme sat $t = 60$ s. In order to clearly show the performances with different schemes, Figure 1b shows the predicted crystal size distributions at the local crystal size. It can be seen that second-order upwind scheme is the worst, which causes both numerical dispersion and diffusion; both the HR-van scheme and the Chebyshev spectral collocation method cause numerical diffusion, but the Chebyshev spectral collocation method causes less numerical diffusion. Table 1 shows the L_1 and L_2 errors with different numerical schemes. The error table clearly shows that the Chebyshev spectral collocation method has the smallest error. Therefore, the Chebyshev spectral collocation method has the highest accuracy among these three methods.

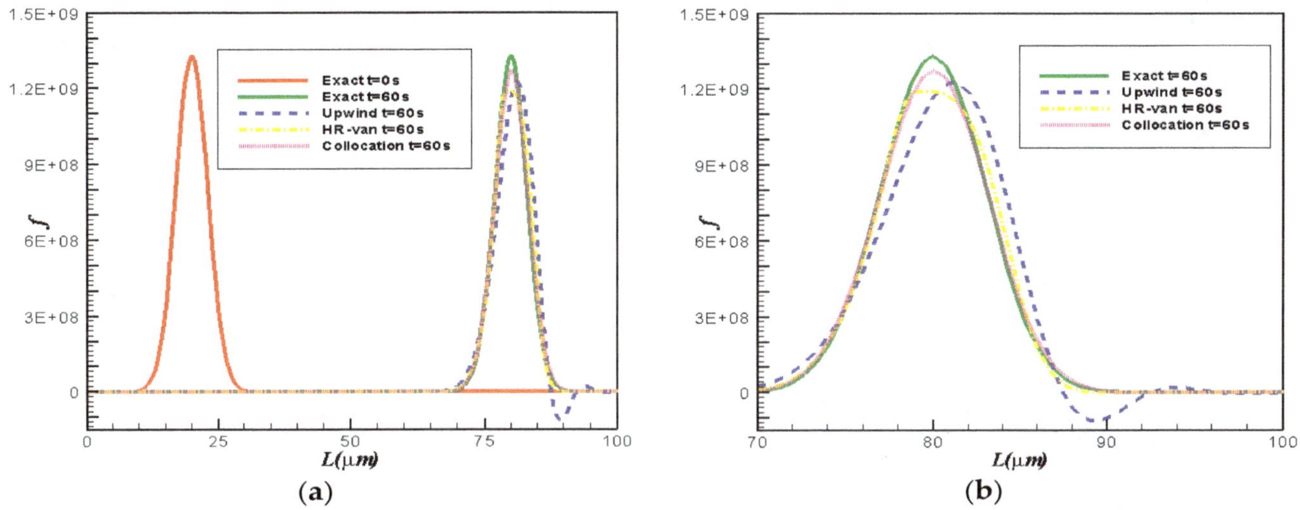

Figure 1. Comparison for final crystal size distribution (CSD) with analytical and numerical results: (**a**) in the whole crystal; (**b**) zoomed figure.

Table 1. L_1 and L_2 errors for different numerical schemes.

Method	Second Upwind	HR-Van	Collocation
L_1-error	0.24941	0.07598	0.04206
L_2-error	0.06334	0.02430	0.01265

We now discuss the effects of the diffusive term. Figure 2 shows the numerical results obtained using the Chebyshev spectral collocation method with different coefficients of diffusive term. As can been seen in Figure 2a, when no diffusive term is added ($\varepsilon = 0$), an oscillation happens near the right boundary at $t = 2$ s. Therefore, in this case, we cannot obtain stable numerical solutions. However, as shown in Figure 2b, when a diffusive term is added ($\varepsilon = 0.002$), a convergent solution is obtained. The CSD function shifts to the right with the velocity of G. Table 2 lists the L_1 and L_2 errors with different coefficients of the diffusive term. When $\varepsilon \leq 0.001$, a convergent numerical solution cannot be obtained. With the increase of ε ($\varepsilon \geq 0.002$), both L_1 and L_2 errors increase. This is consistent with our explanation for the diffusive term; the closer ε is to 0, the more accurate the result is. However, when ε approaches 0, the effect of the diffusive term becomes smaller which leads to an unstable numerical solution. The coefficient of the diffusive term (ε) with $G = 1$ is recommended as 2×10^{-3} to 10^{-2}.

In the size-independent growth case, PBE is a convection equation ($\frac{\partial f}{\partial t} + G\frac{\partial f}{\partial L} = 0$) and displaysa strong hyperbolic property. The diffusive term should be added in order to obtain a stable and convergent numerical solution. Figure 3 shows the relationship between the optimal ε(with minimum L_1 and L_2 errors) and the growth rate G. It is clear that the optimal ε has a linear relationship with the growth rate G. Therefore, the coefficient of diffusive term (ε) is recommended as $2G \times 10^{-3}$ to $G \times 10^{-2}$, where G is the crystal growth rate.

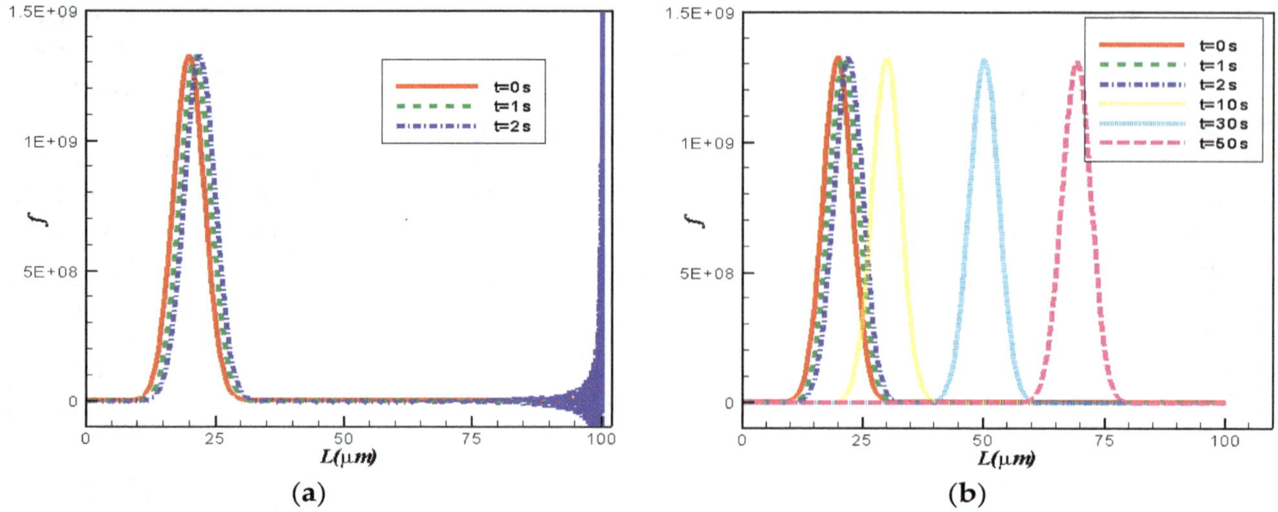

Figure 2. Comparison for numerical results with different coefficients of the diffusive term: (**a**) $\varepsilon = 0$; (**b**) $\varepsilon = 0.002$.

Table 2. L_1 and L_2 errors for different coefficients of the diffusive term.

ε	$\leq$	**0.002**	**0.01**	**0.02**
L_1-error	—	0.01286	0.06111	0.11548
L_2-error	—	0.00389	0.01828	0.03403

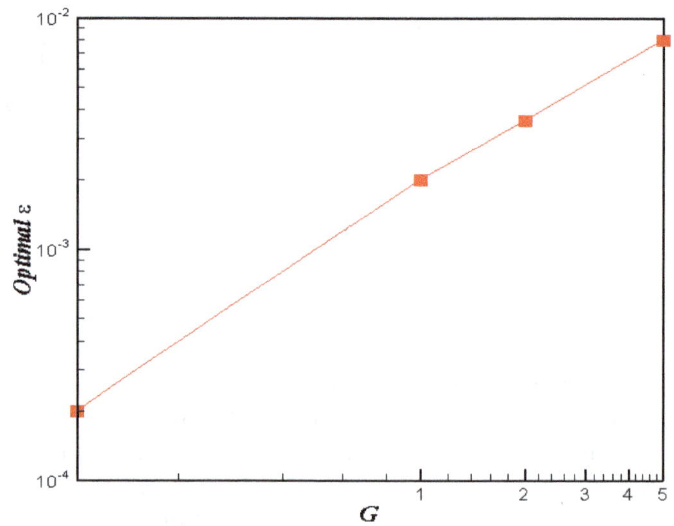

Figure 3. Relationship between optimal ε and G.

3.2. Size-Dependent Growth in a Batch Process

The second example involves crystallization in a batch process with the crystal growth rate G dependent on the crystal size L. Similar to the first example, aggregation, nucleation, and breakage are not taken into account, which means $h(L,t,f) = 0$. Crystal growth rate is supposed as the linear function of crystal size, namely $G(L,t) = G_0 L$, where G_0 is a constant. The initial CSD satisfies the following equation:

$$f(L,0) = \frac{N_0}{\overline{L}} \exp(-\frac{L}{\overline{L}}), \tag{17}$$

where $N_0, \overline{L}$ are constants. The analytical solution for this case is [1] as follows:

$$f(L,t) = \frac{N_0}{\overline{L}} \exp(-\frac{L}{\overline{L}} e^{-G_0 t}) \exp(-G_0 t). \quad (18)$$

Parameters in the simulation are taken as $\overline{L} = 0.01\ \mu m^3, N_0 = 1, G_0 = 0.1\ \mu m^3/s, \varepsilon = 0, N = 200$, $\Delta t = 0.00001\ s$. The number of grids in space used for the second upwind method and HR-van method is 1000.

Figure 4 shows the comparison of analytical solutions with numerical solutions obtained by three numerical schemes at $t = 4\ s$. In Figure 4a, there is no significant difference between the analytical solution and the numerical solutions. Figure 4b gives the comparison of the numerical results near the boundary $L = 0\ \mu m$. From Figure 4b, it is obvious that Chebyshev spectral collocation method performs the best and its result is almost coincident with the analytical solution; however, the second-order upwind scheme and HR-van scheme have some deviations from the analytical solution near the boundary. Table 3 gives the comparison of L_1 and L_2 errors for different numerical schemes. From Table 3, it is clear that the Chebyshev spectral collocation method has the smallest error, whereas the HR-van scheme has a smaller error and the second-order upwind scheme has the biggest error. Thus, the Chebyshev spectral collocation method also has the highest accuracy among these three schemes.

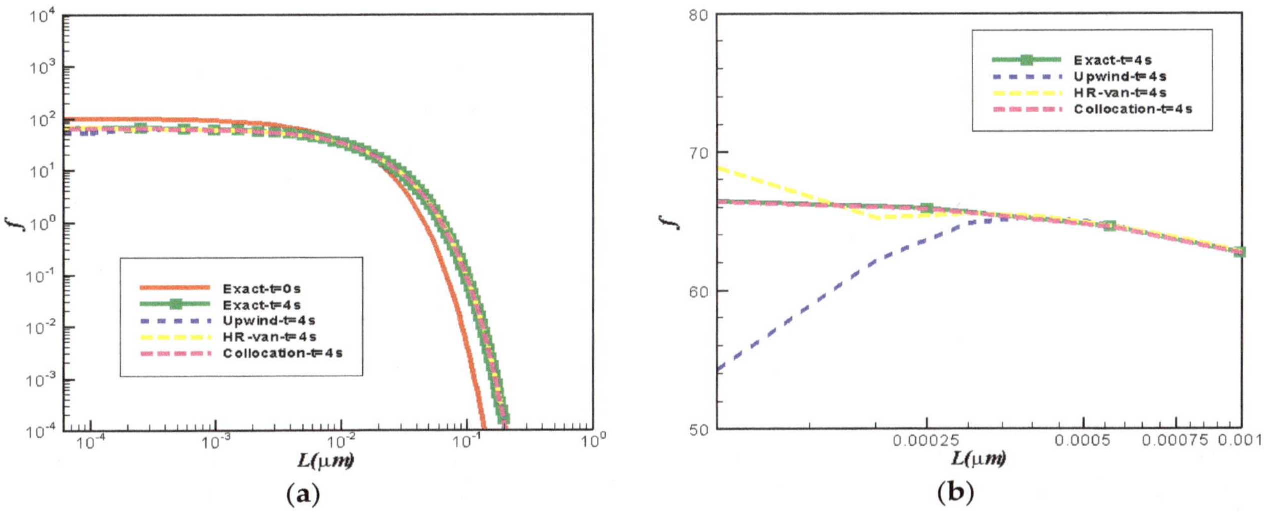

Figure 4. Comparison for final CSD with analytical and numerical results: (**a**) in the whole crystal; (**b**) zoomed figure.

Table 3. L_1 and L_2 errors for different numerical schemes.

Method	Second Upwind	HR-Van	Collocation
L_1−error	0.005036	0.004928	1.7992×10^{-7}
L_2−error	0.352802	0.328718	7.5825×10^{-7}

Figure 5 presents the evolution of CSD obtained using the Chebyshev spectral collocation method. As we can see in Figure 5, there is no oscillation during the computation. Therefore, the diffusive term is not necessary. This is not surprising. In this size-dependent growth case, the PBE is a convection–reaction equation ($\frac{\partial f}{\partial t} + G_0 L \frac{\partial f}{\partial L} + G_0 f = 0$). The reactive term weakens the hyperbolic property of PBE and makes it easy to solve. This is reflected in the decay term "$\exp(-G_0 t)$" of the analytical solution in Equation (18).

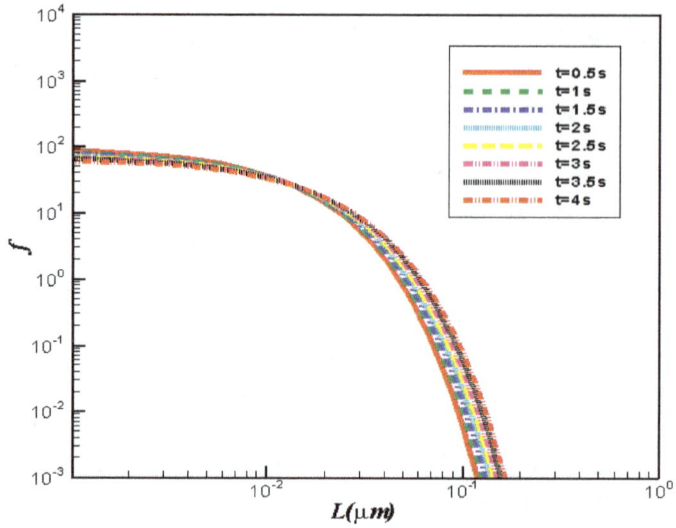

Figure 5. Evolution of CSD obtained using the Chebyshev spectral collocation method.

3.3. *Nucleation and Size-Dependent Growth in a Continuous Process*

The third example is the crystallization in a continuous process with the nucleation and crystal growth rate G dependent on the crystal size L. Here, the source item is assumed to $h(L,t,f) = B_0\delta(L) - \frac{f(L)}{\tau}$, where B_0, τ are constants, and $\delta(L)$ is a Dirac δ function. The relationship between crystal growth rate and crystal size is $G(L) = G_0(1+\gamma L)^z$, where G_0, γ, z are constants. The analytical solution for this case is [1] as follows:

$$\lim_{t\to\infty} f(L) = \frac{B_0}{G_0}(1+\gamma L)^{-z}\exp\left(\frac{1-(1+\gamma L)^{1-z}}{G_0\tau\gamma(1-z)}\right). \tag{19}$$

In order to compare the prediction effects of each numerical method, Equation (19) is given as the initial CSD and the simulation is ended at $t = 10000$ s. In our simulation, parameters are set to $B_0 = 2\times10^{-10}/\mu m^3/s$, $\tau = 100$ s, $G_0 = 0.00168$ $\mu m/s$, $z = 0.3$, $\gamma = 1$, $\varepsilon = 0$, $N = 40$, $dt = 0.001$. The number of grids in space used for the second upwind method and HR-van method is 400.

Figure 6 shows the comparison of analytical solutions with numerical solutions of three numerical schemes. From Figure 6a,b, we can conclude that the result of the second-order upwind scheme is the worst since the predicted result has the biggest deviation from the analytical solution. Results of the HR-van method and the Chebyshev spectral collocation method are satisfactory. Figure 6b shows that the results obtained using the Chebyshev spectral collocation method are slightly better than those of the HR-van method. Table 4 shows the comparison of L_1 and L_2 errors for three different numerical schemes. From the error comparison, we know that the Chebyshev spectral collocation method is the best one, whereas the HR-van scheme is the better one and the second-order upwind scheme is the worst one. Thus, in this case, the Chebyshev spectral collocation method also shows the highest accuracy.

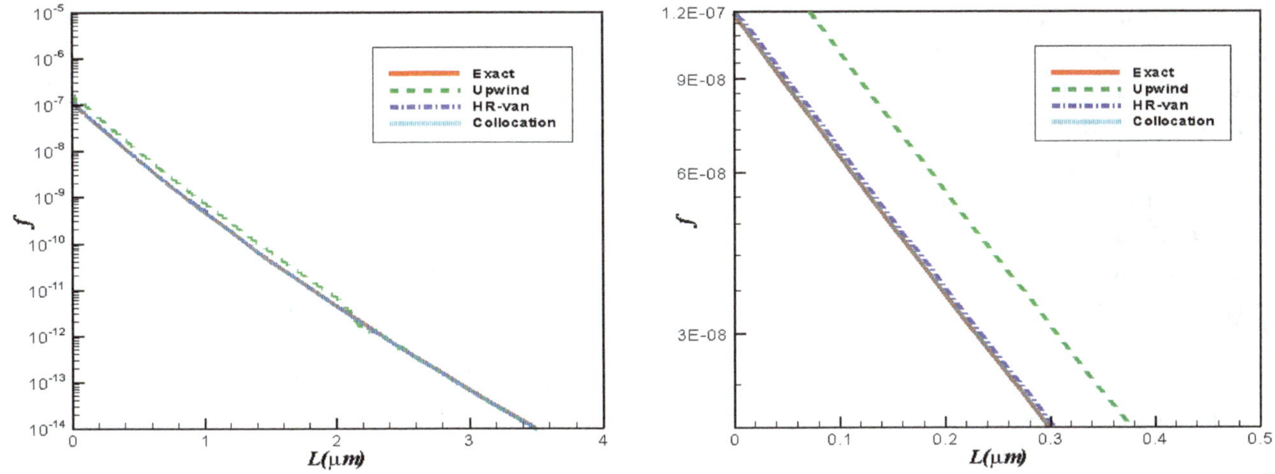

(a) (b)

Figure 6. Comparison for final CSD with analytical and numerical results: (**a**) in the whole crystal; (**b**) zoomed figure.

Table 4. L_1 and L_2 errors for different numerical schemes.

Method	Second Upwind	HR-Van	Collocation
L_1-error	0.513976	0.0285996	3.559151×10^{-7}
L_2-error	0.856945	0.049325	4.422907×10^{-7}

Figure 7 displays the evolution of CSD obtained using the Chebyshev spectral collocation method. It can be seen that it does not result in an unstable solution in the case where no diffusive term is added. In this case, the PBE becomes $\frac{\partial f}{\partial t} + G\frac{\partial f}{\partial L} + f\frac{\partial G}{\partial L} = B_0\delta(L) - \frac{f(L)}{\tau}$. It is a convection–reaction equation. As shown in example 2, the reactive term weakens the hyperbolic property of PBE. Therefore, the diffusive term is not necessary.

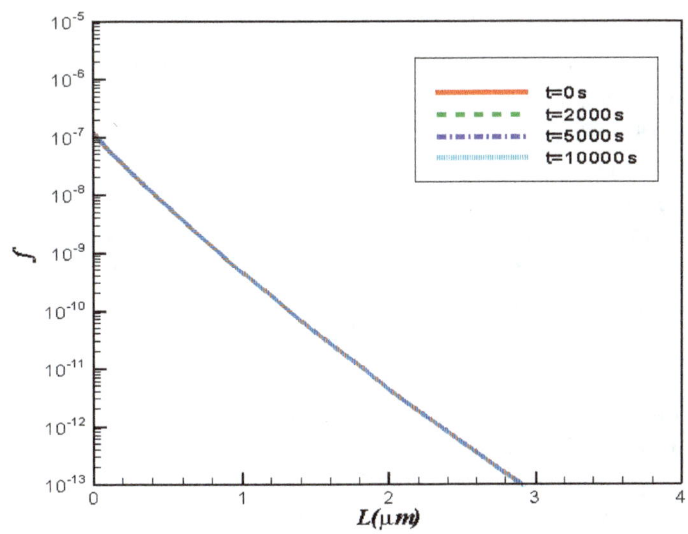

Figure 7. Evolution of CSD obtained using the Chebyshev spectral collocation method.

4. Conclusions

In this paper, the Chebyshev spectral collocation method was introduced to solve the one-dimensional PBE. The detailed implementation and the related techniques of the Chebyshev spectral collocation method are given. Three examples are presented, namely size-independent growth, size-dependent growth in a batch process, and with nucleation, and size-dependent growth in a continuous process. The numerical results showed that the Chebyshev spectral collocation method is effective and highly accurate in solving the PBE. However, in the case of size-independent growth, when PBE is a convection equation, it displays a highly hyperbolic property. A diffusive term should be added in order to obtain a stable solution.

For the more complex crystallization processes, e.g., aggregation and breakage, PBE becomes an integro-differential equation. Other methods, e.g., the cell average method and the fixed pivot method, can be combined with the spectral collocation method to obtain accurate CSD [15]. Moreover, realistic PBE problems involve multidimensional states. The spectral collocation method can be easily implemented for two-dimensional problems, enabling its use in multivariate systems. However, the spectral collocation method as a spectral method has its limitations in solving PBE, i.e., the computational zone should be simple and the distribution of CSD should be sufficiently smooth. Therefore, under such conditions, the Chebyshev spectral collocation method is recommended to solve PBE for which a high accuracy of CSD is needed.

Acknowledgments: Financial support was provided by the Natural Sciences Foundation of China (Nos. 11402078, 51375148, U1304521).

References

1. Gunawan, R.; Fusman, I.; Braatz, R.D. High resolution algorithms for multidimensional population balance equations. *AICHE J.* **2004**, *50*, 2738–2749. [CrossRef]
2. Ramkrishna, D. *Population Balances: Theory and Application to Particulate Systems in Engineering*; Academic Press: New York, NY, USA, 2000.
3. Li, D.Y.; Li, Z.P.; Gao, Z.M. Quadrature-based moment methods for the population balance equation: An algorithm review. *Chin. J. Chem. Eng.* **2019**, in press. [CrossRef]
4. Qamar, S.; Angelov, I.; Elsner, M.P.; Ashfaq, A.; Seidel-Morgenstern, A.; Warnecke, G. Numerical approximations of a population balance model for coupled batch preferential crystallizers. *Appl. Numer. Math.* **2009**, *59*, 739–753. [CrossRef]
5. Qamar, S.; Ashfaq, A.; Angelov, I.A.; Elsner, M.P.; Warnecke, G.; Seidel-Morgenstern, A. Numercial solutions of population balance models in preferential crystallization. *Chem. Eng. Sci.* **2008**, *63*, 1342–11352. [CrossRef]
6. Solsvik, J.; Jakobsen, H.A. Evaluation of weighted residual methods for the solution of a population balance model describing bubbly flows: The least-squares, Galerkin, tau, and orthogonal collocation methods. *Ind. Eng. Chem. Res.* **2013**, *52*, 15988–16013. [CrossRef]
7. Sewerin, F.; Rigopoulos, S. An explicit adaptive grid approach for the numerical solution of the population balance equation. *Chem. Eng. Sci.* **2017**, *168*, 250–270. [CrossRef]
8. Shah, B.H.; Ramkrishna, D.; Borwanker, J.D. Simulation of particulate systems using the concept of the interval of quiescence. *AICHE J.* **1977**, *23*, 897–904. [CrossRef]
9. Gooch, J.R.P.; Hounslow, M.J. Monte Carlo simulation of size-enlargement mechanisms in crystallization. *AICHE J.* **1996**, *42*, 1864–1874. [CrossRef]
10. Kumar, S.; Ramkrishna, D. On the solution of population balance equations by discretization—I. a fixed pivot technique. *Chem. Eng. Sci.* **1997**, *52*, 1311–1332. [CrossRef]
11. Ma, D.L.; Tafti, D.K.; Braatz, R.D. High-resolution simulation of multidimensional crystal growth. *Ind. Eng. Chem. Res.* **2002**, *41*, 6217–6223. [CrossRef]
12. Majumder, A.; Kariwala, V.; Ansumali, S.; Rajendran, A. Lattice Boltzmann method for multi-dimensional population balance models in crystallization. *Chem. Eng. Sci.* **2012**, *70*, 121–134. [CrossRef]
13. Ruan, C.L.; Liang, K.F.; Chang, X.J.; Zhang, L. Weighted Essentially Nonoscillatory method for two-dimensional population balance equations in crystallization. *Math. Prob. Eng.* **2013**, *125128*, 1–11. [CrossRef]
14. Mantzaris, N.V.; Daoutidis, P.; Srienc, F. Numerical simulation of multi-variable cell population balance models. II. Spectral methods. *Compt& Chem. Eng.* **2001**, *25*, 1441–1462.

15. Buck, A.; Klaunick, G.; Kumar, J.; Peglow, M.; Tsotsas, E. Numerical simulation of particulate processes for control and estimation by spectral methods. *AICHE J.* **2012**, *58*, 2309–2319. [CrossRef]
16. Shen, J.; Tang, T.; Wang, L.L. *Spectral Methods: Algorithms, Analysis and Applications*; Springer: Berlin, Germany, 2011.
17. Zhang, X. *Matlab Efficient Solution of Differential Equation: SpectralMethod*; China Machine Press: Beijing, China, 2016.
18. Tian, X.Y.; Li, B.W.; Wu, Y.S.; Zhang, J.K. Chebyshev collocation spectral method simulation for the 2D boundary layer flow and heat transfer in variable viscosity MHD fluid over a stretching plate. *Int. J. Heat. Mass. Trans.* **2015**, *89*, 829–837. [CrossRef]
19. Chen, Y.Y.; Li, B.W.; Zhang, J.K. Spectral collocation method for natural convection in a square porous cavity with local thermal equilibrium and non-equilibrium models. *Int. J. Heat. Mass. Trans.* **2016**, *96*, 84–96. [CrossRef]
20. Fang, J.W.; Wu, B.Y.; Liu, W.J. An explicit spectral collocation method using nonpolynomial basis functions for the time-dependent Schrodinger equation. *Math. Meth. App. Sci.* **2019**, *42*, 186–203. [CrossRef]
21. Zhang, H.; Jiang, X.Y.; Wang, C.; Chen, S.Z. Crank-Nicolson Fourier spectral method for the space fractional nonlinear Schrodinger equation and its parameter estimation. *Int. J. Compt. Math.* **2019**, *96*, 238–263. [CrossRef]
22. Mark, H. *Introduction to Numerical Methods in Differential Equaitons*; Springer: Berlin, Germany, 2011.

9

A Note on the Generalized Relativistic Diffusion Equation

Luisa Beghin *,† **and Roberto Garra** †,*

Dipartimento di Scienze Statistiche, "Sapienza" Università di Roma, P. le A. Moro 5, 00185 Roma, Italy
* Correspondence: luisa.beghin@uniroma1.it (L.B.); roberto.garra@sbai.uniroma1.it (R.G.)
† These authors contributed equally to this work.

Abstract: We study here a generalization of the time-fractional relativistic diffusion equation based on the application of Caputo fractional derivatives of a function with respect to another function. We find the Fourier transform of the fundamental solution and discuss the probabilistic meaning of the results obtained in relation to the time-scaled fractional relativistic stable process. We briefly consider also the application of fractional derivatives of a function with respect to another function in order to generalize fractional Riesz-Bessel equations, suggesting their stochastic meaning.

Keywords: relativistic diffusion equation; Caputo fractional derivatives of a function with respect to another function; Bessel-Riesz motion

MSC: 33E12; 34A08

---

1. Introduction

In recent papers, relativistic diffusion equations have been investigated both from the physical [1] and mathematical [2,3] points of view. It is well known that these kinds of space-fractional equations are strictly related to relativistic stable processes (see, e.g., [2]).

In [2], a fractional relativistic stable process was considered, connected with a time-fractional relativistic diffusion equation involving a derivative in the sense of Caputo. In [3], the Fourier transform of the fundamental solution and the probabilistic interpretation of the solution for the Cauchy problem of a time-fractional relativistic equation was discussed. On the other hand, the fractional derivative of a function with respect to another function is a useful mathematical tool that is recently gaining more interest in relation to models involving time-varying coefficients, see, e.g., [4–6] and the references therein.

The aim of this short paper is to consider the generalization of the time-fractional relativistic diffusion equation by means of a Caputo fractional derivative of a function with respect to another function.

We are able to find the Fourier transform of the fundamental solution for this new class of generalized relativistic diffusion equations. Moreover, we discuss its probabilistic meaning, showing the role of the time-scaling involved in the application of time-fractional operators. A time-scaled fractional tempered stable process is considered in connection to this class of equations.
The time-scaled fractional relativistic stable process is also analyzed by evaluating the covariance and correlation coefficient and the connections with fractional-type equations.

We finally apply fractional derivatives with respect to another function in order to generalize the fractional Bessel-Riesz motion recently considered in [7].

2. Preliminaries on Fractional Relativistic Stable Processes and Fractional Operators

2.1. Fractional Relativistic Diffusion and Relativistic Stable Processes

The tempered stable subordinator (TSS) $\mathcal{T}_{\mu,\alpha}(t)$, $t > 0$, with stability index $\alpha \in (0,1)$ and tempering parameter $\mu > 0$, is a Lévy process with Laplace transform

$$\mathbb{E}(e^{-s\mathcal{T}_{\mu,\alpha}(t)}) = e^{-t((s+\mu)^\alpha - \mu^\alpha)} \tag{1}$$

and transition density

$$f_{\mu,\alpha}(x,t) := e^{-\mu x + \mu^\alpha t} h_\alpha(x,t), \tag{2}$$

where $h_\alpha(x,t) = Pr\{\mathcal{H}_\alpha(t) \in dx\}$ and $\mathcal{H}_\alpha(t)$ is the α-stable subordinator. Moreover, let us denote with $\mathcal{L}_\alpha(t) := \inf\{s : \mathcal{H}_\alpha(s) > t\}, t \geq 0, \alpha \in (0,1)$ the inverse of the α-stable subordinator. For more details about TSS and applications we refer, for example, to [8,9].

The relativistic α-stable process is defined as a Brownian motion (hereafter denoted by $B(t), t \geq 0$) subordinated via an independent TSS, i.e.,

$$\mathcal{X}_{\mu,\alpha}(t) := B(\mathcal{T}_{\mu,\alpha}(t)), \quad t > 0. \tag{3}$$

Following the notation used, for example, in [3], in this paper we generalize the relativistic diffusion equation

$$\frac{\partial u}{\partial t} = H_{\alpha,m} u, \tag{4}$$

where the spatial differential operator

$$H_{\alpha,m} := m - \left(m^{\frac{2}{\alpha}} - \Delta\right)^{\frac{\alpha}{2}}, \quad \alpha \in (0,2),\ m > 0, \tag{5}$$

is a relativistic diffusion operator, i.e., the infinitesimal generator of the relativistic stable process. In [3] the time-fractional generalization has been considered by replacing the first-order time-derivative with a Caputo fractional derivative. The related stochastic process has interesting properties, as discussed in [2,3].

We here briefly recall the notion of fractional tempered stable (TS) process introduced in [2].

Definition 1. *Let $\mathcal{L}_\beta(t)$, $t \geq 0$ be the inverse of the stable subordinator, then the fractional TS process is defined as*

$$\mathcal{T}^\beta_{\mu,\alpha}(t) := \mathcal{T}_{\mu,\alpha}(\mathcal{L}_\beta(t)), \quad t \geq 0, \mu \geq 0, \alpha, \beta \in (0,1) \tag{6}$$

where $\mathcal{L}_\beta$ is independent of the tempered stable subordinator (TSS) $\mathcal{T}_{\mu,\alpha}$.

The density of the fractional TS process $\mathcal{T}^\beta_{\mu,\alpha}(t)$ satisfies the fractional equation (see [2] Theorem 6)

$$D^\beta_t f = \left[\mu^\alpha - \left(\mu + \frac{\partial}{\partial x}\right)^\alpha\right] f, \quad \alpha \in (0,1),\ \mu \geq 0, \tag{7}$$

under initial-boundary conditions

$$\begin{cases} f(x,0) = \delta(x), \\ f(0,t) = 0. \end{cases} \tag{8}$$

We here denote by D^β_t the Caputo fractional derivative of order $\beta \in (0,1)$, i.e.,

$$D^\beta_t f(t) = \frac{1}{\Gamma(1-\beta)} \int_0^t (t-\tau)^{-\beta} \frac{\partial f}{\partial \tau} d\tau. \tag{9}$$

Definition 2. *The fractional relativistic stable process is defined as the time-changed Brownian motion*

$$\mathcal{X}^{\beta}_{\mu,\alpha}(t) := B(\mathcal{T}^{\beta}_{\mu,\alpha}(t)), \quad t \geq 0. \tag{10}$$

The density of the fractional relativistic stable process $\mathcal{X}^{\beta}_{\mu,\alpha}(t)$ coincides with the solution of the time-fractional relativistic diffusion equation ([2], Theorem 16)

$$D^{\beta}_{t} g = \left[\mu^{\alpha} - (\mu - \Delta)^{\alpha}\right] g, \tag{11}$$

under initial and boundary conditions

$$\begin{cases} g(x,0) = \delta(x), \\ \lim_{|x|\to\infty} g(x) = 0, \\ \lim_{|x|\to\infty} \frac{\partial}{\partial x} g(x) = 0. \end{cases} \tag{12}$$

We will here present some generalizations of Equations (7) and (11) obtained by using fractional derivatives of a function with respect to another function.

2.2. Fractional Derivatives of a Function with Respect to Another Function

Fractional derivatives of a function with respect to another function have been considered in the classical monograph by Samko et al. [10] and Kilbas et al. [11] (Section 2.5). Recently they were reconsidered by Almeida in [4] where the Caputo-type regularization of the existing definition and some interesting properties are provided. We recall the basic definitions and properties for the reader's convenience.

Let $\nu > 0$, $f(t) \in C^1([a,t])$ an increasing function such that $f'(t) \neq 0$ in $[a,t]$, the fractional integral of a function $g(t)$ with respect to another function $f(t)$ is given by

$$I^{\nu,f}_{a^+} g(t) := \frac{1}{\Gamma(\nu)} \int_a^t f'(\tau)(f(t) - f(\tau))^{\nu-1} g(\tau) d\tau. \tag{13}$$

Observe that for $f(t) = t^{\beta}$ we recover the definition of Erdélyi-Kober fractional integral recently applied, for example, in connection with the Generalized Grey Brownian Motion [12].

The corresponding Caputo-type evolution operator (according to our notation) is given by

$$\mathcal{O}^{\nu,f}_{t} g(t) := I^{n-\nu,f}_{a^+} \left(\frac{1}{f'(t)} \frac{d}{dt}\right)^n g(t), \tag{14}$$

where $n = [\nu] + 1$. Here we have used the symbol $\mathcal{O}^{\nu,f}_{t}(\cdot)$ in order to underline the generic integro-differential nature of the time-evolution operator, beyond the fractional calculus theory.

A relevant property of Equation (14) is that, if $g(t) = (f(t) - f(a))^{\beta-1}$ with $\beta > 1$, then (see Lemma 1 of [4])

$$\mathcal{O}^{\nu,f}_{t} g(t) = \frac{\Gamma(\beta)}{\Gamma(\beta - \nu)} (f(t) - f(a))^{\beta-\nu-1}. \tag{15}$$

As a consequence, the composite Mittag-Leffler function (see [6,13] for Mittag-Leffler functions)

$$g(t) = E_{\nu}(\lambda(f(t) - f(a))^{\nu}) \tag{16}$$

is an eigenfunction of the operator $\mathcal{O}^{\nu,f}_{t}$. In the next section we will consider $a = 0$.

Hereafter we denote by $\widehat{f}(k)$ the Fourier transform of a given function $f(x)$.

3. On the Generalized Relativistic Diffusion Equation

Theorem 1. *Let $f(t) \in C^2[0,T]$, such that $f(0) = 0$ and $f' \neq 0 \; \forall t \in (0,T]$, then the Fourier transform of the fundamental solution of the generalized time-fractional relativistic diffusion equation*

$$\mathcal{O}_t^{\nu,f} u(x,t) = \left[m - (m^{2/\alpha} - \Delta)^{\alpha/2}\right] u(x,t), \quad \alpha \in (0,2), \nu \in (0,1), \tag{17}$$

is given by

$$\widehat{u}(k,t) = E_\nu(-\theta(|k|)(f(t))^\nu), \tag{18}$$

where

$$\theta(|k|) = (m^{\frac{\alpha}{2}} + |k|^2)^{\alpha/2} - m. \tag{19}$$

Moreover, in the case $\nu = 1$, the fundamental solution of the equation

$$\left(\frac{1}{f'(t)}\frac{\partial}{\partial t}\right) u(x,t) = \left[m - (m^{2/\alpha} - \Delta)^{\alpha/2}\right] u(x,t), \quad \alpha \in (0,2) \tag{20}$$

is given by

$$\widehat{u}(k,t) = e^{-\theta(|k|)f(t)}. \tag{21}$$

Proof. By taking the spatial Fourier transform of Equation (17), we have that

$$\mathcal{O}_t^{\nu,f} \widehat{u}(k,t) = -\{(m^{\frac{\alpha}{2}} + |k|^2)^{\alpha/2} - m\}\widehat{u}(k,t) = -\theta(|k|)\widehat{u}(k,t), \tag{22}$$

under the initial condition $\widehat{u}(k,0) = 1$. Then, by using the fact that

$$\mathcal{O}_t^{\nu,f} E_\nu(-\theta(|k|)(f(t))^\nu) = -\theta(|k|)E_\nu(-\theta(|k|)(f(t))^\nu), \tag{23}$$

we obtain the claimed result.

The case $\nu = 1$ can be checked by simple calculations. □

Remark 1. *Observe that, for $m = 0$, we obtain the Fourier transform of the fundamental solution of the generalized space-time-fractional diffusion equation*

$$\mathcal{O}_t^{\nu,f} u(x,t) = -(-\Delta)^{\alpha/2} u(x,t), \quad \alpha \in (0,2), \nu \in (0,1), \tag{24}$$

while, for $\alpha = 2$, $m = 0$, we have the Fourier transform of the generalized time-fractional heat equation

$$\mathcal{O}_t^{\nu,f} u(x,t) = \Delta u(x,t), \quad \nu \in (0,1). \tag{25}$$

In the latter, some interesting cases have been considered in the literature. In particular

- *if $f(t) = t$, we recover the time-fractional diffusion equation, which is widely studied in the literature, (we refer, for example, to [14,15]).*
- *if $f(t) = (1 - e^{-\beta t})/\beta$, we have the fractional Dodson diffusion equation studied in [6].*
- *if $f(t) = \ln t$, we have the time-fractional diffusion involving a regularized Hadamard fractional derivative.*

Remark 2. *In Theorem 1, we take for simplicity the condition that $a = 0$ in the definition of the fractional operator $\mathcal{O}_t^{\nu,f}$ and we consider the renstriction to functions f such that $f(0) = 0$. In the more general case $a > 0$, $f(a) \neq 0$, we have that the Fourier transform of the fundamental solution for Equation (17) is given by*

$$\widehat{u}(k,t) = E_\nu(-\theta(|k|)(f(t) - f(a))^\nu). \tag{26}$$

Much care should be given, for example, to the case of logarithmic functions, where "a" must be greater than zero.

Remark 3. *Equation* (20) *with* $\alpha = 1$ *can be euristically interpreted as a relativistic Schrödinger-type equation with time-dependent mass.*

4. Time-Scaled Fractional Tempered Stable Process

Let us consider the following equation

$$\mathcal{O}_t^{\beta,f} h = \left[\mu^\alpha - (\mu + \frac{\partial}{\partial x})^\alpha\right] h, \tag{27}$$

under the initial-boundary conditions

$$\begin{cases} h(x,0) = \delta(x), \\ h(0,t) = 0. \end{cases} \tag{28}$$

Let us denote by $\widetilde{f}$ the Laplace transform of a function f, i.e., $\widetilde{f}(\eta) := \int_0^{+\infty} e^{-\eta x} f(x)dx$. We now evaluate the Laplace transform, with respect to the space argument, of Equation (27), which reads

$$\mathcal{O}_t^{\beta,f} \widetilde{h}(\eta,t) = \psi_{\alpha,\mu}(\eta)\widetilde{h}(\eta,t), \qquad \eta > 0,$$

with initial condition $\widetilde{h}(\eta,0) = 1$ and where $\psi_{\alpha,\mu}(\eta) := (\eta+\mu)^\alpha - \mu^\alpha$ (see [2] for details). Thus, we can write its solution as

$$\widetilde{h}(\eta,t) = E_\beta(-\psi_{\alpha,\mu}(\eta)[f(t) - f(0)]^\beta). \tag{29}$$

By comparing Equation (29) with Equation (20) in [2], we can derive the following relationship holding for the corresponding process $\mathcal{T}_{\mu,\alpha}^{\beta,f}(t)$, for any $t > 0$

$$\mathbb{E}e^{-\eta \mathcal{T}_{\mu,\alpha}^{\beta,f}(t)} = \mathbb{E}e^{-\eta \mathcal{T}_{\mu,\alpha}^{\beta}(f(t))} = E_\beta(-\psi_{\alpha,\mu}(\eta)[f(t) - f(0)]^\beta) \tag{30}$$

and, by the unicity of the Laplace transform, we can write the following equality of the one-dimensional distributions

$$\mathcal{T}_{\mu,\alpha}^{\beta,f}(t) \stackrel{d}{=} \mathcal{T}_{\mu,\alpha}^{\beta}(f(t)),$$

which holds for any $t > 0$. Thus the process $\mathcal{T}_{\mu,\alpha}^{\beta,f}$ can be called time-scaled fractional tempered stable process.

Therefore, we have the following:

Theorem 2. *The density of the time-scaled fractional TS process* $\mathcal{T}_{\mu,\alpha}^{\beta}(f(t))$ *satisfies the fractional Equation* (27) *under the initial-boundary Conditions* (28).

From the finiteness of the moments of the standard tempered stable process and by taking into account Equation (6), we can draw the conclusion that the moment generating function of $\mathcal{T}_{\mu,\alpha}^{\beta,f}$ exists, for any f and $\alpha, \beta \in (0,1)$, and is equal to Equation (30), for $\kappa = -\eta$. By differentiating

$$\mathbb{E}e^{\kappa \mathcal{T}_{\mu,\alpha}^{\beta,f}(t)} = E_\beta(-\psi_{\alpha,\mu}(-\kappa)[f(t) - f(0)]^\beta),$$

we then obtain the first and second moments of $\mathcal{T}_{\mu,\alpha}^{\beta,f}$ as follows

$$\mathbb{E}\mathcal{T}_{\mu,\alpha}^{\beta,f}(t) = \frac{\alpha\mu^{\alpha-1}[f(t) - f(0)]^\beta}{\Gamma(\beta+1)} \tag{31}$$

and

$$\mathbb{E}\left[\mathcal{T}^{\beta,f}_{\mu,\alpha}(t)\right]^2 = \frac{\alpha(1-\alpha)\mu^{\alpha-2}[f(t)-f(0)]^{\beta}}{\Gamma(\beta+1)} + \frac{2\alpha^2\mu^{2\alpha-2}[f(t)-f(0)]^{2\beta}}{\Gamma(2\beta+1)}, \tag{32}$$

which, for $\beta = 1$ and $f(t) = t$, for any t, coincide with those of the standard tempered stable process (see [8]). Since, in this case, the process is not Lévy, we also loose the stationarity of increments.

We study the tails' behavior of the distribution of the time-scaled fractional TS process $\mathcal{T}^{\beta,f}_{\mu,\alpha}(t)$, i.e., we obtain an estimate for $P(\mathcal{T}^{\beta,f}_{\mu,\alpha}(t) > x)$, for $x \to +\infty$, by means of the asymptotic result holding for the usual TS subordinator $\mathcal{T}_{\mu,\alpha}(t)$: It is well-known that, for $x \to +\infty$,

$$P(\mathcal{T}_{\mu,\alpha}(t) > x) \sim \frac{e^{-\mu x+\mu^{\alpha}t}x^{-\alpha}\Gamma(1+\alpha)\sin(\pi\alpha)t}{\alpha\pi} \tag{33}$$

(see [16]). Let $f(0) = 0$, for simplicity, and let $l_{\beta}(x,t) := P\{\mathcal{L}_{\beta}(t) \in dx\}$ be the transition density of the inverse β-stable subordinator, then we can write, by considering Equations (6) and (29) together, that

$$\begin{aligned}
P\left(\mathcal{T}^{\beta,f}_{\mu,\alpha}(t) > x\right) &= \int_0^{+\infty} P(\mathcal{T}_{\mu,\alpha}(z) > x)\, l_{\beta}(z, f(t))dz \\
&= [\text{by (33)}] \\
&\sim \frac{e^{-\mu x}x^{-\alpha}\Gamma(1+\alpha)\sin(\pi\alpha)}{\alpha\pi}\int_0^{+\infty} e^{\mu^{\alpha}z} z\, l_{\beta}(z, f(t))dz \\
&= \frac{e^{-\mu x}x^{-\alpha}\Gamma(1+\alpha)\sin(\pi\alpha)}{\alpha\pi}\frac{\partial}{\partial k}\int_0^{+\infty} e^{kz} l_{\beta}(z, f(t))dz\Big|_{k=\mu^{\alpha}} \\
&= \frac{e^{-\mu x}x^{-\alpha}\Gamma(1+\alpha)\sin(\pi\alpha)}{\alpha\pi}\frac{\partial}{\partial k} E_{\beta}(kf(t)^{\beta})\Big|_{k=\mu^{\alpha}} \\
&\sim \frac{e^{-\mu x}x^{-\alpha}\Gamma(1+\alpha)\sin(\pi\alpha)f(t)^{\beta}}{\alpha\pi\beta} E_{\beta,\beta}(\mu^{\alpha}f(t)^{\beta}).
\end{aligned}$$

Taking the derivative out of the integral, in the previous lines, is allowed by the standard arguments for the moment generating function of the inverse stable subordinator

$$\int_0^{+\infty} e^{kz} l_{\beta}(z, f(t))dz = E_{\beta}(kf(t)^{\beta})$$

(see also [17]).

Going back to the problem considered in the previous section, we are now able to provide a stochastic interpretation for the fundamental solution of the generalized fractional relativistic diffusion equation.

Theorem 3. *The fundamental solution of the generalized time-fractional relativistic diffusion Equation* (17) *coincides with the density of the time-scaled fractional relativistic stable process* $\mathcal{X}^{\nu}_{m^{2/\alpha},\frac{\alpha}{2}}(f(t)) = B(\mathcal{T}^{\nu}_{m^{2/\alpha},\frac{\alpha}{2}}(f(t))$.

We omit the complete proof, since it can be easily derived from [2], Corollary 13. Indeed, we here underline that, by using a time-fractional derivative with respect to a function, we obtain a deterministic time-change; therefore the related process is the time-scaled counterpart of the original process.

We can evaluate, by conditioning the expected value and variance of the time-scaled fractional relativistic stable process $\mathcal{X}^{\nu}_{m,\alpha/2}$. Indeed, we can write

$$\mathbb{E}\mathcal{X}^{\nu}_{m^{2/\alpha},\alpha/2}(t) = \mathbb{E}B(\mathcal{T}^{\nu,f}_{\mu,\alpha}(t)) = 0$$

and

$$var\,\mathcal{X}^{\nu}_{m^{2/\alpha},\alpha/2}(t) = \mathbb{E}\left\{\mathbb{E}\left[B(\mathcal{T}^{\nu,f}_{\mu,\alpha}(t))\Big|\,\mathcal{T}^{\nu,f}_{\mu,\alpha}(t)\right]^2\right\} = \mathbb{E}\mathcal{T}^{\nu,f}_{\mu,\alpha}(t) = \frac{\alpha\mu^{\alpha-1}[f(t)-f(0)]^{\nu}}{\Gamma(\nu+1)}.$$

This confirms that the mean-square displacement of the process behaves as $f(t)^\nu$ and thus crucially depends on the choice of f.

We also evaluate the covariance and correlation coefficient of the process $B(\mathcal{T}_{\mu,\alpha}^{\nu,f}(t))$, at least in the special case where f is a sublinear or linear function. Let moreover $f(0) = 0$, for simplicity. We prove, under these assumptions, that the process displays a long-range dependence (LRD). We need the following definition for LRD of non-stationary processes (see [9,18]):

Definition 3. *Let $s > 0$ and $t > s$. If, for a process $Z(t)$, $t \geq 0$, the following asymptotic behavior of the correlation function holds*

$$corr(Z(s), Z(t)) \sim c(s)t^{-d}, \qquad t \to +\infty,$$

where $c(s)$ is a constant (for fixed s) and $d \in (0,1)$, then $Z(t)$ is said to have the LRD property.

We restrict our analysis to the class of increasing, positive functions f, such that $\lim_{t\to+\infty} \frac{f(t)}{t^\rho} = const$, or equivalently $f(t) \sim t^\rho$, for $\rho \in (0,1]$. It corresponds to considering sublinear functions, for which $|f(t-s)| \leq f(t) - f(s)$, for any $t \geq s$, (recalling that the last difference is always positive for an increasing function). The equality holds in the special case of linear functions. We start by recalling that the Brownian motion has stationary increments and the same holds for the tempered stable and the inverse stable subordinators (see, for example, [16,19], respectively). Moreover, it is easy to check that the stationarity of increments is preserved under composition of processes. As a consequence, taking into account Equations (6) and (10), we can write, for $s \leq t$, that

$$\begin{aligned}
& cov(B(\mathcal{T}_{\mu,\alpha}^{\nu,f}(s)), B(\mathcal{T}_{\mu,\alpha}^{\nu,f}(t))) \\
&= \frac{1}{2}\left\{\mathbb{E}\left(B(\mathcal{T}_{\mu,\alpha}^{\nu,f}(s))\right)^2 + \mathbb{E}\left(B(\mathcal{T}_{\mu,\alpha}^{\nu,f}(t))\right)^2 - \mathbb{E}\left(B(\mathcal{T}_{\mu,\alpha}^{\nu,f}(t)) - B(\mathcal{T}_{\mu,\alpha}^{\nu,f}(s))\right)^2\right\} \\
&= \frac{1}{2}\mathbb{E}\left(B(\mathcal{T}_{\mu,\alpha}^{\nu,f}(s))\right)^2 + \frac{1}{2}\mathbb{E}\left(B(\mathcal{T}_{\mu,\alpha}^{\nu,f}(t))\right)^2 - \frac{1}{2}\mathbb{E}\left(B(\mathcal{T}_{\mu,\alpha}^{\nu,f}(t-s))\right)^2 \\
&= \frac{1}{2}\frac{\alpha\mu^{\alpha-1}f(t)^\nu}{\Gamma(\nu+1)} + \frac{1}{2}\frac{\alpha\mu^{\alpha-1}f(s)^\nu}{\Gamma(\nu+1)} - \frac{\alpha\mu^{\alpha-1}f(t-s)^\nu}{2\Gamma(\nu+1)} \\
&\geq \frac{1}{2}\frac{\alpha\mu^{\alpha-1}f(t)^\nu}{\Gamma(\nu+1)}\left[-\sum_{j=1}^{\infty}\binom{\nu}{j}f(t)^{-j}(-f(s))^j + f(s)^\nu f(t)^{-\nu}\right] \\
&= \frac{1}{2}\frac{\alpha\mu^{\alpha-1}f(t)^\nu}{\Gamma(\nu+1)}\left[\frac{\nu f(s)}{f(t)}\sum_{l=0}^{\infty}\binom{\nu-1}{l}\frac{f(t)^{-l}(-f(s))^l}{l+1} + f(s)^\nu f(t)^{-\nu}\right] \\
&= \frac{1}{2}\frac{\alpha\mu^{\alpha-1}f(t)^\nu}{\Gamma(\nu+1)}\left[\frac{\nu f(s)}{f(t)} + f(s)^\nu f(t)^{-\nu} + O(f(t)^{-2})\right] \\
&\sim \frac{1}{2}\frac{\alpha\nu\mu^{\alpha-1}f(s)}{\Gamma(\nu+1)}f(t)^{\nu-1}.
\end{aligned}$$

Thus we easily check that the correlation coefficient is bounded below by the following asymptotic expression $f(s)^{1-\nu/2}f(t)^{(\nu/2)-1} \sim f(s)^{1-\nu/2}t^{\rho(\nu/2)-\rho}$ (for any fixed s and $t \to +\infty$). This means that $B(\mathcal{T}_{\mu,\alpha}^{\nu,f}(t))$ exhibits LRD behavior with $d = \rho(1-\nu/2) \in (\rho/2, 1)$, at least inside the class of sublinear functions (asymptotically behaving as fractional powers of t).

5. An Application to Generalized Fractional Bessel-Riesz Motion

In the recent paper [7], the authors give a stochastic representation for the solution of the following fractional Cauchy problem

$$D_t^\beta p = -\lambda(-\Delta)^{\alpha/2}(I-\Delta)^{\gamma/2}p, \quad ,\gamma \geq 0, \alpha \in (0,2], \beta \in (0,1] \tag{34}$$

$$p(x,0) = \delta(x), \tag{35}$$

where $(-\Delta)^{\alpha/2}$ and $(I-\Delta)^{\gamma/2}$ are the inverses of the Riesz and the Bessel potential respectively. The Fourier transform of the Cauchy Equation (34), has been obtained in the form

$$\widehat{G}(k,t) = E_\beta\left(-\lambda t^\beta \|k\|^\alpha (1+\|k\|^2)^{\gamma/2}\right). \tag{36}$$

In [7], it was proved that the fundamental solution for Equation (34) coincides with the density of the stochastic process

$$X_{BR}(t) = W(L_{\alpha,\gamma}(\mathcal{L}_\beta(t))), \quad t \geq 0 \tag{37}$$

where W is the n-dimensional Brownian motion, $\mathcal{L}_\beta(t)$ is the inverse stable subordinator (according to our notation) and $L_{\alpha,\gamma}$ is the Lévy subordinator with Laplace exponent

$$\Phi(s) = \lambda s^{\alpha/2}(1+s)^{\gamma/2}, s > 0. \tag{38}$$

All these three processes are jointly independent.

Let us consider the generalized fractional counterpart of the Cauchy Equation (34) involving fractional derivatives with respect to another function, i.e.,

$$\mathcal{O}_t^{\nu,f} u(x,t) = -\lambda(-\Delta)^{\alpha/2}(I-\Delta)^{\gamma/2}u(x,t). \tag{39}$$

Assuming that $f(t)$ is a suitable smooth function such that $f(0) = 0$, we have that the fundamental solution of Equation (39) given by

$$u(x,t) = E_\beta\left(-\lambda(f(t))^\beta \|k\|^\alpha (1+\|k\|^2)^{\gamma/2}\right) \tag{40}$$

and therefore it coincides with the density of the time-scaled stochastic processes depending on f

$$X_{BR}(t) = W(L_{\alpha,\gamma}(\mathcal{L}_\beta(f(t)))), \quad t \geq 0 \tag{41}$$

6. Conclusions and Open Problems

We have analyzed here some applications of the time-fractional derivative with respect to a function, in the context of the so-called fractional relativistic diffusion equation. It is well-known that the most interesting case from the physical point of view corresponds to $\alpha = 1$, where the operator $-H_{1,m}$ appearing in Equation (5) represents the free energy of the relativistic Schrödinger equation. A physical discussion about the utility and meaning of the time-fractional generalizations of the relativistic diffusion is still missing. In view of the wide exhisting literature on the time-fractional Schrödinger equation, it is worth investigating this topic for future studies, also for the stochastic interpretation discussed here.

From the mathematical point of view, another interesting topic is the connection between higher order equations and time-fractional equations (see, e.g., [20]). In this case the relation is not trivial and it probably works only for specific values of the real order of the derivative.

Finally, a probabilistic interpretation of the fractional integrals with respect to a function can be developed, following the discussion presented in [21].

Author Contributions: The authors contributed equally to this work.

Acknowledgments: The work of R.G. was carried out in the framework of the activities of the GNFM.

References

1. Baeumer, B.; Meerschaert, M.M.; Naber, M. Stochastic models for relativistic diffusion. *Phys. Rev. E* **2010**, *82*, 011132. [CrossRef] [PubMed]
2. Beghin, L. On fractional tempered stable processes and their governing differential equations. *J. Comput. Phys.* **2015**, *293*, 29–39. [CrossRef]
3. Shieh, N.R. On time-fractional relativistic diffusion equations. *J. Pseudo-Differ. Oper. Appl.* **2012** *3*, 229–237. [CrossRef]
4. Almeida, R. A Caputo fractional derivative of a function with respect to another function. *Commun. Nonlinear Sci. Numer. Simul.* **2017**, *44*, 460–481. [CrossRef]
5. Colombaro, I.; Garra, R.; Giusti, A.; Mainardi, F. Scott-Blair models with time-varying viscosity. *Appl. Math. Lett.* **2018**, *86*, 57–63. [CrossRef]
6. Garra, R.; Giusti, A.; Mainardi, F. The fractional Dodson diffusion equation: A new approach. *Ricerche di Matematica* **2018**, 1–11. [CrossRef]
7. Anh, V.V.; Leonenko, N.N.; Sikorskii, A. Stochastic representation of fractional Bessel-Riesz motion. *Chaos Solitons Fractals* **2017**, *102*, 135–139. [CrossRef]
8. Gajda, J.; Kumar, A.; Wylomańska, A. Stable Lévy process delayed by tempered stable subordinator. *Stat. Probab. Lett.* **2019**, *145*, 284–292. [CrossRef]
9. Kumar, A.; Gajda, J.; Wylomanska, A.; Poloczanski, R. Fractional Brownian motion delayed by tempered and inverse tempered stable subordinators. *Methodol. Comput. Appl. Probab.* **2019**, *21*, 185–202. [CrossRef]
10. Samko, S.G.; Kilbas, A.A.; Marichev, O.I. *Fractional Integrals and Derivatives*; Gordon and Breach Science Publishers: Yverdon-les-Bains, Switzerland, 1993.
11. Kilbas, A.A.A.; Srivastava, H.M.; Trujillo, J.J. *Theory and Applications of Fractional Differential Equations*; Elsevier Science Limited: Amsterdam, The Netherlands, 2006.
12. Pagnini, G. Erdélyi-Kober fractional diffusion. *Fract. Calc. Appl. Anal.* **2012**, *15*, 117–127. [CrossRef]
13. Gorenflo, R.; Kilbas, A.A.; Mainardi, F.; Rogosin, S.V. *Mittag-Leffler Functions, Related Topics and Applications*; Springer: Berlin, Germany, 2014.
14. Mainardi, F. The fundamental solutions for the fractional diffusion-wave equation. *Appl. Math. Lett.* **1996**, *9*, 23–28. [CrossRef]
15. Orsingher, E.; Beghin, L. Fractional diffusion equations and processes with randomly varying time. *Ann. Probab.* **2009**, *37*, 206–249. [CrossRef]
16. Gupta, N.; Kumar, A.; Leonenko, N. Mixtures of tempered stable subordinators. *arXiv* **2019**, arXiv:1905.00192v1.
17. Beghin, L.; Macci, C.; Martinucci, B. Random time-changes and asymptotic results for a class of continuous-time Markov chains on integers with alternating rates. *arXiv* **2018**, arXiv:1802.06434v1.
18. Ayache, A.; Cohen, S.; Lévy Véhel, J. The covariance structure of multifractional Brownian motion, with application to long range dependence. In Proceedings of the IEEE International Conference on Acoustics, Speech, and Signal Processing—ICASSP 2000, Istanbul, Turkey, 5–9 June 2000.
19. Kaj, I.; Martin-Lof, A. Scaling Limit Results for the Sum of Many Inverse Levy Subordinators. *arXiv* **2012**, arXiv:1203.6831
20. D'Ovidio, M. On the fractional counterpart of the higher-order equations. *Stat. Probab. Lett.* **2011**, *81*, 1929–1939. [CrossRef]
21. Tarasov, V.E.; Tarasova, S.S. Probabilistic Interpretation of Kober Fractional Integral of Non-Integer Order. *Prog. Fract. Differ. Appl.* **2019**, *5*, 1–5. [CrossRef]

10

Some Schemata for Applications of the Integral Transforms of Mathematical Physics

Yuri Luchko

Department of Mathematics, Physics, and Chemistry, Beuth University of Applied Sciences Berlin, Luxemburger Str. 10, 13353 Berlin, Germany; luchko@beuth-hochschule.de

Abstract: In this survey article, some schemata for applications of the integral transforms of mathematical physics are presented. First, integral transforms of mathematical physics are defined by using the notions of the inverse transforms and generating operators. The convolutions and generating operators of the integral transforms of mathematical physics are closely connected with the integral, differential, and integro-differential equations that can be solved by means of the corresponding integral transforms. Another important technique for applications of the integral transforms is the Mikusinski-type operational calculi that are also discussed in the article. The general schemata for applications of the integral transforms of mathematical physics are illustrated on an example of the Laplace integral transform. Finally, the Mellin integral transform and its basic properties and applications are briefly discussed.

Keywords: integral transforms; Laplace integral transform; transmutation operator; generating operator; integral equations; differential equations; operational calculus of Mikusinski type; Mellin integral transform

MSC: 45-02; 33C60; 44A10; 44A15; 44A20; 44A45; 45A05; 45E10; 45J05

---

1. Introduction

In this survey article, we discuss some schemata for applications of the integral transforms of mathematical physics to differential, integral, and integro-differential equations, and in the theory of special functions. The literature devoted to this subject is huge and includes many books and reams of papers. For more details regarding this topic we refer the readers to, say, [1–4]. Of course, in a short survey article it is not possible to mention all known integral transforms and their numerous applications. That is why we focus on just some selected integral transforms and their applications that are of general nature and valid for most of the integral transforms in one or another form.

We start with introducing the integral transforms of mathematical physics that possess the inverses in form of the linear integral transforms and can be interpreted as transmutation operators for their generating operators. The integral transforms of mathematical physics, their generating operators, and convolutions are closely related to each other. In particular, the integral transform technique can be employed for derivation of the closed form solutions to some integral equations of convolution type and to the integral, differential, or integro-differential equations with the generating operators.

Another powerful technique for applications of the integral transforms is the Mikusinski-type operational calculi. They can be developed for the left-inverse operators of the generating operators of the integral transforms. A basic element of this construction is the convolutions for the corresponding integral transforms that play the role of multiplication in some rings of functions. This ring is then extended to a field of convolution quotients following the standard procedure. One of the advantages of this extension is that the left-inverse operator $\mathcal{D}$ to the generating operator $\mathcal{L}$ of the given integral transform $\mathcal{T}$ can be then represented as multiplication with a certain field element. Thus, the differential or integro-differential equations with the operator $\mathcal{D}$ are reduced to some algebraic equations in the field of convolution quotients and can be solved in explicit form. The so obtained "generalized" solution can be sometimes represented as a conventional function from the initial ring of functions by using the so-called operational relations.

The general schemata for applications of the integral transforms of mathematical physics mentioned above are demonstrated on an example of the Laplace integral transform. The Laplace integral transform is a simple particular case of the general H-transform that is a Mellin convolution type integral transform with the Fox H-function in the kernel. The general schemata for applications of the integral transforms presented in this article are valid for the H-transform, too. For the theory of the generating operators, convolutions, and operational calculi of the Mikusinski type for the H-transforms we refer the interested readers to [4–8] (see also numerous references therein). In this article, we restrict ourselves to discussion of some fundamental properties of the Mellin integral transform that is a basis for the theory of the Mellin convolution type integral transforms in general and of the H-transform in particular.

The rest of the article is organized as follows: In the second section, general schemata for some applications of the integral transforms to analysis of the integral, differential, and integro-differential equations are presented. In particular, the main ideas behind an operational calculus of Mikusinski type are discussed. The third section illustrates these schemata on the example of the Laplace integral transform. The fourth section deals with the basic properties of the Mellin integral transform.

2. Integral Transforms of Mathematical Physics

The focus of this survey article is on properties of the integral transforms and their applications to different problems of analysis, differential and integral equations, and special functions. Thus, we do not discuss the integral transforms from the viewpoint of functional analysis by considering, say, their mapping properties in some spaces of functions. Instead, we try to illustrate the underlying ideas and procedures both for analysis of the integral transforms and for their applications.

2.1. Applications of the Integral Transforms

The integral transforms of mathematical physics are not arbitrary linear integral operators, but rather those with the known inverse operators and the known generating operators. For the sake of simplicity and clarity, in this article we restrict ourselves to the case of the one-dimensional integral transforms. However, a similar theory can be also developed for the multi-dimensional integral transforms. A one-dimensional integral transform (of mathematical physics) of a function $f : \mathbb{R} \to \mathbb{R}$ at the point $t \in \mathbb{R}$ is defined by the (convergent) integral

$$g(t) = \mathcal{T}\{f(x);t\} = \int_{-\infty}^{+\infty} K(t,x)\, f(x)\, dx\,. \tag{1}$$

Its inverse operator must be also a linear integral transform

$$f(x) = \mathcal{T}^{-1}\{g(t);x\} = \int_{-\infty}^{+\infty} \hat{K}(x,t)\, g(t)\, dt \tag{2}$$

with a known kernel function $\hat{K}$. The kernel functions K and $\hat{K}$ of the integral transforms (1) and (2) satisfy the relation

$$\int_{-\infty}^{+\infty} \hat{K}(x,t)K(t,y)\, dt \;=\; \delta(x-y) \tag{3}$$

with the Dirac δ-function.

Many applications of the integral transforms of mathematical physics are based on the operational relations of the following form:

$$\mathcal{T}\{(\mathcal{L}f)(x);t\} = L(t)\mathcal{T}\{f(x);t\}. \quad (4)$$

The integral transform T satisfying the relation (4) is called a transmutation that translates an operator $\mathcal{L}$ into multiplication by the function L. Following [4,8,9], we call the operator $\mathcal{L}$ the generating operator of the integral transform $\mathcal{T}$. For the general H-transform, one of the important classes of their generating operators can be represented in form of finite compositions of the fractional Erdelyi-Kober integrals and derivatives [4–8]. In the case of the Laplace integral transform, the generating operator is just the first derivative.

Let us now discuss a general schema for applications of the transmutation Formula (4) on an example of the equation

$$P(\mathcal{L})y(x) = f(x), \quad (5)$$

where P is a polynomial and f is a given function. Applying the integral transform (1) to Equation (5) and employing the transmutation Formula (4) lead to the algebraic (in fact, linear) equation

$$P(L(t))\mathcal{T}\{y(x);t\} = \mathcal{T}\{f(x);t\} \quad (6)$$

for the integral transform $\mathcal{T}$ of the unknown function y with a solution in form

$$\mathcal{T}\{y(x);t\} = \frac{\mathcal{T}\{f(x);t\}}{P(L(t))}. \quad (7)$$

In the system theory, the function $1/P(L(t))$ is often called the transfer function. The inversion Formula (2) allows then to represent the solution (7) as follows:

$$y(x) = \mathcal{T}^{-1}\left\{\frac{\mathcal{T}\{f(x);t\}}{P(L(t))};x\right\}. \quad (8)$$

In many applications of the integral transforms of mathematical physics, one deals with the linear differential operators of the form

$$\mathcal{L}(x,\frac{d}{dx})y = \sum_{k=0}^{n} l_k(x)\frac{d^k y}{dx^k}. \quad (9)$$

Let us suppose that $\mathcal{L}$ is a generating operator of the integral transform (1) with the inverse integral transform (2) such that the relation (4) holds true. By

$$\mathcal{L}^T(x,\frac{d}{dx})y = \sum_{k=0}^{n}(-1)^k\frac{d^k}{dx^k}(l_k(x)y) \quad (10)$$

we denote the operator conjugate to the operator $\mathcal{L}$.

Then it is known that the kernel K of the integral transform (1) is an eigenfunction of the operator $\mathcal{L}^T$ and the kernel $\hat{K}$ of the inverse integral transform (2) is an eigenfunction of the operator $\mathcal{L}$ [9]:

$$\mathcal{L}^T(x,\frac{d}{dx})K(t,x) = L(t)K(t,x),$$

$$\mathcal{L}(x,\frac{d}{dx})\hat{K}(x,t) = L(t)\hat{K}(x,t).$$

Let us note that the Formulas (1) and (2) for the integral transform $\mathcal{T}$ of a function f and its inverse integral transform can be put together into the form

$$f(x) = \int_{-\infty}^{+\infty} \hat{K}(x,t)\, \mathcal{T}\{f(x);t\}\, dt$$

and then interpreted as an expansion of the function f by the eigenfunctions of the linear differential operator $\mathcal{L}$.

Thus, the integral transforms of mathematical physics are closely connected with the eigenvalues of some differential operators. However, the eigenvalues of the differential operators are known in explicit form only in a few cases and therefore the amount of the integral transforms of mathematical physics is very restricted.

As a rule, the generating operators of the integral transforms of mathematical physics are differential operators either of the first or of the second order. Examples of the integral transforms with the generating operators in form of the differential operators of the first order are:

(a) the Laplace integral transform with the kernel function $K(t,x) = e^{-xt}$ if $x > 0$ and $K(t,x) = 0$ if $x \le 0$,
(b) the sine- and cosine Fourier integral transforms with the kernel functions $K(t,x) = \sqrt{2/\pi}\sin(xt)$ if $x > 0$ and $K(t,x) = 0$ if $x \le 0$ and $K(t,x) = \sqrt{2/\pi}\cos(xt)$ if $x > 0$ and $K(t,x) = 0$ if $x \le 0$, respectively,
(c) the Fourier integral transform with the kernel function $K(t,x) = e^{-ixt}$,
(d) the Mellin integral transform with the kernel function $K(t,x) = x^{t-1}$ if $x > 0$ and $K(t,x) = 0$ if $x \le 0$.

Following integral transforms possess generating operators in form of the differential operators of the second order:

(a) the Hankel integral transform with the kernel function $K(t,x) = \sqrt{xt}J_\nu(xt)$ (J_ν stands for the Bessel function) if $x > 0$ and $K(t,x) = 0$ if $x \le 0$,
(b) the Meijer integral transform with the kernel function $K(t,x) = \sqrt{xt}K_\nu(xt)$ (K_ν is the Macdonald function) if $x > 0$ and $K(t,x) = 0$ if $x \le 0$,
(c) the Kontorovich- Lebedev integral transform with the kernel function $K(t,x) = K_{it}(x)$ if $x > 0$ and $K(t,x) = 0$ if $x \le 0$,
(d) the Mehler-Fock integral transform with the kernel function $K(t,x) = P^k_{it-1/2}(x)$ (P^μ_ν denotes the Legendre function of the first kind) if $x > 1$ and $K(t,x) = 0$ if $x \le 1$.

As already mentioned, the generating operators of the H-transform are certain compositions of the Erdelyi-Kober fractional integrals and derivatives. This connection allows to solve equations of type (5) with the operator $\mathcal{L}$ in form of a composition of the Erdelyi-Kober fractional integro-differential operators [4–8].

Another operation that plays a very important role in applications of the integral transforms of mathematical physics is a convolution of two functions associated with a certain integral transform.

In general, a convolution on a linear vector space of functions is defined as a bilinear, commutative, and associative operation defined on a direct product of a linear vector space by itself. Together with the usual addition of two elements of the vector space, the convolution thus equips the linear vector space with a structure of a commutative ring.

A convolution $\overset{\mathcal{T}}{*}$ associated with the integral transform $\mathcal{T}$ and defined on a linear functional vector space $\mathcal{X}$ satisfies the relation (convolution theorem)

$$\mathcal{T}\left\{(f \overset{\mathcal{T}}{*} g)(x);t\right\} = \mathcal{T}\{f(x);t\}\,\mathcal{T}\{g(x);t\},\ \forall f,g \in \mathcal{X}. \tag{11}$$

The reader can find many examples of convolutions for different integral transforms of mathematical physics in [4,5].

One of the basic applications of the convolutions is for analysis of the integral equations of convolution type. The convolutions of the integral transforms of mathematical physics are often represented in form of some integrals. In these cases the convolution equations like, e.g.,

$$y(x) - \lambda(y \overset{\mathcal{T}}{*} K)(x) = f(x), \ \lambda \in \mathbb{R} \text{ or } \lambda \in \mathbb{C}, \tag{12}$$

where f and K are some known functions and the function y is unknown, are integral equations.

To solve the integral Equation (12), we apply the integral transform $\mathcal{T}$ to both parts of (12). Then we first get an algebraic (in fact, a linear) equation

$$\mathcal{T}\{y(x);t\} - \lambda\mathcal{T}\{y(x);t\}\,\mathcal{T}\{K(x);t\} = \mathcal{T}\{f(x);t\} \tag{13}$$

with the solution

$$\mathcal{T}\{y(x);t\} = \frac{\mathcal{T}\{f(x);t\}}{1-\lambda\mathcal{T}\{K(x);t\}}. \tag{14}$$

Applying the inverse transform $\mathcal{T}^{-1}$ to (14), the solution of the integral Equation (12) can be then represented in the form

$$y(x) = \mathcal{T}^{-1}\left\{\frac{\mathcal{T}\{f(x);t\}}{1-\lambda\mathcal{T}\{K(x);t\}};x\right\}. \tag{15}$$

In many cases, the right-hand side of (15) has a convolution form and thus the solution to (12) can be rewritten as follows:

$$y(x) = f(x) + \lambda(f \overset{\mathcal{T}}{*} M)(x), \tag{16}$$

where M is a known function.

2.2. Basic Ideas Behind an Operational Calculus of Mikusinski Type

Another useful technique employed for solution of both integral equations of convolution type (12) and differential or integro-differential equations of type (5) is an algebraic approach based on the operational calculi of Mikusinski type [4–6,10–14].

In an operational calculus of Mikusinski type, a close relation between an integral transform, its convolution and its generating operator plays a very essential role as investigated in detail in [15].

Following [15], we first introduce a convolution of a linear operator $\mathcal{L}$. Let $\mathcal{X}$ be a linear vector space and $\mathcal{L} : \mathcal{X} \to \mathcal{X}$ a linear operator defined on the elements of $\mathcal{X}$. A bilinear, commutative, and associative operation $* : \mathcal{X} \times \mathcal{X} \to \mathcal{X}$ is said to be a convolution of the linear operator $\mathcal{L}$ if and only if the relation

$$\mathcal{L}(f * g) = (\mathcal{L}f) * g \tag{17}$$

holds true for all $f, g \in \mathcal{X}$.

As shown in [4], if $\mathcal{L}$ is a generating operator of the integral transform $\mathcal{T}$ (the Formula (4) holds true) and if $\overset{\mathcal{T}}{*}$ is a convolution of $\mathcal{T}$ that satisfies the relation (11), then $\overset{\mathcal{T}}{*}$ is a convolution of the generating operator $\mathcal{L}$ in the sense of the relation (17).

Another important fact is that any of the convolution operators of the type

$$(\mathcal{L}f)(x) = (h \overset{\mathcal{T}}{*} f)(x), \tag{18}$$

where $\overset{\mathcal{T}}{*}$ is a convolution of the integral transform $\mathcal{T}$ and h is a fixed element of $\mathcal{X}$ can be interpreted as a generating operator of the integral transform $\mathcal{T}$, i.e., it satisfies the transmutation relation (4).

The generating operator $\mathcal{L}$ given by (18) is an integral operator that is defined on the functions from the convolution ring $\mathcal{R} = (\mathcal{X}, \overset{\mathcal{T}}{*}, +)$ as multiplication by a fixed element $h \in \mathcal{X}$. It is important to stress that the representations of this type are not possible for the generating operators of the differential type, e.g., for the left-inverse operators of the integral operator (18). However, similar representations of differential operators can be derived on an extension of the convolution ring $\mathcal{R} = (\mathcal{X}, \overset{\mathcal{T}}{*}, +)$ to the field of the convolution quotients. In fact, this extension is a basic element of any operational calculus of Mikusinski type. In the case when the ring $\mathcal{R} = (\mathcal{X}, \overset{\mathcal{T}}{*}, +)$ has no divisors of zero, the extension follows the pattern of the extension of the ring of integer numbers to the field of rational numbers.

If the ring $\mathcal{R} = (\mathcal{X}, \overset{\mathcal{T}}{*}, +)$ has some divisors of zero, the construction of the field of convolution quotients becomes more complicated (see, e.g., [15] for details). A divisor-free convolution ring is usually extended to a field $\mathcal{F}$ of convolution quotients by factorization of the set $\mathcal{X} \times (\mathcal{X} - \{0\})$ with respect to the equivalence relation

$$(f, g) \sim (f_1, g_1) \Leftrightarrow (f \overset{\mathcal{T}}{*} g_1)(x) = (g \overset{\mathcal{T}}{*} f_1)(x). \tag{19}$$

The elements of the field $\mathcal{F}$ are sets of all pairs (f, g), $f, g \in \mathcal{X}$ that are equivalent to each other with respect to the equivalence relation (19). They are often formally denoted as quotients f/g. The addition $+$ and multiplication $\cdot$ operations are defined on $\mathcal{F}$ in a standard way:

$$f/g + f_1/g_1 = (f \overset{\mathcal{T}}{*} g_1 + g \overset{\mathcal{T}}{*} f_1)/(g \overset{\mathcal{T}}{*} g_1), \tag{20}$$

$$f/g \cdot f_1/g_1 = (f \overset{\mathcal{T}}{*} f_1)/(g \overset{\mathcal{T}}{*} g_1). \tag{21}$$

It is an easy exercise in algebra to show that the results of the operations (20) and (21) do not depend on the representatives of the field elements f/g and f_1/g_1 and thus these operations are well defined. Equipped with the operations $+$ and $\cdot$, the set $\mathcal{F}$ becomes a commutative field that is denoted by $(\mathcal{F}, \cdot, +)$.

The ring $\mathcal{R} = (\mathcal{X}, \overset{\mathcal{T}}{*}, +)$ can be embedded into the field $(\mathcal{F}, \cdot, +)$, say, by the map

$$f \in \mathcal{R} \rightarrow (f \overset{\mathcal{T}}{*} h)/h \in \mathcal{F}, \tag{22}$$

where $h \in \mathcal{R}$ is any non-zero element of the ring $\mathcal{R}$. A natural choice for the element $h \in \mathcal{R}$ in the relation (22) is the function from the Formula (18) that defines the generating operator $\mathcal{L}$. In this case, the corresponding operational calculus is constructed for the differential operator $\mathcal{D}$ that is a left-inverse operator to the integral operator (18), i.e., for the operator $\mathcal{D}$ that satisfies the relation

$$\mathcal{D}(\mathcal{L}f) = f, \ \forall f \in \mathcal{X}. \tag{23}$$

On the ring $\mathcal{R} = (\mathcal{X}, \overset{\mathcal{T}}{*}, +)$, the operator $\mathcal{L}$ applied to a function $f \in \mathcal{X}$ is just multiplication of f with a fixed element $h \in \mathcal{X}$. On the other hand, the differential operator $\mathcal{D}$ (a left-inverse operator to the integral operator $\mathcal{L}$) can be represented on the field $(\mathcal{F}, \cdot, +)$ in the form

$$\mathcal{D}f = S \cdot f - S \cdot \mathcal{P}f, \tag{24}$$

where the operator $\mathcal{P} = \text{Id} - \mathcal{L}\mathcal{D}$ is called a projector of the generating operator $\mathcal{L}$ and $S \in \mathcal{F}$ is the element reciprocal to $h \in \mathcal{R} \subset \mathcal{F}$ defined by the Formula (18), i.e.,

$$S = h^{-1} = I/h = h/(h \overset{\mathcal{T}}{*} h) = h/h^2 = \cdots = h^k/h^{k+1}, \ k = 0, 1, 2, \ldots. \tag{25}$$

The element $S \in \mathcal{F}$ is often called an algebraic inverse of the generating operator $\mathcal{L}$ in the field of convolution quotients.

The operational Formula (24) is very important in applications of the constructed operational calculus because it allows a reduction of the linear differential equations with the operator $\mathcal{D}$ to some algebraic equations in the field $(\mathcal{F}, \cdot, +)$ of convolution quotients. The obtained equations can be then often solved in explicit form that leads to the "generalized" solutions that belong to the convolution quotients field $(\mathcal{F}, \cdot, +)$. In some cases, by making use of the embedding (22) and of the so-called operational relations, these generalized solutions can be reduced to the conventional functions from the initial ring $\mathcal{R}$. In particular, the following operational relation plays a very important role in any operational calculus:

$$(S - \rho)^{-1} = I/(S - \rho) = h/(I - \rho h) = h \cdot (I + \rho h + \rho^2 h^2 + \ldots)$$
$$= h(x) + \rho h^2(x) + \rho^2 h^3(x) + \cdots = H(x) \in \mathcal{R},\ \rho \in \mathbb{C},\ h^k = \underbrace{h \overset{\mathcal{T}}{*} h \overset{\mathcal{T}}{*} \cdots \overset{\mathcal{T}}{*} h}_{k}. \tag{26}$$

The general schema for construction of an operational calculus and for its applications that was presented above seems to be not especially complicated. However, in the case of a given generating operator a lot of serious problems can appear while developing the corresponding operational calculus. The main questions are how to construct an appropriate convolution, how to determine its divisors of zero in the corresponding ring of functions (or show that it is divisors-free), how to calculate the projector operator (the projector operator determines the form of the initial conditions for the ordinary or fractional differential equations that can be solved by employing the operational calculus), how to specify the operational relations such as the one given in the Formula (26), etc. For discussions regarding how to overcome all these difficulties for operational calculi for different operators of Fractional Calculus see, e.g., ref. [4] or [5–8].

3. The Laplace Integral Transform

The Laplace integral transform—along with the Fourier integral transform and the Mellin integral transform—is one of the most important classical integral transforms that is widely used in analysis, differential equations, theory of special functions and integral transforms, and for other problems of mathematical physics. For a function f, its Laplace transform at the point $p \in \mathbb{C}$ is defined by the following improper integral (in the case it is a convergent one):

$$\widetilde{f}(p) = \mathcal{L}\{f(t); p\} = \int_0^{\infty} \mathrm{e}^{-pt} f(t)\, dt, \quad \Re(p) > a_f. \tag{27}$$

A sufficient condition for existence of the Laplace integral at the right-hand side of (27) for a function $f \in L^c(0, +\infty)$ is the estimate of the type

$$|f(t)| \leq M_f \mathrm{e}^{a_f t},\ t > T_f, \tag{28}$$

where M_f, a_f, and T_f are some constants depending on the function f. The space of functions $L^c(0, +\infty)$ consists of all real or complex-valued functions of a real variable that are continuous on the open interval $(0, +\infty)$ except, possibly, at a counted number of isolated points, where these functions can tend to infinity and for that the improper Riemann integral absolutely converges on $(0, +\infty)$. In this section, the set of all functions from $L^c(0, +\infty)$ that satisfy the estimate (28) with some constants depending on the functions will be denoted by $\mathcal{O}$. In the following discussions, we always assume that the functions we deal with belong to the space of functions $\mathcal{O}$. For the functions from $\mathcal{O}$, their Laplace transforms $\widetilde{f}(p)$ defined by the right-hand side of (27) are analytic function in the half complex plane $\Re(p) > a_f$. This feature makes the Laplace transform technique very powerful because all methods and ideas elaborated in the well-developed theory of analytical functions can be employed in the Laplace domain.

Let a function f be piecewise differentiable and its Laplace transform exist for $\Re(p) > a_f$. At all points where f is continuous, it can be represented via the inverse Laplace transform

$$f(t) = \mathcal{L}^{-1}\left\{\widetilde{f}(p); t\right\} = \frac{1}{2\pi i}\int_{\gamma-i\infty}^{\gamma+i\infty} e^{pt}\,\widetilde{f}(p)\,dp\,, \quad \Re(p) = \gamma > a_f, \tag{29}$$

where the integral at the right-hand side is understood in the sense of the Cauchy principal value.

Let us mention here that in some cases the bilateral Laplace transform can be useful. It is defined by the formula

$$\mathcal{L}_{bl}\{f(t); p\} = \int_{-\infty}^{\infty} e^{-pt} f(t)\,dt\,, \quad b_f > \Re(p) > a_f. \tag{30}$$

In our article, we assume that the model equations we deal with refer to the causal processes and thus we restrict ourselves to discussion of the Laplace transform.

The Laplace integral transform is treated in many textbooks (see, e.g., [11,16–21]). Here, we demonstrate on the example of the Laplace integral transform how the general schemata for applications of the integral transforms of mathematical physics work. It is worth mentioning that the same constructions can be applied for the general H-transform and its numerous particular cases ([4–8]).

Let the inclusion $f' \in \mathcal{O}$ be valid. The integration by parts formula applied to the Laplace integral leads to following transmutation relation for the Laplace integral transform:

$$\mathcal{L}\left\{\frac{d}{dt}f(t); p\right\} = p\,\mathcal{L}\{f(t); p\} - f(0). \tag{31}$$

This means that the Laplace integral transform is a transmutation operator for the first derivative that translates it into multiplication with the linear factor p.

The Formula (31) and its generalization (for $f^{(n)} \in \mathcal{O}$)

$$\mathcal{L}\left\{f^{(n)}(t); p\right\} = p^n\,\mathcal{L}\{f(t); p\} - p^{n-1}f(0) - \ldots - f^{(n-1)}(0) \tag{32}$$

are the basic formulas for application of the Laplace transform technique to solution of the linear differential equations.

As an example, let us consider the following initial value problem:

$$\begin{cases} x^{(n)}(t) + a_1 x^{(n-1)}(t) + \cdots + a_n x(t) = f(t),\ t > 0, \\ x(0) = x_0,\ \frac{dx}{dt}|_{t=0} = x_1, \ldots, \frac{d^{n-1}x}{dt^{n-1}}|_{t=0} = x_{n-1}. \end{cases} \tag{33}$$

Applying of the Laplace integral transform to Equation (33) to the initial value problem (33) we get an algebraic (in fact, a linear) equation for the Laplace transform of the unknown function x:

$$L(p)\mathcal{L}\{x(t); p\} = \mathcal{L}\{f(t); p\} + M(p), \tag{34}$$

where $L(p) = p^n + a_1 p^{n-1} + \cdots + a_n$ and $M(p) = p^{n-1}x_0 + \cdots + x_{n-1} + a_1(p^{n-2}x_0 + \cdots + x_{n-2}) + \cdots + a_{n-1}x_0$. The polynomial L is known as a characteristic polynomial of Equation (33).

For applicability of this technique, the function f from the left-hand side of Equation (33) must satisfy the condition (28).

The linear Equation (34) can be easily solved:

$$\mathcal{L}\{x(t); p\} = \frac{\mathcal{L}\{f(t); p\} + M(p)}{L(p)}. \tag{35}$$

Thus, the unique solution to (33) can be (formally) obtained by applying the inverse Laplace transform to the right-hand side of (35):

$$x(t) = \mathcal{L}^{-1}\left\{\frac{\mathcal{L}\{f(t);p\} + M(p)}{L(p)};t\right\}. \tag{36}$$

Often, the inverse Laplace transform at the right-hand side of the Formula (36) can be evaluated by means of the Cauchy residue theorem or by employing the tables of the Laplace integral transforms [19,20].

The same procedure as above is applicable for the systems of linear ordinary differential equations with the constant coefficients. A similar method can be applied for the linear ordinary differential equations with the polynomial coefficients ([11]). In the case of the time-dependent partial linear differential equations with the constant coefficients, application of the Laplace integral transform with respect to the time variable transforms them to the stationary partial differential equations of elliptic type with some parameters dependent on the Laplace variable p.

A prominent role in several applications of the Laplace integral transform is played by its convolution that is defined by the well-known formula

$$(f \overset{\mathcal{L}}{*} g)(t) = \int_0^t f(\tau)g(t-\tau)\,d\tau. \tag{37}$$

The Borel convolution theorem states the main property of the Laplace convolution (37). Let the Laplace integral transforms of the functions $f,\ g \in \mathcal{O}$ be well defined for $\Re\,(p) > \gamma$. Then the Laplace convolution (37) also exists for $\Re\,(p) > \gamma$ and the convolution formula

$$\mathcal{L}\left\{(f \overset{\mathcal{L}}{*} g)(t);p\right\} = \mathcal{L}\{f(t);p\} \times \mathcal{L}\{g(t);p\} \tag{38}$$

holds true.

In [4,5], a close relation of the Laplace convolution to the Euler Beta-function was established. It turned out that the Formula (37) for the Laplace convolution follows from the well-known formula

$$B(s,t) = \int_0^1 x^{s-1}(1-x)^{t-1}\,dx = \frac{\Gamma(s)\Gamma(t)}{\Gamma(s+t)}, \tag{39}$$

Γ being the Euler Gamma-function defined as

$$\Gamma(s) = \int_0^\infty e^{-x}x^{s-1}\,dx,\ \Re(s) > 0. \tag{40}$$

Applying the Formula (37) for the Laplace convolution and the Formula (39) for the B-function, convolutions of many other integral transforms including the general H-transform can be constructed [4,5].

The Borel Formula (38) can be employed for solving some integral equations of the Laplace convolution type. As an example, let us consider an integral equation of the second kind in the form

$$x(t) - \lambda\int_0^t k(t-\tau)x(\tau)\,d\tau = f(t). \tag{41}$$

Application of the Laplace integral transform to Equation (41) reduces it to the algebraic (in fact, linear) equation

$$\mathcal{L}\{x(t);p\} - \lambda\mathcal{L}\{x(t);p\}\,\mathcal{L}\{k(t);p\} = \mathcal{L}\{f(t);p\} \tag{42}$$

for the Laplace transform of the unknown function. Its solution is given by the formula

$$\mathcal{L}\{x(t);p\} = \frac{\mathcal{L}\{f(t);p\}}{1-\lambda\mathcal{L}\{k(t);p\}}. \tag{43}$$

To get a solution for the integral Equation (41), the inverse Laplace integral transform can be applied to Equation Lap-8. However, let us follow another approach and get a representation of the solution by using the Formula (38). To do this, let us take a closer look at the expression $\mathcal{L}\{x(t);p\} - \mathcal{L}\{f(t);p\}$. The Formula (43) leads to the representation

$$\begin{aligned} &\mathcal{L}\{x(t);p\} - \mathcal{L}\{f(t);p\} = \tfrac{\mathcal{L}\{f(t);p\}}{1-\lambda\mathcal{L}\{k(t);p\}} - \mathcal{L}\{f(t);p\} \\ &= \mathcal{L}\{f(t);p\}\left(\tfrac{1}{1-\lambda\mathcal{L}\{k(t);p\}} - 1\right) = \lambda\mathcal{L}\{f(t);p\}\tfrac{\mathcal{L}\{k(t);p\}}{1-\lambda\mathcal{L}\{k(t);p\}}. \end{aligned} \tag{44}$$

Because of the Formula (38), we can represent a solution for the integral Equation (41) in the form

$$x(t) = f(t) + \lambda\int_0^t h(t-\tau)f(\tau)\,d\tau. \tag{45}$$

In this formula, the function h is the inverse Laplace transform of the function $\mathcal{L}\{k(t);p\}/(1-\lambda\mathcal{L}\{k(t);p\})$.

The same method can be employed for the initial value problems for the ordinary differential equations in form (33). Let us revisit the Formula (35) and represent the rational functions $1/L$ and M/L as sums of partial fractions

$$\frac{1}{L(p)} = \frac{1}{(p-\lambda_1)^{m_1}\times\ldots\times(p-\lambda_k)^{m_k}} = \sum_{j=1}^{k}\sum_{r=1}^{m_j}\frac{c_{jr}}{(p-\lambda_j)^r},\ \lambda_j,\, c_{jr}\in\mathbb{C} \tag{46}$$

$$\frac{M(p)}{L(p)} = \sum_{j=1}^{m}\sum_{r=1}^{s_j}\frac{d_{jr}}{(p-\eta_j)^r},\ \eta_j,\, d_{jr}\in\mathbb{C}. \tag{47}$$

The operational formula

$$\mathcal{L}\{t^n\,e^{at};p\} = \int_0^\infty t^n\,e^{-t(p-a)}\,dt = \frac{\Gamma(n+1)}{(p-a)^{n+1}} = \frac{n!}{(p-a)^{n+1}},\ \Re(p-a)>0 \tag{48}$$

and the Borel convolution Formula (38) allow us to represent the solution of the initial value problem (33) in the form

$$x(t) = \int_0^t f(t-\tau)\left(\sum_{j=1}^{k}\sum_{r=1}^{m_j}\frac{c_{jr}}{(r-1)!}\tau^{r-1}\,e^{\lambda_j\tau}\right)d\tau + \sum_{j=1}^{m}\sum_{r=1}^{s_j}\frac{d_{jr}}{(r-1)!}t^{r-1}\,e^{\eta_j t}. \tag{49}$$

It is worth mentioning that the Formula (48) is valid not only for the power functions in the form $t^n, n\in\mathbb{N}$, but also for the arbitrary power functions with the exponents $\alpha > -1$:

$$\mathcal{L}\{t^\alpha\,e^{at};p\} = \int_0^\infty t^\alpha\,e^{-t(p-a)}\,dt = \frac{\Gamma(\alpha+1)}{(p-a)^{\alpha+1}},\ \alpha>-1,\ \Re(p-a)>0. \tag{50}$$

In order to solve the linear ordinary differential equations with the polynomial coefficients, a slightly different procedure compared to the method demonstrated above is often employed. Its main element consists in representation of an unknown solution to a differential equation in the form similar to the form of the inverse Laplace transform

$$x(t) = \int_C \phi(p)\,e^{pt}\,dp \tag{51}$$

with an unknown contour C. In the process of solution, the contour C must be appropriately chosen [22].

To illustrate this technique, let us consider another—a little bit exotic—application of the Laplace integral transform technique, namely, for solving a functional equation. The Euler Gamma-function (40) is a generalization of the factorial function. It satisfies the functional equation

$$F(s+1) = s\,F(s),\ \Re(s) > 0,\ F(1) = 1. \tag{52}$$

Let us show that the functional Equation (52) has a unique solution, namely, the Gamma-function in the functional space of smooth functions. To solve the functional Equation (52), the Laplace transform method mentioned above is employed. We look for the solutions F of (52) in form

$$F(s) = \int_{-\infty}^{\infty} \phi(p)\, e^{-ps}\, dp. \tag{53}$$

Then the relation

$$s\,F(s) = s\int_{-\infty}^{\infty} \phi(p)\, e^{-ps}\, dp = -\phi(p)\, e^{-ps}|_{-\infty}^{\infty} + \int_{-\infty}^{\infty} \phi'(p)\, e^{-ps}\, dp \tag{54}$$

holds valid if the unknown function ϕ is a smooth function on $\mathbb{R}$ and the limits $\lim_{p\to\pm\infty} \phi(p)\, e^{-ps}$ exist and are finite. Moreover, let the relations $\lim_{p\to\pm\infty} \phi(p)\, e^{-ps} = 0$ hold true. At this stage we just suppose that the unknown function ϕ satisfies the conditions above. However, after the functional Equation (52) will be solved, these conditions can be directly verified.

Using the representation (that directly follows from (53))

$$F(s+1) = \int_{-\infty}^{\infty} \phi(p)\, e^{-ps}\, e^{-p}\, dp,$$

and the Formula (54), the functional Equation (52) is reduced to a differential equation for the unknown function ϕ:

$$\phi(p)\, e^{-p} = \phi'(p).$$

The general solution formula of the above equation can be derived by separating the variables p and ϕ:

$$\phi(p) = -C\, e^{-e^{-p}},\ C \in \mathbb{R}. \tag{55}$$

Evidently, the function ϕ defined by the right-hand side of (55) satisfies the conditions $\lim_{p\to\pm\infty} \phi(p)\, e^{-ps} = 0$ for $\Re(s) > 0$. Substituting (55) into (53) and by the variables substitution $x = \exp(-p)$, we get the solution formula

$$F(s) = C\int_{0}^{\infty} e^{-x}\, x^{s-1}\, dx,\ \Re(s) > 0. \tag{56}$$

The initial condition from (52) leads to a unique value of the constant C:

$$1 = F(1) = C\int_{0}^{\infty} e^{-x}\, dx = C.$$

Thus, the solution Formula (56) coincides with the integral representation of the Gamma-function and the Gamma-function is the only smooth solution to the functional Equation (52) with the initial condition $F(1) = 1$.

In our last example of this section, we prove the Formula (39) for the Beta-function by employing the Laplace transform technique. We start with an observation that the Beta-function B is the Laplace convolution of two power functions evaluated at the point $x = 1$:

$$B(s,t) = (\tau^{s-1} \overset{\mathcal{L}}{*} \tau^{t-1})(x)|_{x=1} = \int_{0}^{x} \tau^{s-1}\, (x-\tau)^{t-1}\, d\tau|_{x=1}.$$

Let us define an auxiliary function

$$\beta(s,t,x) = (\tau^{s-1} \overset{\mathcal{L}}{*} \tau^{t-1})(x) = \int_0^x \tau^{s-1}\,(x-\tau)^{t-1}\,d\tau,\ x>0.$$

Its Laplace integral transform is given by the formula

$$\mathcal{L}\{\beta(x);p\} = \frac{\Gamma(s)}{p^s}\frac{\Gamma(t)}{p^t} = \frac{\Gamma(s)\Gamma(t)}{p^{s+t}} \tag{57}$$

because of the operational relation (50) with $a=0$ and the Borel convolution formula.

Again using the operational relation (50), we determine the function β with the Laplace transform given by (57) in the form:

$$\beta(s,t,x) = \frac{\Gamma(s)\Gamma(t)}{\Gamma(s+t)}\,x^{s+t-1}. \tag{58}$$

Specifying the Formula (58) for $x=1$, we receive the well-known representation of the Beta-function in terms of the Gamma-function:

$$B(s,t) = \int_0^x \tau^{s-1}\,(x-\tau)^{t-1}\,d\tau|_{x=1} = \beta(s,t,1) = \frac{\Gamma(s)\Gamma(t)}{\Gamma(s+t)}.$$

As to the Mikusinski-type operational calculus associated with the Laplace integral transform, we start with the Volterra integral operator that is one of the generating operators for the Laplace integral transform. Indeed, this operator can be represented as the Laplace convolution of the functions f and $\{1\}$ ($\{1\}$ is the function that is identically equal to 1):

$$(\mathcal{V} f)(t) = (f \overset{\mathcal{L}}{*} \{1\})(t) = \int_0^t f(\tau)\,d\tau. \tag{59}$$

Evidently, the Volterra integral operator is a linear operator on $C[0,\infty)$ with the Laplace convolution as its convolution in the sense of Formula (17).

As stated by the well-known Titchmarsh theorem [23], the space of functions $C[0,\infty)$ equipped with the operations $+$ and $\overset{\mathcal{L}}{*}$ is a commutative ring without divisors of zero. The basic idea behind the classical Mikusinski operational calculus is an extension of this ring to a field of convolution quotients according to the schema presented in the previous section. The so constructed Mikusinski operational calculus is closely related with the first derivative $\frac{d}{dt}$ that is a left-inverse operator to the Volterra integral operator (59). The projector of the Volterra integral operator is given by the expression

$$\mathcal{P} f = f - \mathcal{V}\frac{df}{dt} = f(0). \tag{60}$$

It is worth mentioning that the projector (60) determines the form of the initial conditions for the differential equations that can be solved by applying the Mikusinski operational calculus.

4. The Mellin Integral Transform

In this section, some basic definitions and formulas for the Mellin integral transform are presented. The Mellin integral transform is one of the main tools for a treatment of the integral transforms of the Mellin convolution type, their convolutions and generating operators. More details regarding the Mellin integral transform, its properties and particular cases can be found in [4,24–29].

The Mellin integral transform of a sufficiently well-behaved function f at the point $s \in \mathbb{C}$ is defined as

$$\mathcal{M}\{f(t);s\} = f^*(s) = \int_0^{+\infty} f(t)t^{s-1}dt, \tag{61}$$

and the inverse Mellin integral transform as

$$f(t) = \mathcal{M}^{-1}\{f^*(s);t\} = \frac{1}{2\pi i}\int_{\gamma-i\infty}^{\gamma+i\infty} f^*(s)t^{-s}ds,\ t>0,\ \gamma=\Re(s), \tag{62}$$

where the integral is understood in the sense of the Cauchy principal value.

The Mellin integral transform can be interpreted as the Fourier integral transform changed by a variables substitution and by a rotation of the complex plane:

$$\mathcal{M}\{f(t);s\} = \int_0^{+\infty} f(t)t^{s-1}dt = \int_{-\infty}^{+\infty} f(e^t)e^{it(-is)}dt = \mathcal{F}\{f(e^t);-is\}.$$

The integral at the right-hand side of the Formula (61) is well defined, say, for the functions $f \in L^c(\epsilon, E)$, $0<\epsilon<E<\infty$ that are continuous on the intervals $(0,\epsilon]$, $[E,+\infty)$ and satisfy the estimates $|f(t)| \le Mt^{-\gamma_1}$ for $0<t<\epsilon$, $|f(t)| \le Mt^{-\gamma_2}$ for $t>E$, where M is a constant and $\gamma_1<\gamma_2$. Under these conditions, the Mellin transform f^* exists and is an analytical function in the vertical strip $\gamma_1<\Re(s)<\gamma_2$.

The Formula (62) for the inverse Mellin integral transform holds true at all points where the function f is continuous if f is piecewise differentiable, $f(t)\,t^{\gamma-1} \in L^c(0,+\infty)$, and its Mellin integral transform f^* is given by (61).

For the reader's convenience, we present in this section some important theorems concerning the Mellin integral transform (for the proofs see e.g., [29]).

Theorem 1. *Let f be a function of bounded variation in the neighborhood of a point $t=x$, $t^{\gamma-1}f(t) \in L(0,\infty)$, and*

$$F(s) = \mathcal{M}\{f(t),s\} = \int_0^{+\infty} f(t)t^{s-1}dt,\ s=\gamma+i\tau. \tag{63}$$

Then

$$\frac{f(x+0)+f(x-0)}{2} = \frac{1}{2\pi i}\lim_{T\to\infty}\int_{\gamma-iT}^{\gamma+iT} F(s)x^{-s}ds. \tag{64}$$

Theorem 2. *Let $F(s)$, $s=\gamma+i\tau$ be a function of bounded variation in the neighborhood of a point $\tau=x$, $F\in L(-\infty,+\infty)$, and*

$$f(t) = \mathcal{M}^{-1}\{F(s);t\} = \frac{1}{2\pi i}\int_{\gamma-i\infty}^{\gamma+i\infty} F(s)t^{-s}ds. \tag{65}$$

Then

$$\frac{F(\gamma+i(x+0))+F(\gamma+i(x-0))}{2} = \lim_{\lambda\to\infty}\int_{1/\lambda}^{\lambda} f(t)t^{\gamma+ix-1}dt. \tag{66}$$

To formulate other theorems, some special spaces of functions are first introduced. By $L_p(\mu(t);\mathbb{R}_+)$, $p\ge 1$ the space of functions summable in the Lebesgue sense on the interval $(0,+\infty)$ to the power p and with the weight $\mu(t)>0$, $t>0$ is denoted. The norm of the space $L_p(\mu(t);\mathbb{R}_+)$ is defined by

$$\|f\|_{L_p(\mu(t);\mathbb{R}_+)} = \left\{\int_0^{\infty}\mu(t)|f(t)|^p dt\right\}^{1/p} < \infty. \tag{67}$$

In particular, when $\mu(t)\equiv 1$, $t>0$, the space $L_p(\mu(t);\mathbb{R}_+)$ is reduced to the usual L_p-space. While estimating the integrals in L_p, the Hölder inequality

$$\int_0^{\infty}|f(t)g(t)|dt \le \|f\|_{L_p(\mathbb{R}_+)}\|g\|_{L_q(\mathbb{R}_+)}, \tag{68}$$

where $\frac{1}{p} + \frac{1}{q} = 1$ and the Minkowski inequality

$$\left\{\int_0^\infty dx \left|\int_0^\infty f(x,y)dy\right|^p\right\}^{1/p} \leq \int_0^\infty dy \left\{\int_0^\infty |f(x,y)|^p dx\right\}^{1/p} \tag{69}$$

are often used.

Theorem 3. *Let $f \in L_2(t^{2\gamma-1}; \mathbb{R}_+)$. Then the function*

$$f^*(s,\lambda) = \int_{1/\lambda}^{\lambda} f(t)t^{\gamma+i\tau-1}dt, \; s = \gamma + i\tau \tag{70}$$

converges in the norm of $L_2(\gamma - i\infty, \gamma + i\infty)$ to a function f^* and the function

$$f(t,\lambda) = \frac{1}{2\pi i}\int_{\gamma-i\lambda}^{\gamma+i\lambda} f^*(s)t^{-s}ds \tag{71}$$

converges in the norm of $L_2(t^{2\gamma-1}; \mathbb{R}_+)$ to the function f, i.e.,

$$\lim_{\lambda\to\infty}\int_0^\infty |f(t) - f(t,\lambda)|^2 t^{2\gamma-1}dt = 0. \tag{72}$$

Moreover, the Parseval equality

$$\int_0^\infty |f(t)|^2 t^{2\gamma-1}dt = \frac{1}{2\pi}\int_{-\infty}^{+\infty} |f^*(\gamma+i\tau)|^2 d\tau \tag{73}$$

holds true.

Theorem 4. *Let $f \in L_2(t^{2\gamma-1}; \mathbb{R}_+)$, $g \in L_2(t^{1-2\gamma}; \mathbb{R}_+)$ and f^*, g^* be their Mellin integral transforms, respectively. Then the Mellin-Parseval equality*

$$\int_0^\infty f(t)g(t)dt = \frac{1}{2\pi i}\int_{\gamma-i\infty}^{\gamma+i\infty} f^*(s)g^*(1-s)ds \tag{74}$$

holds true.

The Mellin convolution

$$(f \overset{\mathcal{M}}{*} g)(x) = \int_0^{+\infty} f(x/t)g(t)\frac{dt}{t} \tag{75}$$

is a very essential element of the integral transforms of the Mellin convolution type. Following [29], we formulate the following important theorem:

Theorem 5. *Let $f(t)\,t^{\gamma-1} \in L(0,\infty)$ and $g(t)\,t^{\gamma-1} \in L(0,\infty)$. Then the Mellin convolution $h = (f \overset{\mathcal{M}}{*} g)$ given by (75) is well defined, $h(x)\,x^{\gamma-1} \in L(0,\infty)$ and the convolution formula*

$$\mathcal{M}\left\{(f \overset{\mathcal{M}}{*} g)(x); s\right\} = \mathcal{M}\left\{f(t); s\right\} \times \mathcal{M}\left\{g(t); s\right\} \tag{76}$$

holds true along with the Parseval equality

$$\int_0^{+\infty} f(x/t)g(t)\frac{dt}{t} = \frac{1}{2\pi i}\int_{\gamma-i\infty}^{\gamma+i\infty} f^*(s)g^*(s)x^{-s}ds. \tag{77}$$

In particular, the Parseval equality (77) can be employed while treating integrals of the Fourier *cos*- and *sin*-transforms type for $x > 0$:

$$I_c(x) = \frac{1}{\pi} \int_0^\infty f(t) \cos(t\,x)\,dt,$$

$$I_s(x) = \frac{1}{\pi} \int_0^\infty f(t) \sin(t\,x)\,dt.$$

Indeed, the integrals I_c and I_s can be interpreted as Mellin convolutions (75) of the function f and the functions

$$g_c(t) = \frac{1}{\pi\,x\,t} \cos\left(\frac{1}{t}\right),\; g_s(t) = \frac{1}{\pi\,x\,t} \sin\left(\frac{1}{t}\right),$$

respectively, evaluated at the point $1/x$.

The Mellin integral transforms of the functions g_c, g_s are well known ([26] or [27]):

$$g_c^*(s) = \mathcal{M}\{g_c(t);s\} = \frac{\Gamma(1-s)}{\pi\,x} \sin\left(\frac{\pi s}{2}\right),\; 0 < \Re(s) < 1,$$

$$g_s^*(s) = \mathcal{M}\{g_s(t);s\} = \frac{\Gamma(1-s)}{\pi\,x} \cos\left(\frac{\pi s}{2}\right),\; 0 < \Re(s) < 2.$$

Thus the integrals I_c and I_s can be represented by means of the Parseval equality (77) as follows:

$$I_c(x) = \frac{1}{\pi\,x}\frac{1}{2\pi i} \int_{\gamma-i\infty}^{\gamma+i\infty} f^*(s)\,\Gamma(1-s)\,\sin\left(\frac{\pi\,s}{2}\right)\,x^s\,ds,\; x > 0,\; 0 < \gamma < 1,$$

$$I_s(x) = \frac{1}{\pi\,x}\frac{1}{2\pi i} \int_{\gamma-i\infty}^{\gamma+i\infty} f^*(s)\,\Gamma(1-s)\,\cos\left(\frac{\pi\,s}{2}\right)\,x^s\,ds,\; x > 0,\; 0 < \gamma < 2.$$

In [4], the method presented above was employed to introduce the notion of the generalized H-transform.

In the applications, elementary properties of the Mellin integral transform are often employed. They are presented in the rest of this section.

Let us denote by $\overset{\mathcal{M}}{\leftrightarrow}$ the juxtaposition of a function f with its Mellin transform f^*. The basic Mellin transform rules are as follows:

$$f(at) \overset{\mathcal{M}}{\leftrightarrow} a^{-s} f^*(s),\; a > 0, \tag{78}$$

$$t^p f(t) \overset{\mathcal{M}}{\leftrightarrow} f^*(s+p), \tag{79}$$

$$f(t^p) \overset{\mathcal{M}}{\leftrightarrow} \frac{1}{|p|} f^*(s/p),\; p \neq 0, \tag{80}$$

$$f^{(n)}(t) \overset{\mathcal{M}}{\leftrightarrow} \frac{\Gamma(n+1-s)}{\Gamma(1-s)} f^*(s-n) \quad \text{if } \lim_{t\to 0} t^{s-k-1} f^{(k)}(t) = 0,\; k = 0,1,\ldots,n-1, \tag{81}$$

$$\left(t\frac{d}{dt}\right)^n f(t) \overset{\mathcal{M}}{\leftrightarrow} (-s)^n f^*(s), \tag{82}$$

$$\left(\frac{d}{dt}t\right)^n f(t) \overset{\mathcal{M}}{\leftrightarrow} (1-s)^n f^*(s). \tag{83}$$

In [24,26,27], the Mellin transforms of the elementary and many of the special functions are given. Here we list just some basic Mellin transform formulas that are often used in applications.

$$e^{-t^p} \overset{\mathcal{M}}{\leftrightarrow} \frac{1}{|p|}\Gamma(\frac{s}{p}) \text{ if } \Re(\frac{s}{p}) > 0, \tag{84}$$

$$\frac{(1-t^p)_+^{\alpha-1}}{\Gamma(\alpha)} \overset{\mathcal{M}}{\leftrightarrow} \frac{\Gamma(\frac{s}{p})}{|p|\Gamma(\frac{s}{p}+\alpha)} \text{ if } \Re(\alpha) > 0,\ \Re(\frac{s}{p}) > 0, \tag{85}$$

$$\frac{(t^p-1)_+^{\alpha-1}}{\Gamma(\alpha)} \overset{\mathcal{M}}{\leftrightarrow} \frac{\Gamma(1-\alpha-\frac{s}{p})}{|p|\Gamma(1-\frac{s}{p})} \text{ if } 0 < \Re(\alpha) < 1 - \Re(\frac{s}{p}), \tag{86}$$

$$\Gamma(\rho)(1+t)^{-\rho} \overset{\mathcal{M}}{\leftrightarrow} \Gamma(s)\Gamma(\rho - s) \text{ if } 0 < \Re(s) < \Re(\rho), \tag{87}$$

$$\frac{1}{\pi(1-t)} \overset{\mathcal{M}}{\leftrightarrow} \frac{\Gamma(s)\Gamma(1-s)}{\Gamma(s+1/2)\Gamma(1/2-s)} \text{ if } 0 < \Re(s) < 1, \tag{88}$$

$$\frac{\sin(2\sqrt{t})}{\sqrt{\pi}} \overset{\mathcal{M}}{\leftrightarrow} \frac{\Gamma(s+1/2)}{\Gamma(1-s)} \text{ if } |\Re(s)| < 1/2, \tag{89}$$

$$\sqrt{\pi}\,\mathrm{erf}\,(\sqrt{t}) \overset{\mathcal{M}}{\leftrightarrow} \frac{\Gamma(s+1/2)\Gamma(-s)}{\Gamma(1-s)} \text{ if } -1/2 < \Re(s) < 0, \tag{90}$$

$$J_\nu(2\sqrt{t}) \overset{\mathcal{M}}{\leftrightarrow} \frac{\Gamma(s+\frac{\nu}{2})}{\Gamma(1+\frac{\nu}{2}-s)} \text{ if } -\Re(\frac{\nu}{2}) < \Re(s) < \frac{3}{4}, \tag{91}$$

$$2K_\nu(2\sqrt{t}) \overset{\mathcal{M}}{\leftrightarrow} \Gamma(s+\frac{\nu}{2})\Gamma(s-\frac{\nu}{2}) \text{ if } \Re(s) > \frac{|\Re(\nu)|}{2}, \tag{92}$$

$$\frac{\Gamma(a)}{\Gamma(c)}{}_1F_1(a;c;-t) \overset{\mathcal{M}}{\leftrightarrow} \frac{\Gamma(s)\Gamma(a-s)}{\Gamma(c-s)} \text{ if } 0 < \Re(s) < \Re(a), \tag{93}$$

$$|1-t|^{\mu/2}P_\nu^\mu(\sqrt{t}) \overset{\mathcal{M}}{\leftrightarrow} \frac{\Gamma(s)\Gamma(s+\frac{1}{2})\Gamma(\frac{1+\nu-\mu}{2}-s)\Gamma(-\frac{\mu+\nu}{2}-s)}{\pi 2^{\mu+1}\Gamma(1-\mu+\nu)\Gamma(-\mu-\nu)} \tag{94}$$
$$\text{if } 0 < \Re(s) < \min\{\frac{1+\Re(\nu-\mu)}{2}, -\frac{\Re(\nu+\mu)}{2}\},$$

$$\frac{\Gamma(a)\Gamma(b)}{\Gamma(c)}{}_2F_1(a,b;c;-t) \overset{\mathcal{M}}{\leftrightarrow} \frac{\Gamma(s)\Gamma(a-s)\Gamma(b-s)}{\Gamma(c-s)} \tag{95}$$
$$\text{if } 0 < \Re(s) < \min\{\Re(a), \Re(b)\},$$

$$\frac{(1-t)_+^{c-1}}{\Gamma(c)}{}_2F_1(a,b;c;1-t) \overset{\mathcal{M}}{\leftrightarrow} \frac{\Gamma(s)\Gamma(s+c-a-b)}{\Gamma(s+c-a)\Gamma(s+c-b)} \text{ if } 0 < \Re(s), \tag{96}$$
$$0 < \Re(c),\ 0 < \Re(s+c-a-b),$$

$${}_pF_q((a)_p;(b)_q;-t) \overset{\mathcal{M}}{\leftrightarrow} \frac{\prod_{j=1}^q \Gamma(b_j)}{\prod_{j=1}^p \Gamma(a_j)} \frac{\prod_{j=1}^p \Gamma(a_j - s)\Gamma(s)}{\prod_{j=1}^q \Gamma(b_j - s)} \tag{97}$$
$$\text{if } 0 < \Re(s) < \min_{1\le j\le p} \Re(a_j),$$
$$b_k \neq 0, -1, \ldots,\ 1 \le k \le q$$
$$\text{and}$$
$$(1)\ q = p-1 \text{ or}$$
$$(2)\ q = p \text{ or}$$
$$(3)\ q = p+1 \text{ and}$$
$$\Re(s) < \frac{1}{4} - \frac{1}{2}\left(\Re\left(\sum_{j=1}^p a_j - \sum_{j=1}^q b_j\right)\right),$$

where J_ν is the Bessel function, K_ν is the Macdonald function, P_ν^μ denotes the Legendre function of the first kind, and ${}_pF_q((u)_p;(b)_q;z)$ stands for the generalized hypergeometric function.

As to applications of the Mellin integral transform, we mention here its applications in the theory of the integral transform of the Mellin convolution type [4], for evaluation of improper integrals [26,27], in the theory of special functions of the hypergeometric type [26], for construction of the operational calculi of Mikusinski type for the compositions of the fractional Erdelyi-Kober derivatives [5], for derivation of the fundamental solutions to the space-time fractional diffusion equation [30,31], for analysis of the multi-dimensional fractional diffusion-wave equations [32], for derivation of the subordination principles for the multi-dimensional space-time-fractional diffusion-wave equation [33], and for several other important problems in Fractional Calculus [34].

5. Conclusions

In this survey article, we considered some elements of theory and applications of the integral transforms of mathematical physics. These integral transforms are not arbitrary integral transforms but those possessing well defined inverse integral transforms and generating operators. The basic constructions for most of applications of these integral transforms are their convolutions and generating operators. They lead to some simple and efficient solution methods for the corresponding integral, differential, and integro-differential equations. Another important technique for applications of the integral transforms is the Mikusinski-type operational calculi that were also discussed in the article.

The general schemata for applications of the integral transforms of mathematical physics were illustrated in detail by considering the Laplace integral transform. Similar, but more complicated constructions and solution methods are valid for the general H-transform as a generalization of the Laplace integral transform. In this case, the "integral" generating operators are in form of the compositions of the fractional Erdelyi-Kober right- and left-hand sided fractional integrals and derivatives. Their left-inverse "differential" operators are in form of certain compositions of the fractional Erdelyi-Kober left- and right-hand sided fractional derivatives and integrals. The convolutions of the general H-transform can be constructed in explicit form as some multiple integrals.

In the article, some basic elements of the Mellin integral transform were discussed, too. The Mellin integral transform is a foundation for the theory of the Mellin convolution type integral transforms in general and of the H-transform in particular. For details we refer the interested readers to [4] or [5–8].

References

1. Davies, B. *Integral Transforms and Their Applications*; Springer: New York, NY, USA, 2002.
2. Debnath, L.; Bhatta, D. *Integral Transforms and Their Applications*, 3rd ed.; Chapman and Hall/CRC: Boca Raton, FL, USA, 2014.
3. Tranter, C.J. *Integral Transforms in Mathematical Physics*; Methuen: York, UK, 1971.
4. Yakubovich, S.B.; Luchko, Y.F. *The Hypergeometric Approach to Integral Transforms and Convolutions*; Kluwer: Amsterdam, The Netherlands, 1994.
5. Luchko, Y.F. Some Operational Relations for the H-Transforms and Their Applications. Ph.D. Thesis, Belarusian State University, Minsk, Belarus, 1993. (In Russian)
6. Luchko, Y. Operational method in fractional calculus. *Fract. Calc. Appl. Anal.* **1999**, *2*, 463–489.
7. Luchko, Y. Operational rules for a mixed operator of the Erdélyi-Kober type. *Fract. Calc. Appl. Anal.* **2004**, *7*, 339–364.
8. Luchko, Y. Integral transforms of the Mellin convolution type and their generating operators. *Integr. Transf. Spec. Funct.* **2008**, *19*, 809–851. [CrossRef]
9. Fedoryuk, M.V. Integral transforms. *Itogi Nauki i Techniki* **1986**, *13*, 211–253. (In Russian)
10. Berg, L. Operational calculus in linear spaces. *Stud. Math.* **1961**, *20*, 1–18.
11. Ditkin, V.A.; Prudnikov, A.P. *Operational Calculus*; Nauka: Moscow, Russia, 1975. (In Russian)

12. Erdelyi, A. *Operational Calculus and Generalized Functions*; Holt, Rinehart and Winston: New York, NY, USA, 1962.
13. Mikusinski, J. *Operational Calculus*; Oxford University Press: Oxford, UK, 1959.
14. Yosida, K. *Operational Calculus. A Theory of Hyperfunctions*; Springer: Berlin, Germany; New York, NY, USA, 1984.
15. Dimovski, I.H. *Convolutional Calculus*; Publ. House of the Bulgarian Academy of Sciences: Sofia, Bulgaria, 1982.
16. Churchill, R.V. *Operational Mathematics*; McGraw-Hill: New York, NY, USA, 1958.
17. Doetsch, G. *Handbuch der Laplace-Transformation*; Bd. I-IV; Birkhauser Verlag: Basel, Switzerland, 1956.
18. Oberhettinger, F. *Tables of Laplace Transforms*; Springer: New York, NY, USA, 1973.
19. Prudnikov, A.P.; Brychkov, Y.A.; Marichev, O.I. *Integrals and Series*: *Direct Laplace Transforms*; Gordon and Breach: New York, NY, USA, 1992; Volume 4.
20. Prudnikov, A.P.; Brychkov, Y.A.; Marichev, O.I. *Integrals and Series*: *Inverse Laplace Transforms*; Gordon and Breach: New York, NY, USA, 1992; Volume 5.
21. Widder, D.V. *The Laplace Transform*; Oxford University Press: Oxford, UK, 1946.
22. Evgrafov, M.A. Series and integral representations. *Itogi Nauki i Techniki* **1986**, *13*, 5–92. (In Russian)
23. Titchmarsh, E.C. The zeros of certain integral functions. *Proc. Lond. Math. Soc.* **1926**, *25*, 283–302. [CrossRef]
24. Erdelyi, A.; Magnus, W.; Oberhettinger, F.; Tricomi, F.G. *Tables of Integral Transforms*; McGraw-Hill: New York, NY, USA, 1954; Volumes 1–2.
25. Gradshteyn, I.H.; Ryzhik, I.M. *Tables of Integrals, Series, and Products*, 6th ed.; Academic Press: San Diego, CA, USA, 2000.
26. Marichev, O.I. *Handbook of Integral Transforms of Higher Transcendental Functions, Theory and Algorithmic Tables*; Ellis Horwood: Chichester, UK, 1983.
27. Prudnikov, A.P.; Brychkov, Y.A.; Marichev, O.I. *Integrals and Series*: *More Special Functions*; Gordon and Breach: New York, NY, USA, 1989; Volume 3.
28. Prudnikov, A.P.; Brychkov, Y.A.; Marichev, O.I. Evaluation of Integrals and the Mellin Transform. *Itogi Nauki i Tekhniki Seriya Matematicheskii Analiz* **1989**, *27*, 3–146. (In Russian)
29. Titchmarsh, E.C. *Introduction to theory of Fourier integrals*; Oxford University Press: Oxford, UK, 1937.
30. Gorenflo, R.; Iskenderov, A.; Luchko, Y. Mapping between solutions of fractional diffusion-wave equations. *Fract. Calc. Appl. Anal.* **2000**, *3*, 75–86.
31. Mainardi, F.; Luchko, Y.; Pagnini, G. The fundamental solution of the space-time fractional diffusion equation. *Fract. Calc. Appl. Anal.* **2001**, *4*, 153–192.
32. Boyadjiev, L.; Luchko, Y. Mellin integral transform approach to analyze the multidimensional diffusion-wave equations. *Chaos Solitons Fract.* **2017**, *102*, 127–134. [CrossRef]
33. Luchko, Y. Subordination principles for the multi-dimensional space-time-fractional diffusion-wave equation. *Theory Probab. Math. Stat.* **2018**, *98*, 121–141.
34. Luchko, Y.; Kiryakova, V. The Mellin integral transform in fractional calculus. *Fract. Calc. Appl. Anal.* **2013**, *16*, 405–430. [CrossRef]

11

Shift Scheduling with the Goal Programming Method

Özlem Kaçmaz, Haci Mehmet Alakaş and Tamer Eren *

Department of Industrial Engineering, Kirikkale University, Kirikkale 71450, Turkey; ozlemkacmazz@hotmail.com (Ö.K.); hmalagas@gmail.com (H.M.A.)

* Correspondence: teren@kku.edu.tr.

Abstract: Nowadays, resource utilization and management are very important for businesses. They try to make a profit by providing high levels of efficiency from available sources. Their labor force is one of these sources. Therefore, scheduling based on personnel satisfaction has become an important problem in recent years. In this study, a case study was carried out in a glass factory in Ankara which has 7 department and 80 personnel. The aim of the study is to provide better service by distributing personnel to shifts in a fair and balanced manner. Assignment points are different for the departments where the personnel will work. Every personnel member is assigned to the department as best as possible. A goal programming method was used, and the results were better than those obtained using other methods.

Keywords: shift schedule; goal programming; labor; assignment; personnel

---

1. Introduction

Today, resource utilization and management is very important for businesses. They try to make a profit by providing high levels of efficiency from available sources. Their labor force is one of these sources. The human factor has different demands and expectations than other sources, and it increases the level of interest in this field. The shift scheduling problem, which is a sub-problem of personnel scheduling, is the most frequently studied problem from past to present. The shift scheduling problem, which is encountered in industries, enterprises, hospitals and many other places, is an important problem. To solve this problem, we aimed to provide better service by distributing personnel to shifts in a fair and balanced manner. In addition, the shift scheduling problem provides more income for institutions and organizations by improving employee wages, overtime and break times [1].

The problem becomes complicated if the solution to the shift scheduling problems is about increasing the satisfaction level of the employee and the enterprise. In terms of business, the seniority levels, knowledge and skills of the personnel are taken into consideration, and from the point of view of the personnel, expectations are met. Being one of the conditions that the parties want to provide for each other, it becomes difficult to solve the problem manually. Goal Programming is a mathematical programming method which aims to simultaneously provide multiple and conflicting goal constraints. As with other mathematical programming methods, the optimal solution is not the result. It works to minimize the deviation variables added to the goal constraints. So, the results are displayed as the closest to the targeted value or values.

In this study, 80 personnel and 7 departments (cutting (1), sanding (2), grinding (3), tempering (4), laminating (5), double glazing (6), shipment (7)) in a glass factory were used. In practice, the shifts of personnel working in the factory are scheduled. In the table consisting of two shifts, besides the provision of the number of personnel needed for the shift, the knowledge, skills and requests of the personnel were taken into consideration. The question was asked which personnel should be assigned to which shift for which department. Improvements were made according to the table prepared in the current situation.

In the second part of the study, shift scheduling is examined; in the third part, goal programming (GP) is discussed. The fourth part comprises a literature review, the fifth part a case study, and the last part presents the results.

2. Shift Scheduling

One situation in which staff scheduling problems are commonly encountered is shift scheduling. During the planning period, personnel perform certain activities such as work, rest, eating and taking tea breaks, week holidays and annual leave. These activities should occur within the framework of certain rules and laws. The equitable and balanced distribution of the personnel to the shifts is called shift scheduling. Businesses and institutions can make a profit by providing a good chart and a high yield.

Shift scheduling problems have been very widely discussed in the literature. The first integer mathematical model for shift scheduling was developed by George Dantzig in 1954, and the second by Elbridge Keith in 1979. In the academic literature, the model developed by Dantzig on shift scheduling and that developed by Keith in commercial studies have attracted the most attention [2].

In shift schedules, it is sometimes desirable to employ personnel mainly in the fields of business or departments where they are experts. Considering both the shift scheduling of these problems and the employment of the personnel in the fields where they are experts, it takes a long time to find an optimal solution. In addition, it is very difficult to deal with this problem manually, even by a skilled person. Developing a mathematical model for such problems is of benefit in every respect [3].

3. Goal Programming Method

Goal programming is a type of multi-purpose programming model. Model constraints are written by adding deviation variables to the targeted constraints, except for the indispensable constraints when creating the model. The aim is to minimize deviation variables in goal constraints. It is expected that multiple objectives will be provided at the same time in the models created using goal programming. At the same time, deviations of constraints can be minimized by converting these objectives into constraints and ranking them according to their importance [1].

It is not always possible to achieve every goal determined by the target programming method. Optimal results are selected from the among the most satisfactory. Targets are created for the selected solution. The priorities for the created target are determined. By performing these steps, the model is provided in general. Finally, the solution is determined.

The mathematical representation of the goal programming is as follows [4].

$$Min\ Z = \sum_{i=1}^{k}\left(d_i^+ + d_i^-\right), i = 1, \ldots, k \tag{1}$$

$$\sum_{j=1}^{n} a_{ij}X_j + d_i^- - d_i^+ = b_i, i = 1, \ldots, k, j = 1, \ldots, n \tag{2}$$

$$X_j, d_i^+, d_i^- \geq 0, i = 1, \ldots, k, j = 1, \ldots, n \tag{3}$$

Variables

X_j : j. decision variable, $j = 1, \ldots, n$
a_{ij} : coefficients of ith goal in variable j, $i = 1, \ldots, k$ $j = 1, \ldots, n$
b_i : desired goal value of the ith goal, $i = 1, \ldots, m$
d_i^+ : the deviation values in the positive directions from the ith goal, $i = 1, \ldots, m$
d_i^- : the deviation values in the negative directions from the ith goal, $i = 1, \ldots, m$

4. Literature Review

In the literature, there are many comprehensive studies on shift scheduling and staff/personnel scheduling. In [5], Gungor proposes an integer linear model for nursing scheduling in a hospital that is open 24 h a day, 7 days a week, where all nurses were staffed and worked for 40 h per week. The model consists of two stages. First, the minimum number of nurses that need to be fulfilled, and how many of them could be a student nurse is determined; then, a work and holiday schedule for a period of 2 weeks is designed. Bard et al. [6] modeled the tour scheduling problem in the United States Postal Service using integer programming. They added the restrictions set by a trade union agreement to the model. They presented scenarios that aimed to reduce the size of the workforce by producing solutions in 1 h. Wong and Chun [7] examined the nursing scheduling problem using a probability-based technique. As a result of their work, they brought solutions to the problem in a short time and presented appropriate charts. Azaiez and Sharif [8] developed a model for a computerized 0-1 Goal Programming method for nurse scheduling. This model is adapted to a hospital program in Saudi Arabia. The model prevents unnecessary overtime costs when considering hospital goals and nurse preferences. It is also implemented over a six-month period. Seckiner et al. [9] presented the hierarchical workforce scheduling problem with an integer programming model. When they compared the results between the current model and the solved model, they found that worker costs had been reduced. Personnel assigned to a single shift in the previous model can be assigned to alternate shifts by the proposed model. The model provides flexibility to the decision maker. Sungur, et al. [2], using the integer programming method, aimed to reduce labor costs and enabled employers to assign shifts in the most appropriate way. Castillo et al. [10] aimed to minimize labor costs by including service quality in their work at a call center, as well as achieving optimal staff scheduling. They put forward a multi-criteria paradigm. Olive [11] developed a mathematical model for the manufacturing industry to address personnel scheduling problems; to this end, he used mixed integer programming. The aim of this study is to meet the required number of work hours in different time periods. Topaloglu [12] discussed the problem of scheduling medical assistants in a hospital. Assistants are classified according to their seniority levels. The model is solved and scheduled for six months with goal programming. When the results are examined, better performance is shown compared to manual methods. Heimerl and Kolisch [13] discussed the labor planning problem in multiple projects using mixed integer linear programming. The aim of integer linear programming is to minimize labor costs. In terms of solution results, it has received better results than simple heuristic methods. Karaatli [14] studied the health sector and worked on the nursing scheduling problem in which work hours were 24/7. Fuzzy linear programming has benefited from genetic and heuristic algorithms; the results differed according to the methods. Koruca [15] developed a simulation-supported shift planning module in a small-scale enterprise manufacturing central heating boilers. In order to determine the situation, he undertook study and data collection. As a result of the study, four different shift plan scenarios suitable for possible crisis environments were presented. Atmaca et al. [16] determined the number of nurses who needed to be assigned to shifts in a hospital while increasing hospital efficiency, ensuring customer satisfaction and minimizing costs. Then, they compared the results obtained from current methods and the proposed model. Bag et al. [17] addressed the problem of nurse scheduling in a state hospital in Kirikkale using 0-1 goal programming and the Analytical Network Process (ANP) method. They solved the model with 0-1 goal programming and used ANP to determine the weight in goal programming. They used 5 goals in the study. They compared the solution results to the previous situation. Bektur and Hasgul [18] applied this technique to restaurant taking into account the staff, seniority levels and system staff. The objective function of the model is to minimize deviations from loose constraints according to seniority levels. The proposed model gave better results than current methods. Desert [19] examined the problem of labor recruitment using the goal programming method. A schedule for employees in a restaurant, seniority levels, preferences and shifts in accordance with the preferences of the day was made. Unal and Eren [20] addressed the problem of scheduling personnel in a government agency. They used the weighted goal programming method by considering the demands of the personnel, and solved the model with the GAMS 22.5 package program. As a result, the personnel were assigned according to their seniority levels, and most importantly, they were assigned shifts on their preferred days. More effective charts were prepared in a shorter time with the model. Ozcan et al. [4], in a large-scale hydroelectric power plant in Turkey, set a goal programming model by using real data. In the installed model, a 91% improvement was achieved in production stoppages caused by operator errors, taking into account the performance of employees and the requirements of the work. In this study, 3 goal constraints were determined. Ozder et al. [21] provided the best possible cleaning service to a 24-hour university hospital, proposing a monthly chart for 70 staff. They benefited from goal programming as a method. In [3], Varli devised the monthly working schedules of formen working in the bearing sector by considering different scenarios. Goal programming was used as the method. Attempts were made to take the wishes of the formen into account with the least possible level of deviation. The shifts, days and sections used in the scenarios are the same. The different formen numbers are special constraints and goal constraints. In [1], Varli and Eren studied the seniority levels of the workers in a factory using the Analytic Hierarchical Process (AHP) method. They then developed a model with goal programming to meet the number of employees needed for each shift and to make a distribution in a balanced and fair manner. Five goal constraints were used in the study. The same authors [22] aimed to ensure that the supervisory appointments of research assistants in the Faculty of Engineering of Kirikkale University were made in the most appropriate way during the final period. Goal programming was used as the method. Seventy four research assistants were assigned to 741 exams. They also [23] discussed the staff scheduling problem for nurses working in the internal medicine and endocrine departments of a hospital. For the monthly schedule to be created, hospital rules and special permission requests of the nurses were taken into consideration. Once again, the goal programming method was used. As a result of improvements and solutions, service quality is expected to increase. Bedir [24] aimed at reducing production downtime costs from personnel at a hydroelectric power plant by suggesting the use of the 0-1 Priority goal programming model considering personnel competencies. Competencies may be prioritized with the PROMETHEE method. The criteria affecting personnel competences were weighted using the AHP method. As a result of the solution, an 86% improvement was achieved for August 2017, i.e., when the plant was operating most intensively. The studies conducted in the literature have similarities with the present study. In our study, attention was given to the prioritization of personnel competencies. Gur and Eren [25] examined scheduling and planning problems using the goal programming method. As a result of the examination, they categorized the problems with a detailed analysis. Koc [26] aimed at minimizing all direct and indirect costs related to workforce scheduling. The method made use of integer programming. Koctepe [27] created a model for a meeting organization using the 0-1 integer programming method. In a model where personnel competencies are taken into consideration, planning was devised for 2 shifts and 80 personnel for 7 days. As a result of that study, it was determined that personnel satisfaction had increased. Tapkan [28] discussed the task scheduling problem of the Kayseri rail transportation system. Again, the multipurpose 0-1 mixed integer model was used. In the objective function of the mathematical model, the largest difference between the number of staff, the weekly statutory working time of the weekly working hours of the staff, the sum of the overrun periods and the average rest period and the rest period were taken into account. Ozder et al. [29] changed the shifts of personnel using ANP and goal programming methods in a natural gas combined cycle power plant. In order to incorporate the personnel skills into the model, they calculated the seniority levels with the ANP method. They identified 4 levels of seniority among 80 staff members for 3 shifts. The aforementioned studies are presented in Table 1.

Table 1. Literature review.

Author	Type	Methods
Varlı and Eren [1]	Shift scheduling	AHP, Goal Programming
Sungur [2]	Shift scheduling	Fuzzy Integer Programming
Varlı [3]	Shift scheduling	AHP, Goal Programming
Ozcan et al. [4]	Shift scheduling	Goal Programming
Güngör [5]	Personnel Scheduling	Integer Programming
Bard et al. [6]	Personnel Scheduling	Integer programming
Wong and Chun [7]	Personnel Scheduling	Constraint programming and Heuristic Method
Azaiez and Al Sharif [8]	Personnel Scheduling	0-1 Goal Programming
Seçkiner et al. [9]	Shift scheduling	Integer programming
Castillo et al. [10]	Shift scheduling	labor scheduling paradigm
Olive [11]	Shift scheduling	Mathematical model
Topaloglu [12]	Shift scheduling	Multiple objective programming
Heimerl and Kolisch [13]	Project scheduling	Integer programming
Karaatlı [14]	Personnel Scheduling	Linear Programming, Genetic and Heuristic Algorithm
Koruca [15]	Shift scheduling	simulation
Atmaca et al. [16]	Personnel Scheduling	0-1 linear goal programming
Bag et al. [17]	Personnel Scheduling	0-1 goal programming, ANP method
Bektur and Hasgul [18]	Shift scheduling	Goal Programming
Çöl [19]	Shift scheduling	Goal Programming
Unal and Eren [20]	Shift scheduling	Goal Programming, Multiple-Objective Decision Making
Ozder et al. [21]	Personnel Scheduling	Goal Programming
Varlı et al. [22]	Personnel Scheduling	Goal Programming
Varlı et al. [23]	Personnel Scheduling	Goal Programming
Bedir [24]	Shift scheduling	Goal Programming, AHP, PROMETHEE
Gur and Eren [25]	Literature Review	Research
Koc [26]	Shift scheduling	Integer Programming
Koctepe et al. [27]	Fair Scheduling	Integer Programming
Tapkan et al. [28]	Personnel Scheduling	0-1 mixed integer model
Ozder et al. [29]	Shift scheduling	Goal Programming, ANP

5. A Case Study in a Glass Factory in Ankara Province

The present study was carried out in a glass factory operating in the facade sector in Ankara in order to ensure the optimal appointment of personnel to shifts and departments. The factory workday comprises two shifts, i.e., 08:00-18:00 and 22:00-08:00. Personnel change shifts at one week intervals; the factory is closed on Sundays. Seven sections and 80 personnel, in which production goes on continuously, are discussed. When evaluating the existing system, it was assumed that the department and staff were working at full capacity. The relevant sections are cutting (1), sanding (2), grinding (3), tempering (4), laminating (5), double glazing (6) and shipment (7).

Orders from customers are numbered according to the type of product, in other words, according to the last process before the merchandise is made ready for shipment. Materials progress through the work order between the processes. The product range includes flat glass, colored glass, solar and temperature controlled glass, tempered glass, laminated glass and bulletproof glass.

5.1. Product Type

- Flat Glass: Flat glass has high light transmittance due to its transparency.
- Colored Glass: Colored glass is obtained by adding colorants to the glass paste; available in green, smoked, bronze and blue.
- Solar and Heat Controlled Glass: This is a type of glass with different aesthetics and designs that can save energy.
- Tempered Glass: A type of glass whose durability and resistance to thermal stresses are 5 times higher than those of flat glass. Areas of application are generally glass railings and doors, walk-in showers, intermediate compartments, glass furniture, refrigerator and oven windows, and side and rear windows of automobiles.

- Laminated Glass: Two or more glass plates are produced by combining special binder polyvinyl butyral (PVB) layers under heat and pressure. This process minimizes the risk of glass breakage by keeping the pieces in place in such an event. It contributes to sound insulation.
- Bullet Proof Glass: Bulletproof glass is aimed at preventing crime and facilitating the capture of the criminal after the action. Areas of use are banks, police stations, museums, military buildings and other official organizations, psychiatric wards, jewelers and so on. This category comprises polyvinyl butyral (PVB) or polycarbonate interlayer laminated glass.

5.2. Production Rotation

The order number is determined according to the process after which the glass will be ready for dispatch. Routes are created according to the following order numbers. The routes to be determined are cutting, grinding, tempering, laminating and double glazing. Glass is usually prepared by following these routes, except for cases of special orders. The following routes are shown in a flow chart. Routes of the products are shown in the flow charts given in Figure 1 for cutting, Figure 2 for grinding, Figure 3 for temper, Figure 4 for laminate and Figure 5 for double glazing glass.

Cutting Section Route; Cutting-Shipment

Figure 1. Cutting flow chart.

Grinding Section Route; Cutting- Grinding-Shipment

Figure 2. Grinding flow chart.

Temper Section Route; Cutting-Sanding Or Grinding-Tempering-Shipment

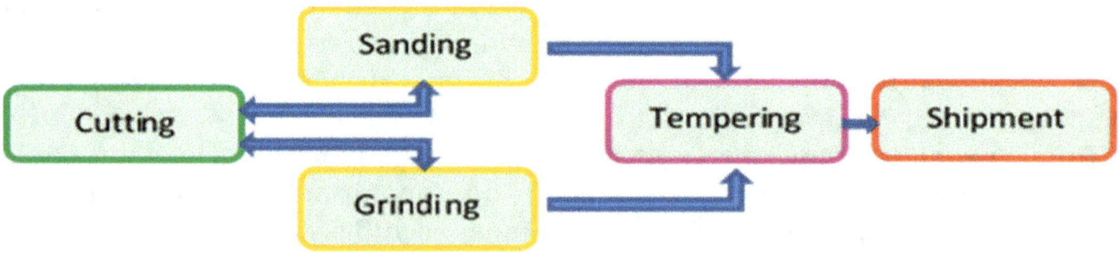

Figure 3. Tempering flow chart.

Laminating Section Route; Cutting-Sanding Or Grinding-Tempering-Laminated-Shipment

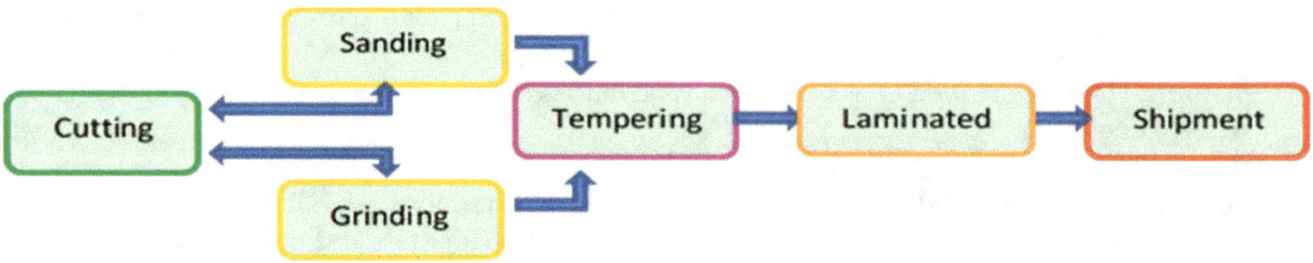

Figure 4. Laminating flow chart.

Double Glazing Section Route; Cutting-Sanding Or Grinding-Tempering-Double Glazing-Shipment

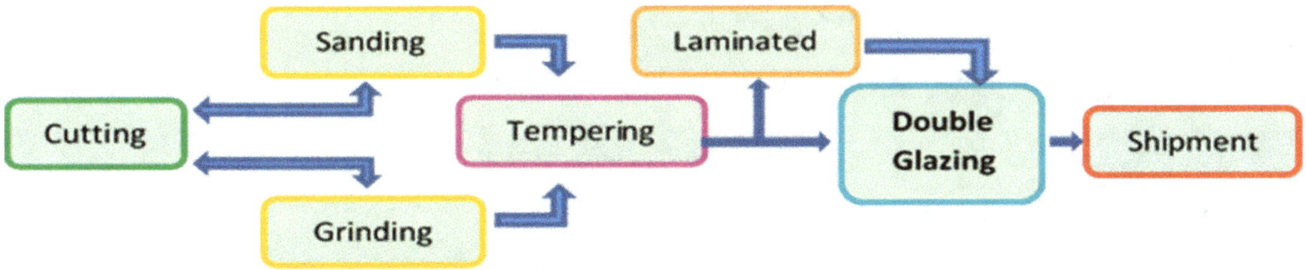

Figure 5. Double glazing flow chart.

5.3. *List of Personnel*

The factory has a total of 80 personnel. In practice, each personnel member receives points for each department. These scores are determined according to the opinions of experts. There scale is 1 to 3 points, according to the sections to be assigned. In the determined section, we tried to restricted to system to a maximum of 5 points. The model uses the minimum objective function; 1 point indicates that the competence level is greater than 2, while 3 points indicates that an employee is not competent in that section. The points obtained according to the personnel list and the sections are given in Table A1 in Appendix A section.

5.4. *Mathematical Model*

5.4.1. Parameters

n: number of personnel working in the factory, $n = 80$
m: number of days, $m = 30$
s: Number of sections in the factory, $s = 7$
t: Number of shifts, $t = 2$
i: Personnel index, $i = 1,2, \ldots ,n$
j: Day index, $j = 1,2, \ldots ,m$
k: Section index, $k = 1,2, \ldots ,s$
l: Shift index, $l = 1,2, \ldots ,t$

5.4.2. Decision variables

$$X_{ijkl} = \begin{cases} 1, \text{if shift, chapter and day is chosen for personnel} \\ 0, \text{otherwise} \end{cases} ,i = 1,2,\ldots,n, j = 1,2,\ldots,m, k = 1,2,\ldots,s, l = 1,2,\ldots,t \quad (4)$$

$$h_{ij} = \begin{cases} 1, \text{if vacation for personnel} \\ 0, \text{otherwise} \end{cases} ,i = 1,2,\ldots,n, j = 1,2,\ldots,m \quad (5)$$

5.4.3. Constraints

1-To meet the daily personnel needs of the departments:
Number of personnel needed for each shift in the cutting(1) section.

$$\Sigma_{i=1}^{n}\left(X_{ij1l}\right) = 3, j = 1,2,\ldots,m, l = 1,2 \quad (6)$$

Number of personnel needed for each shift in the sanding(2) section.

$$\Sigma_{i=1}^{n}\left(X_{ij2l}\right) = 4, j = 1,2,\ldots,m, l = 1,2 \quad (7)$$

Number of personnel needed for each shift in the grinding(3) section.

$$\sum_{i=1}^{n}\left(X_{ij3l}\right)=4, j=1,2,\ldots,m, l=1,2 \tag{8}$$

Number of personnel needed for each shift in the tempering (4) section

$$\sum_{i=1}^{n}\left(X_{ij4l}\right)=4, j=1,2,\ldots,m, l=1,2 \tag{9}$$

Number of personnel needed for each shift in the laminating (5) section

$$\sum_{i=1}^{n}\left(X_{ij5l}\right)=6, j=1,2,\ldots,m, l=1,2 \tag{10}$$

Number of personnel needed for each shift in the double glazing (6) section

$$\sum_{i=1}^{n}\left(X_{ij6l}\right)=8, j=1,2,\ldots,m, l=1,2 \tag{11}$$

Number of personnel needed for each shift in the shipment (7) section

$$\sum_{i=1}^{n}\left(X_{ij7l}\right)=5, j=1,2,\ldots,m, l=1,2 \tag{12}$$

2- Only one shift per personnel per day:

$$\sum_{l=1}^{t}\sum_{k=1}^{s}\left(X_{ijkl}\right)\leq 1, i=1,2,\ldots,n, j=1,2,\ldots,m \tag{13}$$

3- personnel not working on the day of leave:

$$\sum_{l=1}^{t}\sum_{k=1}^{s}\left(X_{ijkl}\right)\leq\left(1-h_{ij}\right), i=1,2,\ldots,n, j=1,2,\ldots,m \tag{14}$$

4- Each personnel member has a minimum of 1 and a maximum of 2 days a week.:

$$h_{ij}+h_{i(j+1)}+h_{i(j+2)}+h_{i(j+3)}+h_{i(j+4)}+h_{i(j+5)}+h_{i(j+6)}\leq 2, i=1,2,\ldots,n, j=1,2,\ldots,m-6 \tag{15}$$

$$h_{ij}+h_{i(j+1)}+h_{i(j+2)}+h_{i(j+3)}+h_{i(j+4)}+h_{i(j+5)}+h_{i(j+6)}\geq 1, i=1,2,\ldots,n, j=1,2,\ldots,m-6 \tag{16}$$

5-Upper limit for each personnel to work on 1 and 2 shifts:

$$\sum_{j=1}^{m}\sum_{k=1}^{s}\left(X_{ijk1}\right)\leq 12, i=1,2,\ldots,n \tag{17}$$

$$\sum_{j=1}^{m}\sum_{k=1}^{s}\left(X_{ijk2}\right)\leq 12, i=1,2,\ldots,n \tag{18}$$

6- Lower limit restrictions for each personnel on 1 and 2 shifts:

$$\sum_{j=1}^{m}\sum_{k=1}^{s}\left(X_{ijk1}\right)\geq 10, i=1,2,\ldots,n \tag{19}$$

$$\sum_{j=1}^{m}\sum_{k=1}^{s}\left(X_{ijk2}\right)\geq 10, i=1,2,\ldots,n \tag{20}$$

7- If an employee were assigned to the night shift on a given day, the next day's shift in the morning shift would be limited:

$$\sum_{k=1}^{s}\left(X_{ijk2}\right)+\left(X_{i(j+1)k1}\right)\leq 1, i=1,2,\ldots,n, j=1,2,\ldots,m-1 \tag{21}$$

5.4.4. Goal Constraints

Goal 1: Goal constraint where personnel are asked to minimize the assignment as day of leave-workday-leave when being assigned shifts:

$$h_{ij} + \Sigma_{k=1}^{s}\Sigma_{l=1}^{t}\left(X_{i(j+1)kl}\right) + h_{i(j+2)} + d_{1ij}^{-} - d_{1ij}^{+} = 2, i = 1,2,\ldots,n, j = 1,2,\ldots,m-2 \quad (22)$$

Goal 2: Goal constraint where personnel are asked to minimize the assignment of working day-tracking-working day when being assigned to shifts:

$$\Sigma_{k=1}^{s}\sum_{l=1}^{t}\left(X_{ijkl}\right) + h_{i(j+1)} + \Sigma_{k=1}^{s}\Sigma_{l=1}^{t}\left(X_{i(j+1)kl}\right) + d_{2ij}^{-} - d_{2ij}^{+} = 2, i = 1,2,\ldots,n, j = 1,2,\ldots,m-2 \quad (23)$$

Goal 3: Goal constraint on which the total number of vacancies for which each personnel is assigned is intended to be as equal as possible:

$$\Sigma_{j=1}^{m}\Sigma_{k=1}^{s}\Sigma_{l=1}^{t}\left(X_{ijkl}\right) + d_{3i}^{-} - d_{3i}^{+} = 22, i = 1,2,\ldots,n \quad (24)$$

Goal 4: Personnel assigned to the departments in each shift will provide the required sum of points as a qualification:

$$\Sigma_{i=1}^{n}\left(X_{ij1l}\right) * (1) + d_{4jl}^{-} - d_{4jl}^{+} = 3, j = 1,2,\ldots,m, l = 1,2 \quad (25)$$

$$\Sigma_{i=1}^{n}\left(X_{ij2l}\right) * (2) + d_{5jl}^{-} - d_{5jl}^{+} = 4, j = 1,2,\ldots,m, l = 1,2 \quad (26)$$

$$\Sigma_{i=1}^{n}\left(X_{ij3l}\right) * (3) + d_{6jl}^{-} - d_{6jl}^{+} = 4, j = 1,2,\ldots,m, l = 1,2 \quad (27)$$

$$\Sigma_{i=1}^{n}\left(X_{ij4l}\right) * (4) + d_{7jl}^{-} - d_{7jl}^{+} = 4, j = 1,2,\ldots,m, l = 1,2 \quad (28)$$

$$\Sigma_{i=1}^{n}\left(X_{ij5l}\right) * (5) + d_{8jl}^{-} - d_{8jl}^{+} = 6, j = 1,2,\ldots,m, l = 1,2 \quad (29)$$

$$\Sigma_{i=1}^{n}\left(X_{ij6l}\right) * (6) + d_{9jl}^{-} - d_{9jl}^{+} = 8, j = 1,2,\ldots,m, l = 1,2 \quad (30)$$

$$\Sigma_{i=1}^{n}\left(X_{ij7l}\right) * (7) + d_{10jl}^{-} - d_{10jl}^{+} = 5, j = 1,2,\ldots,m, l = 1,2 \quad (31)$$

5.4.5. Objective Function

$$\text{MinZ} = \Sigma_{i=1}^{n}\Sigma_{j=1}^{m}\left(d_{1ij}^{-} + d_{1ij}^{+}\right) + (d_{2ij}^{-} + d_{2ij}^{+}) + (d_{3i}^{-} + d_{3i}^{+}) + \Sigma_{j=1}^{m}\Sigma_{l=1}^{t}d_{4jl}^{+} + d_{5jl}^{+} + d_{6jl}^{+} + d_{7jl}^{+} + d_{8jl}^{+} + d_{9jl}^{+} + d_{10jl}^{+} \quad (32)$$

5.5. Model Solution

First of all, general constraints and goal constraints are determined for the mathematical model. The current system was examined and the number of personnel needed by the departments was changed. Personnel are also allowed to leave on Sundays. Deviations in goal constraints were minimized and the objective function was created. The solutions used in the solution of the mathematical models were obtained using ILOG version 12.6.2. As a result of the solution, 30-day work schedules for 80 personnel were created. The suggested data is given in Table A2 in Appendix B. The solution produced using the proposed model was compared to that produced using current methods. The total competence score of the personnel working in the departments was divided by the number of people required in that department, and the shift labor force was created. The current system is given in Table 2 and the proposed system is given in Table 3.

Table 2. Average workforce in the current system.

Current System														
Section	**Cutting**		**Sanding**		**Grinding**		**Tempering**		**Laminated**		**Double Glazing**		**Shipment**	
Day \ Shift	1	2	1	2	1	2	1	2	1	2	1	2	1	2
1	2.75	3	2.5	2.75	2.75	2.75	3.2	2.6	3	3.14	2.9	2.9	2.67	3.17
2	2.75	3	2.5	2.75	2.75	2.75	3.2	2.6	3	3.14	2.9	2.9	2.67	3.17
3	2.75	3	2.5	2.75	2.75	2.75	3.2	2.6	3	3.14	2.9	2.9	2.67	3.17
4	2.75	3	2.5	2.75	2.75	2.75	3.2	2.6	3	3.14	2.9	2.9	2.67	3.17
5	2.75	3	2.5	2.75	2.75	2.75	3.2	2.6	3	3.14	2.9	2.9	2.67	3.17
6	2.75	3	2.5	2.75	2.75	2.75	3.2	2.6	3	3.14	2.9	2.9	2.67	3.17
7	**Staff on Leave**													
8	3	2.75	2.75	2.5	2.75	2.75	2.6	3.2	3.14	3	2.9	2.9	3.17	2.67
9	3	2.75	2.75	2.5	2.75	2.75	2.6	3.2	3.14	3	2.9	2.9	3.17	2.67
10	3	2.75	2.75	2.5	2.75	2.75	2.6	3.2	3.14	3	2.9	2.9	3.17	2.67
11	3	2.75	2.75	2.5	2.75	2.75	2.6	3.2	3.14	3	2.9	2.9	3.17	2.67
12	3	2.75	2.75	2.5	2.75	2.75	2.6	3.2	3.14	3	2.9	2.9	3.17	2.67
13	3	2.75	2.75	2.5	2.75	2.75	2.6	3.2	3.14	3	2.9	2.9	3.17	2.67
14	**Staff on Leave**													
15	2.75	3	2.5	2.75	2.75	2.75	3.2	2.6	3	3.14	2.9	2.9	2.67	3.17
16	2.75	3	2.5	2.75	2.75	2.75	3.2	2.6	3	3.14	2.9	2.9	2.67	3.17
17	2.75	3	2.5	2.75	2.75	2.75	3.2	2.6	3	3.14	2.9	2.9	2.67	3.17
18	2.75	3	2.5	2.75	2.75	2.75	3.2	2.6	3	3.14	2.9	2.9	2.67	3.17
19	2.75	3	2.5	2.75	2.75	2.75	3.2	2.6	3	3.14	2.9	2.9	2.67	3.17
20	2.75	3	2.5	2.75	2.75	2.75	3.2	2.6	3	3.14	2.9	2.9	2.67	3.17
21	**Staff on Leave**													
22	3	2.75	2.75	2.5	2.75	2.75	2.6	3.2	3.14	3	2.9	2.9	3.17	2.67
23	3	2.75	2.75	2.5	2.75	2.75	2.6	3.2	3.14	3	2.9	2.9	3.17	2.67
24	3	2.75	2.75	2.5	2.75	2.75	2.6	3.2	3.14	3	2.9	2.9	3.17	2.67
25	3	2.75	2.75	2.5	2.75	2.75	2.6	3.2	3.14	3	2.9	2.9	3.17	2.67
26	3	2.75	2.75	2.5	2.75	2.75	2.6	3.2	3.14	3	2.9	2.9	3.17	2.67
27	3	2.75	2.75	2.5	2.75	2.75	2.6	3.2	3.14	3	2.9	2.9	3.17	2.67
28	**Staff on Leave**													
29	2.75	3	2.5	2.75	2.75	2.75	3.2	2.6	3	3.14	2.9	2.9	2.67	3.17
30	2.75	3	2.5	2.75	2.75	2.75	3.2	2.6	3	3.14	2.9	2.9	2.67	3.17

In the current system, it is assumed that the factory was operating at full capacity. Staff are only allowed a day off on Sundays. Persons are generally requested to work in the department in which they are employed. A task assignment is made to meet needs rather than special talents; that is why the average labor force is constant for each department, every week and for every shift.

For a sample average workforce calculation from the table, the 8th day shift of the sanding section will be considered; on that day, 2, 18, 34, 75 staff members were assigned. The total score of the assigned personnel for the sanding section was 2 + 2 + 3 + 3 = 10. The score obtained was divided by the number of staff required by the department, and the average workforce was found. This is equal to 10/4 = 2.5.

Table 3. Average workforce in the proposed model.

Recommended Model														
Section	**Cutting**		**Sanding**		**Grinding**		**Tempering**		**Laminated**		**Double Glazing**		**Shipment**	
Shift / Day	1	2	1	2	1	2	1	2	1	2	1	2	1	2
1	1.67	2.33	1	1.5	2	1	2	1	1.83	1.67	1.38	1.75	1	1
2	1	3	1.5	1.75	1.5	1.75	2	1	2.67	1.83	1.5	2	1.2	2.8
3	1	3.67	1.5	1.25	1.25	1.5	2	1.5	1.67	1.67	1.25	1.25	1	1
4	1	2.33	1.5	1.25	1	3.75	1.5	1.5	1.33	2.33	1.5	2.5	1	1
5	2.33	1	1.25	1.25	2.25	1.25	1	2	1.17	1.67	1.5	1.5	1.4	1
6	2.33	2.33	1	1	1.75	1.25	1.25	1	1.17	1.67	2	1.38	1.8	2
7	3.67	1.67	1.25	1	1.5	2	2	1.5	1.33	2.17	2	2.25	1.6	1.4
8	1	2.33	1	1	1.25	1.25	2	1.5	1.67	1.33	2	1	1	1.2
9	1	2.33	1.5	1.25	1.5	1	2	1	1.67	1.67	1.5	1.88	1	1.8
10	1	2.33	1	1.25	1.5	1	2	2.25	1.83	2	1.25	2	1.2	1
11	2.33	1.67	1.5	1	2.5	2	2	1.25	1.33	2.67	1.75	1.13	1.4	1
12	3.67	1	1.25	1	1.5	1.5	2	1.75	2	1	1.75	2.13	1	1.6
13	3	2.33	1.25	1	1.5	1.5	1	1	1.17	1.33	2.75	1.5	2	1.4
14	3.67	1.67	1.25	1	1	1	1.5	1	1.67	1.67	1	1.25	1	1.6
15	2.33	1	1.25	1	1.5	1	2.5	1	1	1.67	1.5	1.63	1	1.4
16	3.67	1	1.75	1.25	1	3	1	2	1.33	1.5	2	2.13	1.2	1.8
17	1.67	2.33	1.25	2.25	1.5	1	2.25	1.5	1.33	1	1.5	1.38	1	1.2
18	1.67	1.33	1	1.25	1.5	2	1.25	1	1.5	1	1.75	1	1.4	1
19	1.67	2	1.25	1.25	2	2	1	2.25	1.33	1.33	1.63	2.13	1.4	1
20	2.33	2.33	1.5	1.25	2.25	1.25	2	2	1.33	1.33	1.5	1.5	1.2	1.4
21	3	2.33	1.5	1	1.5	1	1.25	2	2	1.83	1.5	1.25	1.2	1.6
22	3	2.33	1.25	2.25	1.25	3	2	2	1.33	1.33	1.5	1.25	1.2	2
23	1	2.33	1	1	2	2.25	1	2.5	1.33	1	2	1.5	1.2	1
24	1.67	2.33	1	1.25	1.25	2.5	1	1.5	1.83	1.67	1.25	1.75	1.8	1.8
25	2.33	1	1	1	1	1.25	1	1	1	1.33	1	1.25	1	1
26	1	1.67	1	1.25	1	1.5	1	1	1.33	1.33	2.38	1	1	1
27	1	2.33	1.25	1	1	1.25	3	2.5	1.5	1.83	2	2	1.4	1.4
28	3	1	1.5	1	1	1.5	1.5	2	1	1.33	2	2	1.8	1.8
29	4.33	2.33	1.25	1.25	1.5	1.25	1.5	2.5	1.33	1	1.25	1.75	1.8	1.8
30	1.67	1.67	1.75	1.25	1	1	1	2	1.67	2	1.5	1.5	1	1

With the proposed model, specific qualification scores have been defined for each department. Personnel can be assigned to different departments to provide a qualification score. The flexibility of work on different parts by the person provided flexibility for the day of leave. Thanks to our model, the shortage of authorized personnel was addressed. Thus, production can continue without stopages. The average workforce changes daily for each shift.

6. Conclusions

In the factory, we aimed to determine morning and evening shifts for 7 personnel and 80 personnel. Firstly, the number of personnel needed by each departments is emphasized. The number of personnel changed by taking into consideration that the right employee needs to be assigned to the appropriate department. In the current system, 4 staff are required for cutting, sanding, and grinding, 5 are needed for tempering, 7 for laminating, 10 for heaters and 6 for shipments. In the proposed system, these numbers are 3 for cutting, sanding, grinding, 4 for tempering, 6 for laminating, 8 for heaters and 5 for shipments. On Sunday, the personnel permits were distributed. Thus, the factory, which is in need of production, was allowed to operate on Sundays. In terms of the average labor force, only 5 of the 420 shifts did not achieve the desired result within a monthly planning period. In other shifts, the result was quite successful compared to the current situation. Every personnel member was assigned to only one shift during the day and no appointment was made on that employee's day of leave. Personnel were rated between 1 and 3 according to the departments to be assigned. One point indicates that the employee is more than competent for that part, 2 points indicates that he/she is moderately competent, and 3 points indicates insufficient competence. The assignment was restricted by giving 5 points for cases where personnel appointments were not requested. The upper and lower limit numbers can be determined and the distribution is equal for the morning and evening shifts. The personnel who worked a night shift were prevented from being assigned to the day shift the next day, as this would cause require more than 24 h of work withour a rest. In addition to this, it was desireable that the total number of shifts assigned to the personnel during the one-month planning period be the same, and that these appointments should be sequential around the day of leave. Finally, in order to achieve a certain level of points among the personnel assigned to the departments, a specific score for each department was given. The targets were gathered in one place and the objective function was established. Deviations from the target values determined by the goal programming method were very small. In this way, the most suitable shifts and departments could be assigned on the appropriate days by taking into consideration the talents of the personnel. The aim of the goal programming method is to perform multiple goals simultaneously.

In the existing system used in the factory, personnel are divided into two separate teams; each shift team is divided into 7 sections. The teams change at one week intervals. Each team should comprise only from competent people. Generally, a sufficient number of qualified persons are assigned to each section and the remaining personnel are used to complete the task; however, in this way, full efficiency cannot be achieved. The competence of the departments was improved with the proposed new model. Personnel were used more effectively being better assigned, and the satisfaction levels of the personnel increased. As a result, it was observed that the new system offers better results.

Author Contributions: Ö.K. conceptualized the study, prepared dataset, conducted analyses, contributed to writing the manuscript, and provided modeling processes. H.M.A. supported the solution process of the mathematical model and was a contributor to writing the manuscript. T.E. provided overall guidance and expertise in conducting the analysis and was a contributor to writing the manuscript. All authors read and approved the final manuscript.

Appendix A

Table A1. The points of the personnel according to the sections.

	Cutting(1)	Sanding(2)	Grinding(3)	Tempering(4)	Laminated(5)	Double Glazing(6)	Shipment(7)
P1	1	1	1	5	5	5	2
P2	5	2	2	1	5	5	3
P3	5	1	1	3	5	3	5
P4	5	1	2	2	5	5	3
P5	3	3	5	1	5	5	5
P6	2	5	5	5	1	5	3
P7	5	1	1	2	3	5	5
P8	5	3	3	2	5	1	3
P9	3	2	2	5	1	5	5
P10	5	2	3	5	1	5	1

Table A1. *Cont.*

	Cutting(1)	Sanding(2)	Grinding(3)	Tempering(4)	Laminated(5)	Double Glazing(6)	Shipment(7)
P11	5	1	2	3	3	5	5
P12	5	1	1	5	5	5	3
P13	1	1	2	2	5	5	3
P14	5	1	2	1	5	5	5
P15	5	1	2	5	1	5	3
P16	5	2	3	1	5	5	1
P17	5	2	3	5	1	5	2
P18	3	2	5	2	5	1	5
P19	5	2	5	2	5	5	1
P20	5	1	2	5	5	2	1
P21	5	2	1	3	5	5	3
P22	1	1	1	5	3	5	2
P23	5	1	3	1	5	5	3
P24	5	2	3	2	5	1	5
P25	5	2	5	1	5	3	1
P26	3	1	2	5	3	5	3
P27	3	2	1	5	3	5	5
P28	5	5	5	3	1	5	1
P29	5	2	5	3	2	1	5
P30	5	2	1	3	5	5	2
P31	3	3	2	5	1	5	5
P32	1	5	5	3	2	3	1
P33	2	1	5	5	5	1	2
P34	5	3	1	5	5	3	5
P35	5	1	2	5	2	5	1
P36	3	5	5	1	3	1	1
P37	5	1	2	1	5	5	2
P38	1	2	3	3	5	5	3
P39	3	1	2	5	3	5	5
P40	5	2	1	5	5	5	1
P41	5	2	3	1	5	1	5
P42	5	1	1	1	3	5	5
P43	3	1	2	5	5	3	2
P44	3	1	1	5	5	3	5
P45	5	2	1	5	5	1	3
P46	5	2	1	3	3	5	5
P47	3	2	3	1	5	1	5
P48	5	2	3	2	1	5	5
P49	5	2	3	1	3	5	5
P50	5	2	1	5	5	1	2
P51	3	2	3	1	5	5	5
P52	5	1	2	3	5	3	5
P53	5	2	2	5	2	1	5
P54	5	1	2	5	1	5	5
P55	5	5	5	2	2	5	1
P56	3	2	2	1	5	5	3
P57	5	1	1	3	3	5	5
P58	3	1	1	5	3	5	2
P59	1	3	2	5	2	5	3
P60	3	5	5	2	5	5	1
P61	1	1	5	3	3	5	3
P62	5	2	5	5	1	3	5
P63	5	2	2	1	3	5	5
P64	5	2	3	5	3	1	1
P65	5	5	5	1	5	3	1
P66	3	1	1	5	5	5	3
P67	5	2	5	1	3	1	5
P68	5	2	3	5	1	5	2

Table A1. *Cont.*

	Cutting(1)	Sanding(2)	Grinding(3)	Tempering(4)	Laminated(5)	Double Glazing(6)	Shipment(7)
P69	5	2	3	1	5	1	5
P70	5	2	2	5	1	3	5
P71	1	5	3	2	3	3	1
P72	3	3	5	5	1	3	5
P73	5	1	2	2	5	5	3
P74	5	3	5	2	1	1	5
P75	5	3	3	1	1	5	5
P76	5	2	2	5	5	1	1
P77	5	2	5	5	5	3	1
P78	3	2	3	1	5	5	5
P79	5	2	5	2	5	5	1
P80	5	1	2	5	5	1	2

Appendix B

Table A2. Monthly chart of factory personnel.

	Monday	Tuesday	Wednesday	Thursday	Friday	Saturday	Sunday	Day	Night
P1	1(1)	X	1(1)	5(2)	1(2)	1(2)	7(2)	14	11
	X	1(1)	1(1)	1(1)	1(2)	2(2)	3(2)		
	X	7(1)	3(1)	7(1)	1(1)	1(2)	1(2)		
	X	1(1)	1(1)	1(1)	1(1)	1(2)	1(2)		
	X	1(1)							
P2	X	4(1)	4(1)	2(1)	4(1)	4(1)	4(2)	12	13
	X	4(2)	4(2)	4(2)	7(2)	1(2)	4(2)		
	X	7(2)	4(2)	4(2)	2(2)	4(2)	4(2)		
	X	4(1)	4(1)	1(1)	4(1)	4(1)	X		
	2(1)	4(1)							
P3	X	5(2)	1(2)	6(2)	6(2)	7(2)	6(2)	14	10
	X	3(1)	2(1)	2(1)	1(1)	1(1)	3(1)		
	X	3(1)	6(1)	2(1)	6(2)	2(2)	X		
	3(1)	6(1)	6(1)	2(1)	3(2)	6(2)	X		
	X	3(1)							
P4	X	2(2)	1(2)	5(2)	2(2)	1(2)	5(2)	11	14
	X	2(1)	2(1)	6(1)	2(1)	2(2)	2(2)		
	X	2(1)	2(1)	2(1)	2(1)	2(1)	2(1)		
	X	2(2)	7(2)	2(2)	2(2)	2(2)	2(2)		
	X	2(1)							
P5	4(1)	4(1)	4(1)	4(1)	X	4(1)	4(2)	13	13
	4(2)	7(2)	5(2)	1(2)	X	6(1)	4(1)		
	6(1)	4(1)	4(2)	4(2)	X	4(2)	4(2)		
	4(2)	4(2)	1(2)	4(2)	X	7(1)	4(1)		
	4(1)	4(1)							
P6	5(1)	5(2)	5(2)	5(2)	5(2)	5(2)	X	13	12
	5(1)	4(1)	5(1)	5(1)	5(2)	5(2)	X		
	5(1)	5(1)	5(1)	1(2)	1(2)	5(2)	X		
	5(1)	5(1)	5(1)	5(1)	5(2)	X	5(1)		
	5(2)	X							
P7	X	2(2)	3(2)	1(2)	2(2)	2(2)	3(2)	13	12
	X	2(1)	2(1)	3(1)	3(1)	7(1)	2(1)		
	X	5(2)	3(2)	2(2)	4(2)	1(2)	2(2)		
	X	3(1)	3(1)	3(1)	2(1)	3(1)	3(1)		
	X	3(1)							

Table A2. *Cont.*

	Monday	Tuesday	Wednesday	Thursday	Friday	Saturday	Sunday	Day	Night
P8	X	5(1)	6(1)	6(1)	6(1)	5(2)	6(2)	13	12
	X	6(1)	5(1)	6(1)	6(1)	6(2)	6(2)		
	X	6(2)	6(2)	6(2)	6(2)	6(2)	6(2)		
	X	3(1)	6(1)	6(1)	6(2)	6(2)	X		
	6(1)	6(1)							
P9	2(2)	5(2)	X	5(2)	5(2)	5(2)	5(2)	12	14
	5(2)	2(2)	X	2(1)	5(1)	5(1)	2(1)		
	3(1)	2(2)	X	5(1)	5(1)	5(1)	3(1)		
	3(1)	4(2)	X	5(1)	5(1)	5(2)	5(2)		
	5(2)	5(2)							
P10	7(1)	2(1)	3(2)	5(2)	X	5(1)	5(1)	13	13
	5(1)	5(1)	5(2)	X	5(1)	5(2)	5(2)		
	5(2)	4(2)	5(2)	X	5(1)	5(1)	5(1)		
	5(2)	5(2)	1(2)	X	5(1)	5(1)	7(1)		
	5(2)	7(2)							
P11	2(1)	2(1)	2(1)	6(1)	2(1)	X	2(1)	14	12
	2(2)	2(2)	2(2)	2(2)	2(2)	X	2(2)		
	2(2)	2(2)	2(2)	2(2)	X	3(1)	1(1)		
	2(1)	2(1)	2(1)	2(2)	X	5(1)	2(1)		
	2(1)	5(2)							
P12	3(2)	X	2(1)	2(1)	2(1)	7(1)	2(2)	13	12
	2(2)	X	3(1)	3(2)	2(2)	3(2)	2(2)		
	2(2)	X	3(1)	2(1)	2(1)	3(1)	2(1)		
	4(2)	X	7(1)	2(1)	3(1)	2(2)	3(2)		
	2(2)	X							
P13	X	5(1)	3(1)	1(1)	1(1)	1(1)	1(2)	12	13
	X	1(2)	6(2)	2(2)	4(2)	6(2)	1(2)		
	X	1(1)	1(1)	1(1)	1(1)	2(2)	1(2)		
	X	3(1)	2(1)	1(2)	1(2)	1(2)	1(2)		
	X	1(1)							
P14	4(2)	3(2)	X	2(1)	3(1)	4(1)	3(1)	12	14
	4(2)	2(2)	X	2(1)	4(1)	4(1)	4(2)		
	4(2)	6(2)	X	4(1)	2(1)	4(1)	4(1)		
	2(2)	3(2)	X	2(1)	4(2)	1(2)	2(2)		
	4(2)	4(2)							
P15	5(2)	5(2)	5(2)	3(2)	3(2)	3(2)	X	12	14
	6(1)	5(1)	5(2)	5(2)	5(2)	5(2)	X		
	5(1)	1(1)	5(1)	5(1)	5(1)	X	5(1)		
	5(1)	5(2)	5(2)	5(2)	5(2)	X	5(1)		
	5(1)	5(1)							
P16	7(2)	3(2)	7(2)	X	2(1)	3(1)	7(1)	14	12
	7(1)	7(1)	7(2)	X	2(1)	4(1)	1(1)		
	7(1)	7(1)	7(1)	X	7(1)	7(1)	7(2)		
	7(2)	1(2)	X	7(1)	7(2)	7(2)	7(2)		
	7(2)	2(2)							
P17	5(1)	3(1)	5(1)	5(1)	X	5(1)	5(1)	12	14
	4(1)	5(2)	5(2)	5(2)	X	5(1)	5(1)		
	5(1)	5(1)	5(2)	5(2)	X	1(1)	5(2)		
	2(2)	5(2)	3(2)	5(2)	X	5(2)	5(2)		
	1(2)	5(2)							
P18	6(2)	6(2)	6(2)	X	6(1)	6(2)	3(2)	14	11
	6(2)	6(2)	X	6(1)	6(1)	6(1)	6(1)		
	6(1)	X	6(1)	6(1)	6(1)	4(1)	6(2)		
	6(2)	X	6(1)	6(1)	6(1)	6(1)	7(2)		
	7(2)	X							

Table A2. *Cont.*

	Monday	Tuesday	Wednesday	Thursday	Friday	Saturday	Sunday	Day	Night
P19	3(1)	7(1)	7(1)	7(1)	7(2)	7(2)	X	14	12
	7(1)	7(1)	7(1)	7(1)	7(1)	7(2)	X		
	7(2)	3(2)	7(2)	3(2)	3(2)	7(2)	X		
	7(1)	7(1)	7(1)	7(1)	7(1)	7(2)	X		
	7(2)	7(2)							
P20	6(1)	7(1)	7(1)	7(1)	7(2)	6(2)	X	14	12
	3(2)	6(2)	7(2)	6(2)	6(2)	X	7(1)		
	6(2)	6(2)	6(2)	X	6(1)	7(1)	7(1)		
	2(1)	3(1)	3(1)	X	6(1)	7(1)	7(1)		
	7(2)	7(2)							
P21	3(1)	4(1)	3(1)	4(2)	X	3(1)	2(1)	14	12
	3(1)	3(2)	3(2)	3(2)	X	3(1)	3(1)		
	3(1)	3(1)	3(2)	3(2)	X	3(2)	3(2)		
	3(2)	3(2)	3(2)	3(2)	X	3(1)	3(1)		
	3(1)	3(1)							
P22	X	5(1)	4(1)	1(1)	1(1)	7(2)	1(2)	14	11
	X	1(2)	3(2)	5(2)	1(2)	3(2)	1(2)		
	X	6(1)	2(1)	1(1)	2(2)	4(2)	3(2)		
	X	1(1)	7(1)	7(1)	3(1)	1(1)	X		
	3(1)	3(1)							
P23	4(2)	4(2)	X	4(1)	4(1)	2(1)	4(1)	12	13
	2(1)	4(2)	X	2(2)	4(2)	2(2)	4(2)		
	X	4(1)	3(1)	3(1)	4(1)	2(1)	2(2)		
	X	2(1)	4(1)	4(2)	3(2)	4(2)	3(2)		
	X	2(2)							
P24	6(1)	6(1)	X	2(1)	6(1)	6(1)	6(1)	14	11
	6(2)	X	6(1)	6(1)	6(1)	6(2)	6(2)		
	6(2)	X	4(1)	6(1)	6(1)	6(1)	6(2)		
	6(2)	X	6(1)	6(2)	6(2)	6(2)	6(2)		
	6(2)	X							
P25	7(2)	7(2)	7(2)	7(2)	X	7(1)	7(1)	13	13
	7(1)	7(1)	4(1)	7(1)	X	7(2)	7(2)		
	7(2)	7(2)	7(2)	7(2)	X	3(1)	7(1)		
	3(2)	7(2)	3(2)	X	7(1)	7(1)	2(1)		
	7(1)	7(1)							
P26	2(2)	X	2(1)	2(2)	2(2)	2(2)	2(2)	12	13
	2(2)	X	4(1)	7(1)	2(1)	2(1)	5(1)		
	4(1)	X	1(1)	5(1)	1(1)	3(2)	7(2)		
	7(2)	X	5(1)	2(1)	2(2)	2(2)	2(2)		
	2(2)	X							
P27	3(2)	X	X	3(1)	3(1)	6(1)	3(1)	11	14
	3(2)	3(2)	X	4(1)	6(2)	3(2)	3(2)		
	3(2)	3(2)	X	1(1)	3(1)	2(1)	1(1)		
	3(1)	5(1)	X	3(2)	3(2)	3(2)	3(2)		
	3(2)	3(2)							
P28	5(1)	5(1)	5(1)	5(1)	5(1)	X	4(2)	14	12
	4(2)	5(2)	5(2)	7(2)	5(2)	X	4(1)		
	4(1)	7(1)	5(2)	5(2)	5(2)	X	5(1)		
	1(1)	5(1)	5(2)	7(2)	5(2)	X	5(1)		
	1(1)	7(1)							
P29	6(1)	2(1)	6(1)	6(1)	3(1)	5(1)	X	13	13
	6(2)	6(2)	6(2)	6(2)	4(2)	6(2)	X		
	6(2)	6(2)	6(2)	6(2)	6(2)	6(2)	X		
	6(1)	6(1)	6(1)	6(1)	6(2)	X	4(1)		
	4(1)	6(1)							

Table A2. *Cont.*

	Monday	Tuesday	Wednesday	Thursday	Friday	Saturday	Sunday	Day	Night
P30	3(1)	3(1)	X	4(1)	7(1)	3(2)	3(2)	11	14
	7(2)	3(2)	X	1(1)	4(1)	3(1)	3(2)		
	3(2)	3(2)	X	7(1)	3(1)	7(2)	7(2)		
	3(2)	4(2)	X	3(1)	3(1)	6(2)	3(2)		
	3(2)	X							
P31	5(2)	5(2)	5(2)	5(2)	5(2)	X	5(1)	12	14
	3(1)	5(1)	4(2)	5(2)	5(2)	X	5(1)		
	5(1)	5(1)	5(1)	5(1)	5(1)	X	5(1)		
	5(1)	5(1)	6(2)	5(2)	X	3(2)	5(2)		
	3(2)	1(2)							
P32	1(2)	1(2)	1(2)	7(2)	1(2)	X	1(1)	12	14
	1(1)	1(1)	1(1)	X	4(1)	5(1)	7(1)		
	1(1)	7(1)	1(1)	X	7(1)	1(1)	6(2)		
	1(2)	1(2)	7(2)	X	7(2)	7(2)	7(2)		
	4(2)	7(2)							
P33	5(2)	6(2)	6(2)	X	6(2)	6(2)	2(2)	14	12
	6(2)	6(2)	6(2)	X	6(1)	6(1)	6(1)		
	6(1)	3(2)	6(2)	X	6(1)	6(1)	6(1)		
	6(1)	6(1)	6(2)	X	6(1)	4(1)	6(1)		
	6(1)	6(1)							
P34	6(2)	7(2)	X	3(1)	3(2)	3(2)	6(2)	12	13
	1(2)	3(2)	X	6(1)	5(1)	6(1)	3(1)		
	3(1)	2(1)	X	6(1)	6(1)	6(1)	3(1)		
	3(1)	X	3(2)	6(2)	3(2)	3(2)	4(2)		
	4(2)	X							
P35	7(2)	5(2)	7(2)	2(2)	7(2)	7(2)	X	14	12
	7(1)	7(1)	7(1)	7(1)	7(1)	7(2)	X		
	7(1)	5(1)	7(1)	5(1)	7(2)	7(2)	X		
	7(1)	7(1)	7(1)	7(2)	7(2)	5(2)	X		
	7(1)	7(1)							
P36	6(1)	6(1)	6(2)	6(2)	6(2)	X	6(1)	13	13
	6(1)	6(1)	6(1)	6(1)	7(1)	X	6(1)		
	6(1)	6(1)	2(2)	6(2)	1(2)	X	6(1)		
	2(2)	6(2)	1(2)	6(2)	1(2)	X	6(1)		
	6(2)	6(2)							
P37	X	7(1)	5(1)	4(2)	4(2)	4(2)	2(2)	11	14
	X	4(1)	7(1)	5(2)	7(2)	4(2)	X		
	2(1)	4(2)	7(2)	2(2)	2(2)	7(2)	X		
	7(1)	2(1)	2(1)	4(1)	4(1)	6(1)	X		
	2(2)	4(2)							
P38	5(1)	1(1)	1(1)	1(2)	X	1(1)	6(1)	13	13
	1(1)	1(1)	1(2)	1(2)	X	1(1)	1(1)		
	1(1)	1(2)	4(2)	2(2)	X	4(1)	1(1)		
	1(1)	1(2)	7(2)	1(2)	X	4(2)	1(2)		
	6(2)	1(2)							
P39	6(2)	7(2)	X	5(1)	6(1)	3(1)	2(1)	12	13
	2(1)	X	2(1)	2(2)	2(2)	5(2)	5(2)		
	2(2)	X	2(1)	2(1)	5(1)	5(1)	3(1)		
	X	2(2)	2(2)	2(2)	5(2)	5(2)	2(2)		
	X	5(1)							
P40	7(1)	2(2)	2(2)	7(2)	7(2)	X	4(1)	13	13
	5(1)	5(1)	7(1)	7(2)	7(2)	X	7(1)		
	7(1)	7(2)	3(2)	7(2)	7(2)	X	7(1)		
	4(1)	7(2)	7(2)	7(2)	X	3(1)	7(1)		
	7(1)	7(1)							

Table A2. *Cont.*

	Monday	Tuesday	Wednesday	Thursday	Friday	Saturday	Sunday	Day	Night
P41	6(1)	6(1)	6(1)	6(1)	6(1)	X	6(1)	14	12
	6(1)	6(1)	6(1)	6(2)	3(2)	X	6(1)		
	6(1)	6(2)	6(2)	6(2)	6(2)	X	6(2)		
	6(2)	6(2)	2(2)	6(2)	6(2)	X	6(1)		
	6(1)	6(1)							
P42	4(2)	3(2)	3(2)	X	4(1)	2(1)	4(2)	12	13
	4(2)	3(2)	X	5(1)	3(1)	4(1)	4(1)		
	4(2)	4(2)	X	3(2)	4(2)	5(2)	3(2)		
	4(2)	X	5(1)	4(1)	2(1)	4(1)	4(1)		
	3(1)	X							
P43	2(2)	6(2)	2(2)	X	7(1)	2(1)	7(1)	12	14
	2(1)	2(1)	6(2)	X	6(1)	2(1)	7(2)		
	2(2)	2(2)	2(2)	X	7(1)	7(1)	5(2)		
	7(2)	2(2)	2(2)	X	6(1)	2(1)	6(1)		
	6(2)	2(2)							
P44	6(1)	2(1)	3(1)	6(2)	X	6(1)	3(1)	14	12
	3(1)	3(1)	3(1)	5(2)	X	2(1)	2(2)		
	3(2)	2(2)	6(2)	3(2)	X	6(1)	6(1)		
	6(1)	2(1)	3(1)	2(2)	X	2(2)	6(2)		
	3(2)	3(2)							
P45	6(2)	6(2)	X	3(1)	6(1)	6(1)	6(1)	13	13
	6(1)	6(2)	X	6(1)	3(2)	6(2)	6(2)		
	6(2)	6(2)	X	6(1)	2(1)	6(1)	6(1)		
	6(2)	6(2)	X	6(1)	6(1)	6(1)	6(2)		
	6(2)	6(2)							
P46	3(1)	3(1)	3(1)	3(1)	3(2)	X	6(1)	13	13
	3(2)	5(2)	3(2)	3(2)	3(2)	X	3(1)		
	2(1)	1(1)	3(1)	3(1)	3(2)	X	3(1)		
	5(2)	3(2)	5(2)	3(2)	2(2)	X	3(1)		
	5(1)	3(2)							
P47	6(2)	6(2)	6(2)	6(2)	X	6(1)	6(1)	14	12
	6(1)	3(1)	6(1)	2(1)	X	6(1)	6(2)		
	6(2)	6(2)	6(2)	6(2)	X	6(1)	5(1)		
	6(1)	6(1)	6(1)	6(1)	X	6(1)	6(2)		
	6(2)	6(2)							
P48	5(2)	6(2)	5(2)	3(2)	5(2)	X	5(1)	12	14
	5(2)	5(2)	4(2)	4(2)	X	5(1)	5(1)		
	5(1)	5(1)	5(1)	5(2)	X	5(1)	4(1)		
	5(1)	5(1)	5(2)	5(2)	X	2(1)	5(1)		
	5(2)	5(2)							
P49	4(2)	4(2)	4(2)	2(2)	4(2)	4(2)	X	13	13
	6(1)	4(1)	4(1)	4(1)	5(1)	2(1)	X		
	4(1)	4(1)	4(2)	4(2)	5(2)	2(2)	X		
	4(1)	4(1)	4(2)	5(2)	4(2)	X	2(1)		
	3(1)	5(1)							
P50	X	6(1)	6(2)	6(2)	6(2)	6(2)	6(2)	11	14
	X	6(1)	2(2)	6(2)	6(2)	6(2)	6(2)		
	X	3(1)	6(1)	6(1)	6(2)	6(2)	X		
	6(1)	6(1)	6(1)	6(1)	6(2)	4(2)	X		
	1(1)	6(1)							
P51	X	4(2)	5(2)	6(2)	4(2)	4(2)	1(2)	14	10
	X	6(1)	4(1)	4(1)	4(1)	4(2)	4(2)		
	X	2(1)	4(1)	4(2)	4(2)	X	2(1)		
	2(1)	4(1)	4(1)	4(1)	4(1)	X	X		
	4(1)	4(1)							

Table A2. *Cont.*

	Monday	Tuesday	Wednesday	Thursday	Friday	Saturday	Sunday	Day	Night
P52	2(1)	4(1)	6(1)	X	6(2)	2(2)	6(2)	14	12
	2(2)	2(2)	6(2)	X	6(1)	6(1)	2(1)		
	2(1)	2(1)	6(1)	X	6(1)	6(2)	2(2)		
	6(2)	2(2)	2(2)	X	2(1)	2(1)	2(1)		
	6(1)	2(2)							
P53	6(2)	X	6(1)	6(1)	1(1)	6(1)	3(1)	13	12
	6(2)	X	6(2)	6(2)	6(2)	6(2)	6(2)		
	X	5(1)	6(1)	6(1)	4(2)	6(2)	6(2)		
	X	6(1)	5(1)	6(1)	6(1)	6(1)	6(2)		
	X	6(2)							
P54	2(1)	5(1)	5(1)	5(1)	5(2)	5(2)	X	14	12
	5(1)	5(1)	5(1)	5(1)	5(2)	5(2)	X		
	1(1)	5(2)	5(2)	5(2)	5(2)	5(2)	X		
	5(1)	5(1)	X	5(1)	5(1)	5(1)	5(2)		
	5(2)	5(2)							
P55	5(1)	7(1)	7(1)	7(2)	X	4(1)	7(1)	13	13
	7(1)	7(1)	5(1)	7(2)	X	7(1)	7(1)		
	7(2)	5(2)	7(2)	7(2)	X	7(1)	1(2)		
	1(2)	3(2)	4(2)	X	6(1)	5(1)	7(2)		
	7(2)	7(2)							
P56	4(1)	6(1)	4(2)	4(2)	X	7(1)	4(1)	12	14
	4(1)	4(2)	4(2)	4(2)	X	6(1)	4(1)		
	4(2)	4(2)	2(2)	X	4(1)	2(1)	4(1)		
	4(1)	4(1)	X	4(2)	4(2)	7(2)	4(2)		
	4(2)	5(2)							
P57	3(2)	3(2)	4(2)	2(2)	X	3(2)	3(2)	11	14
	5(2)	X	3(1)	3(1)	2(1)	3(1)	5(1)		
	6(2)	X	5(1)	3(1)	3(2)	3(2)	3(2)		
	X	6(1)	3(1)	3(1)	2(2)	3(2)	5(2)		
	X	2(1)							
P58	3(2)	1(2)	2(2)	X	3(2)	2(2)	5(2)	14	12
	3(2)	5(2)	2(2)	X	3(1)	7(1)	2(1)		
	2(1)	3(1)	3(2)	X	3(1)	3(1)	7(1)		
	2(1)	7(1)	5(2)	X	3(1)	3(1)	1(1)		
	1(1)	3(2)							
P59	1(1)	1(1)	2(1)	1(1)	5(1)	X	5(2)	11	13
	1(2)	X	1(2)	1(2)	1(2)	1(2)	1(2)		
	1(2)	X	1(2)	1(2)	1(2)	1(2)	5(2)		
	1(2)	X	1(1)	1(1)	1(1)	1(1)	1(1)		
	X	2(1)							
P60	7(1)	7(1)	7(1)	7(1)	7(1)	7(1)	X	12	13
	7(2)	7(2)	7(2)	7(2)	X	1(1)	X		
	7(1)	7(1)	7(2)	7(2)	7(2)	X	7(2)		
	7(2)	7(2)	7(2)	7(2)	7(2)	X	7(1)		
	7(1)	1(1)							
P61	1(2)	X	1(1)	1(2)	1(2)	1(2)	6(2)	12	13
	1(2)	X	1(1)	1(1)	1(1)	X	1(1)		
	1(2)	1(2)	1(2)	1(2)	X	1(1)	5(1)		
	5(1)	1(1)	1(1)	X	1(1)	1(1)	6(2)		
	1(2)	1(2)							
P62	4(1)	5(1)	5(1)	X	5(1)	5(1)	5(1)	12	14
	5(1)	5(1)	5(1)	X	5(1)	6(1)	5(1)		
	5(2)	5(2)	5(2)	X	5(2)	5(2)	5(2)		
	5(2)	5(2)	5(2)	X	5(2)	5(2)	5(2)		
	5(2)	6(2)							

Table A2. *Cont.*

	Monday	Tuesday	Wednesday	Thursday	Friday	Saturday	Sunday	Day	Night
P63	4(1)	X	4(1)	4(1)	4(1)	1(1)	4(1)	13	12
	4(1)	X	5(2)	4(2)	4(2)	4(2)	5(2)		
	4(2)	X	2(1)	6(1)	4(1)	4(2)	4(2)		
	4(2)	X	4(1)	4(1)	4(1)	4(2)	4(2)		
	X	4(2)							
P64	6(2)	X	6(1)	7(1)	6(1)	6(2)	7(2)	13	12
	6(2)	X	6(1)	6(1)	1(1)	3(1)	6(2)		
	6(2)	X	6(2)	6(2)	6(2)	6(2)	6(2)		
	6(2)	X	6(1)	6(1)	5(1)	6(1)	1(1)		
	6(1)	X							
P65	7(1)	X	7(1)	3(2)	7(2)	7(2)	7(2)	11	14
	7(2)	X	7(1)	3(2)	7(2)	7(2)	6(2)		
	5(2)	X	7(1)	7(1)	7(2)	6(2)	7(2)		
	X	7(1)	7(1)	7(1)	7(1)	7(1)	7(2)		
	X	7(1)							
P66	1(1)	3(1)	3(2)	X	3(1)	3(1)	2(1)	14	11
	3(1)	3(1)	3(2)	X	3(2)	2(2)	3(2)		
	3(2)	6(2)	X	3(1)	3(2)	3(2)	4(2)		
	7(2)	X	2(1)	3(1)	6(1)	6(1)	3(1)		
	2(1)	X							
P67	1(2)	X	6(1)	6(1)	6(1)	6(1)	1(1)	14	11
	X	6(1)	6(1)	3(1)	6(2)	6(2)	X		
	6(1)	6(1)	6(1)	6(2)	6(2)	6(2)	X		
	6(1)	6(2)	6(2)	6(2)	6(2)	6(2)	X		
	6(1)	6(1)							
P68	2(2)	7(2)	X	5(1)	5(1)	5(1)	7(2)	14	12
	5(2)	5(2)	X	5(1)	5(2)	3(2)	5(2)		
	5(2)	5(2)	X	5(1)	7(1)	5(1)	2(1)		
	5(2)	5(2)	X	5(1)	5(1)	5(1)	5(1)		
	5(1)	5(1)							
P69	6(1)	6(1)	X	6(1)	6(2)	6(2)	6(2)	11	14
	6(2)	1(2)	X	3(1)	6(1)	4(1)	6(1)		
	6(2)	X	4(1)	4(1)	6(1)	6(1)	6(2)		
	6(2)	X	6(2)	6(2)	6(2)	6(2)	4(2)		
	6(2)	X							
P70	5(1)	6(2)	5(2)	X	5(1)	5(1)	5(2)	13	13
	5(2)	6(2)	1(2)	X	6(1)	5(1)	5(2)		
	5(2)	5(2)	5(2)	X	5(1)	5(1)	5(2)		
	5(2)	5(2)	X	5(1)	5(1)	4(1)	6(1)		
	5(1)	5(1)							
P71	7(1)	1(1)	7(2)	X	7(1)	6(1)	5(1)	12	14
	1(1)	7(2)	7(2)	X	7(1)	1(2)	7(2)		
	1(2)	1(2)	X	7(1)	7(1)	4(1)	6(1)		
	6(1)	7(2)	X	1(2)	1(2)	5(2)	6(2)		
	1(2)	6(2)							
P72	5(2)	2(2)	6(2)	3(2)	4(2)	5(2)	X	13	13
	5(1)	2(1)	6(1)	5(1)	5(1)	5(2)	X		
	5(1)	6(1)	5(1)	5(2)	5(2)	5(2)	X		
	1(1)	6(1)	5(1)	5(2)	5(2)	6(2)	X		
	5(1)	5(1)							
P73	2(1)	1(2)	2(2)	X	2(1)	2(1)	7(1)	12	14
	2(1)	6(2)	2(2)	X	2(2)	7(2)	7(2)		
	7(2)	7(2)	X	4(1)	2(2)	2(2)	2(2)		
	2(2)	6(2)	X	7(1)	2(1)	2(1)	6(1)		
	2(1)	6(1)							

Table A2. *Cont.*

	Monday	Tuesday	Wednesday	Thursday	Friday	Saturday	Sunday	Day	Night
P74	6(1)	6(1)	6(1)	6(1)	5(1)	X	6(2)	12	14
	6(2)	6(2)	6(2)	6(2)	6(2)	X	6(1)		
	6(1)	6(1)	6(1)	6(1)	X	6(1)	6(1)		
	3(2)	6(2)	6(2)	6(2)	X	6(2)	6(2)		
	6(2)	6(2)							
P75	5(2)	X	5(1)	5(1)	5(2)	5(2)	5(2)	11	13
	5(2)	X	5(1)	5(1)	3(1)	5(1)	5(2)		
	5(2)	X	1(2)	5(2)	5(2)	5(2)	5(2)		
	5(2)	X	5(1)	5(1)	7(1)	5(1)	X		
	5(1)	X							
P76	X	6(1)	6(2)	6(2)	6(2)	6(2)	X	14	11
	6(1)	6(1)	6(1)	6(2)	7(2)	X	6(1)		
	6(1)	6(1)	7(1)	7(1)	6(1)	X	6(1)		
	6(2)	6(2)	6(2)	6(2)	6(2)	X	6(1)		
	6(1)	6(1)							
P77	7(2)	X	7(2)	7(2)	X	6(2)	7(2)	12	13
	7(2)	7(2)	7(2)	7(2)	X	7(1)	7(1)		
	7(1)	6(1)	7(1)	X	3(1)	7(1)	7(1)		
	7(1)	7(2)	6(2)	X	7(1)	7(1)	7(1)		
	2(2)	4(2)							
P78	X	4(2)	4(2)	4(2)	2(2)	4(2)	X	14	11
	4(1)	4(1)	3(1)	4(1)	6(2)	4(2)	X		
	4(1)	4(1)	4(1)	4(1)	4(1)	X	4(1)		
	4(1)	4(2)	4(2)	4(2)	4(2)	X	4(1)		
	4(1)	4(1)							
P79	7(2)	7(2)	X	7(1)	7(1)	7(1)	1(1)	11	14
	7(2)	7(2)	X	7(1)	7(1)	7(1)	7(2)		
	7(2)	7(2)	X	7(2)	7(2)	7(2)	X		
	7(1)	7(1)	4(2)	7(2)	7(2)	7(2)	X		
	7(1)	2(1)							
P80	6(2)	6(2)	6(2)	6(2)	6(2)	X	6(1)	12	14
	6(1)	6(1)	6(2)	6(2)	6(2)	X	6(1)		
	3(1)	6(1)	6(1)	6(2)	6(2)	X	6(1)		
	6(1)	6(2)	6(2)	3(2)	X	6(1)	6(1)		
	6(1)	6(2)							

Notes: *l(k): l*: shift; *k*: section assigned personnel.

References

1. Eren, T.; Varli, E. Shift Scheduling Problems and A Case Study. *Int. J. Inform. Technol.* **2017**, *10*, 185–197.
2. Sungur, B. Integer Programming Model for Fuzzy Shift Scheduling Problems. *Erciyes Univ. J. Fac. Econ. Adm. Sci.* **2008**, *30*, 211–227.
3. Varli, E. Solution of the Shift Scheduling Problem for the Foremans in the Manufacturing Sector with Ahp-Goal Programming. Master's Thesis, Kirikkale University, Kirikkale, Turkey, 2017.
4. Özcan, E.C.; Varli, E.; Eren, T. Goal Programming Approach for Shift Scheduling Problems in Hydroelectric Power Plants. *J. Inf. Technol.* **2017**, *10*, 363–370.
5. Güngör, İ. A Model Proposal for Nursing Assignment and Scheduling. *Süleyman Demirel Univ. J. Fac. Econ. Adm. Sci.* **2002**, *7*, 77–94.
6. Bard, J.F.; Binici, C. Staff Scheduling at the United States Postal Service. *Comput. Oper. Res.* **2003**, *30*, 745–771. [CrossRef]
7. Wong, G.Y.C.; Chun, A.H.W. Constraint-Based Rostering Using Meta-Level Reasoning and Probability-Based Ordering. *Eng. Appl. Artif. Intell.* **2004**, *17*, 599–610. [CrossRef]

8. Azaiez, M.N.; Al Sharif, S.S. A 0-1 Goal Programming Model for Nurse Scheduling. *Comput. Oper. Res.* **2005**, *32*, 491–507. [CrossRef]
9. Seçkiner, S.U.; Gökçen, H.; Kurt, M. An Integer Programming Model for Hierarchical Workforce Scheduling Problem. *Eur. J. Oper. Res.* **2007**, *183*, 694–699. [CrossRef]
10. Castillo, I.; Joro, T.; Li, Y.Y. Workforce Scheduling with Multiple Objectives. *Eur. J. Oper. Res.* **2009**, *196*, 162–170. [CrossRef]
11. Olive, J.T. A Proposed Mathematical Model for the Personnel Scheduling Problem in a Manufacturing Company. Master's Thesis, Istanbul Technical University, İstanbul, Turkey, 2009.
12. Topaloğlu, S. A Shift Scheduling Model for Employees with Different Seniority Levels and an Application in Healthcare. *Eur. J. Oper. Res.* **2009**, *198*, 943–957. [CrossRef]
13. Heimerl, C.; Kolisch, R. Scheduling and Staffing Multiple Projects with a Multi-Skilled Workforce. *OR Spectr.* **2010**, *32*, 343–368. [CrossRef]
14. Karaatli, M. Multi Purpose Manpower Scheduling in Fuzzy Environment: An Application for Nurses. Ph.D. Thesis, Süleyman Demirel University, Isparta, Turkey, 2010.
15. Koruca, H.İ. Development of A Simulation-Based Shift Planning Module. *J. Fac. Eng. Archit. Gazi Univ.* **2010**, *25*, 469–482.
16. Atmaca, E.; Pehlivan, C.; Aydoğdu, B.; Yakici, M. Nurse Scheduling Problem and Application. *Erciyes Univ. J. Inst. Sci. Technol.* **2012**, *28*, 351–358.
17. Bağ, N.; Özdemir, N.M.; Eren, T. 0-1 Goal Programming and ANP Method and Nurse Scheduling Problem Solution. *Int. J. Eng. Res. Dev.* **2012**, *4*, 2–6.
18. Bektur, G.; Hasgül, S. Workforce Scheduling Problem According to Seniority Levels: An Application For A Service System. *Afyon Kocatepe Univ. J. Fac. Econ. Adm. Sci.* **2013**, *15*, 385–402.
19. Çöl, G. Goal Programming Approach to Workforce Scheduling Problem with Different Seniority Levels. Master's Thesis, Eskişehir Osmangazi University, Eskişehir, Turkey, 2013.
20. Ünal, F.M.; Eren, T. The Solution of Shift Scheduling Problem by Using Goal Programming. *Acad. Platf. J. Eng. Sci.* **2016**, *4*, 28–37.
21. Özder, E.H.; Varli, E.; Eren, T. A Model Suggestion for Cleaning Staff Scheduling Problem with Goal Programming Approach. *Karadeniz J. Sci.* **2017**, *7*, 114–127.
22. Varli, E.; Alağaş, H.M.; Eren, T.; Özder, E.H. Goal Programming Solution of the Examiner Assignment Problem. *Bilge Int. J. Sci. Technol. Res.* **2017**, *1*, 105–118.
23. Varli, E.; Ergişi, B.; Eren, T. Nurse Scheduling Problem with Special Constraints: Goal Programming Approach. *Erciyes Univ. J. Fac. Econ. Adm. Sci.* **2017**, *49*, 189–206.
24. Bedir, N. Shift Scheduling Problems Solution with Combined Ahppromethee and Goal Programming Methods: An Aplication in Hydroelectric Power Plant. Master's Thesis, Kirikkale University, Kirikkale, Turkey, 2018.
25. Gür, Ş.; Eren, T. Scheduling and Planning in Service Systems with Goal Programming: Literature Review. *Mathematics* **2018**, *6*, 265. [CrossRef]
26. Koç, E. An Exact Approach for a Dynamic Workforce Scheduling Problem. Master's Thesis, Özyeğin University, İstanbul, Turkey, 2018.
27. Koçtepe, S.; Bedir, N.; Eren, T.; Gür, Ş. Solution of Personnel Scheduling Problem for Organization Officers With 0-1 Integer Programming. *J. Econ. Bus. Manag.* **2018**, *2*, 25–46.
28. Tapkan, P.Z.; Özbakir, L.; Kulluk, S.; Telcioğlu, B. Modelling and solving railway crew rostering problem. *J. Fac. Eng. Archit. Gazi Univ.* **2018**, *33*, 953–965.
29. Özder, E.H.; Özcan, E.; Eren, T. Staff Task-Based Shift Scheduling Solution with an ANP and Goal Programming Method in a Natural Gas Combined Cycle Power Plant. *Mathematics* **2019**, *7*, 192. [CrossRef]

12

Modeling Heavy Metal Sorption and Interaction in a Multispecies Biofilm

Berardino D'Acunto, Luigi Frunzo, Vincenzo Luongo * and Maria Rosaria Mattei

Department of Mathematics and Applications, University of Naples Federico II, 80126 Naples, Italy
* Correspondence: vincenzo.luongo@unina.it.

Abstract: A mathematical model able to simulate the physical, chemical and biological interactions prevailing in multispecies biofilms in the presence of a toxic heavy metal is presented. The free boundary value problem related to biofilm growth and evolution is governed by a nonlinear ordinary differential equation. The problem requires the integration of a system of nonlinear hyperbolic partial differential equations describing the biofilm components evolution, and a systems of semilinear parabolic partial differential equations accounting for substrates diffusion and reaction within the biofilm. In addition, a semilinear parabolic partial differential equation is introduced to describe heavy metal diffusion and sorption. The biosoption process modeling is completed by the definition and integration of other two systems of nonlinear hyperbolic partial differential equations describing the free and occupied binding sites evolution, respectively. Numerical simulations of the heterotrophic-autotrophic interaction occurring in biofilm reactors devoted to wastewater treatment are presented. The high biosorption ability of bacteria living in a mature biofilm is highlighted, as well as the toxicity effect of heavy metals on autotrophic bacteria, whose growth directly affects the nitrification performance of bioreactors.

Keywords: multispecies biofilm; biosorption; free boundary value problem; heavy metals toxicity; method of characteristics

---

1. Introduction

Most of the living microbial communities organize themselves in complex structures where the interaction between different species leads to advantageous environmental conditions for their growth [1,2]. These structures, known as multispecies biofilms, include different components, such as living cells, inert materials and extracellular polymeric substances (EPS) [3–6]. Their structural organization confers to these biological systems enhanced mechanical characteristics and adaptive features to many environmental conditions [7–9]. For instance, the protective self-secreted EPS matrix can strongly affect the dynamics of substances within the biofilm and it can also serve as a source of nutrients for bacteria [10,11].

These aspects are highly relevant in many applications as biofilms result in being more resistant than individual planktonic cells to toxic substances such as heavy metals, antibiotics, chlorine and detergents, due to the presence of natural diffusion barriers [12]. In recent years, biofilms have been widely used as biosorption technologies for metal immobilization and sequestration [13,14]. Biosorption is a combination of complex phenomena leading to the entrapment of a substance onto the surface of a living/dead organism or EPS. The mechanisms involved (complexation, precipitation, ion exchange, adsorption) are strongly affected by several biotic and abiotic parameters, such as pH, temperature, binding site density and affinity, which in turn influence the biosorption efficiency. Significant applications of biofilm technology to biosorption have been presented in the field of groundwater purification and mining industry wastewater treatment.

The sorption properties of various components constituting a biofilm (i.e., microorganisms, EPS and inert materials) depend on the different affinity of each specific component to heavy metals. It is known, for instance, that the cell membrane of many microorganisms allows for heavy metals accumulation due to the presence of surface functional groups [15]. These act as binding agents removing heavy metals during biofilm growth. On the other hand, heavy metals can be highly toxic compounds for a wide range of bacteria, i.e., autotrophic microorganisms, as they can act as inhibiting agent when significant metal concentrations are reached in bioreactors [16–19].

Many experimental studies demonstrated the possibility of using bacteria to govern heavy metal mobility in different aquatic ecosystems [20–22], but additional efforts are still required to completely understand the complex dynamics and interactions occurring between biofilms and heavy metals. In this context, mathematical modeling represents an appropriate tool to provide basic information on specific biosorption phenomena and stimulate further research on the multiplicity of mechanisms regulating biosorption process by biofilms [2]. For instance, multidimensional models can be implemented for specific applications when micro-scale outputs are required [7]. The spatial distribution of diffusing compounds and microbial species within the biofilm, and the physical structure of the biofilm at a micro-scale level can be investigated by using complex 2D and 3D mathematical models [23–26]. If a macro-scale output is required, as in the case of engineering biofilm reactors, 1D formulations have been recognized as efficient tools to analyze bioreactor performances in terms of biomass accumulation and degradation of substrates [27].

To this aim, a 1D mathematical model reproducing a biosorption phenomenon occurring in a typical biofilm reactor devoted to wastewater treatment has been proposed. The model is presented in its general form and then applied to a relevant case in wastewater treatment field. Specifically, the case study accounts for the coexistence of two different microbial species performing nitrification and organic carbon degradation. A continuum approach was used to describe biomass growth and decay within the biofilm [28]. The model accounts for the diffusion–reaction of substrates and the diffusion–biosorption of heavy metals within the biofilm [29,30]. More precisely, in this work, heterotrofic bacteria have been characterized by a high specific number of binding sites on their cell wall allowing heavy metal sequestration during biofilm evolution. On the other hand, the kinetics of autotrophic bacteria, which are usually more sensitive to toxic compounds then heterotrophic species, have been supposed to be negatively affected by the heavy metal concentration, which acts as an inhibiting agent and affects the efficiency of the nitrification process.

The main objective is to apply the knowledge of recently introduced mathematical approaches for biosorption in multispecies biofilms to highlight the effects of heavy metals in a traditional biofilm system for wastewater treatment. The work elucidates different ecological aspects of biofilms/heavy metals interaction, such as spatial distribution of biofilm components over time, substrate and heavy metals dynamics, and effects of heavy metals contamination. Numerical simulations remarked on the consistency of the model and showed the effect of toxic heavy metals on different microbial species coexisting in a multispecies biofilm.

2. Statement of the Problem

The effect of an inhibiting agent diffusing in a multispecies biofilm and the related biosorption interactions are discussed in the following sections. The specific case study concerns the competition for oxygen of heterotrophic and autotrophic microorganisms performing organic carbon degradation and nitrification, respectively. This is a typical situation occurring in the biological treatment units of municipal wastewater treatment plants.

According to [29], the biofilm dynamics were modeled as a free boundary problem essentially hyperbolic, where the free boundary is represented by the biofilm thickness. Its evolution is dictated by the growth of the microbial species constituting the biofilm $X_i(z,t)$ and the exchange fluxes between the biofilm and the bulk liquid. The biofilm growth is catalyzed by the availability of substrates $S_j(z,t)$, which diffuse from the bulk liquid into the biofilm where they are consumed by microbial species.

The model considers biofilm as constituted of four different components $X_i, i = 1, ..., 4$ (green and gray in Figure 1), which can accumulate/growth and decrease/decay during time. The biofilm growth and development is governed by the availability of substrates $S_j, j = 1, ..., 3$ (blue in Figure 1) within the biofilm, which regulate the microbial metabolism and interactions. These components include heterotrophic bacteria $X_1 = \rho_1 f_1$, autotrophic bacteria $X_2 = \rho_2 f_2$, inert material $X_3 = \rho_3 f_3$ and EPS $X_4 = \rho_4 f_4$, with f_i denoting the volume fraction of each biofilm component i and ρ_i the corresponding density. Specifically, EPS production was taken into account according to the approach proposed by [31]. Three different substrates, ammonium S_1, organic carbon S_2, and oxygen S_3 were taken into account as they are involved in metabolic pathways. The active microbial biomasses X_1 and X_2 naturally decrease via respiration and decay processes, producing residual inert biomass X_3. Contextually, they produce EPS as a metabolic byproduct during their growth. The autotrophs X_2 are nitrifying bacteria that grow by consuming ammonium S_1 and oxygen S_3. On the other hand, the heterotrophic bacteria simultaneously uptake organic carbon S_2 and oxygen S_3 for their growth. The two species compete for space and oxygen in multispecies biofilms [28].

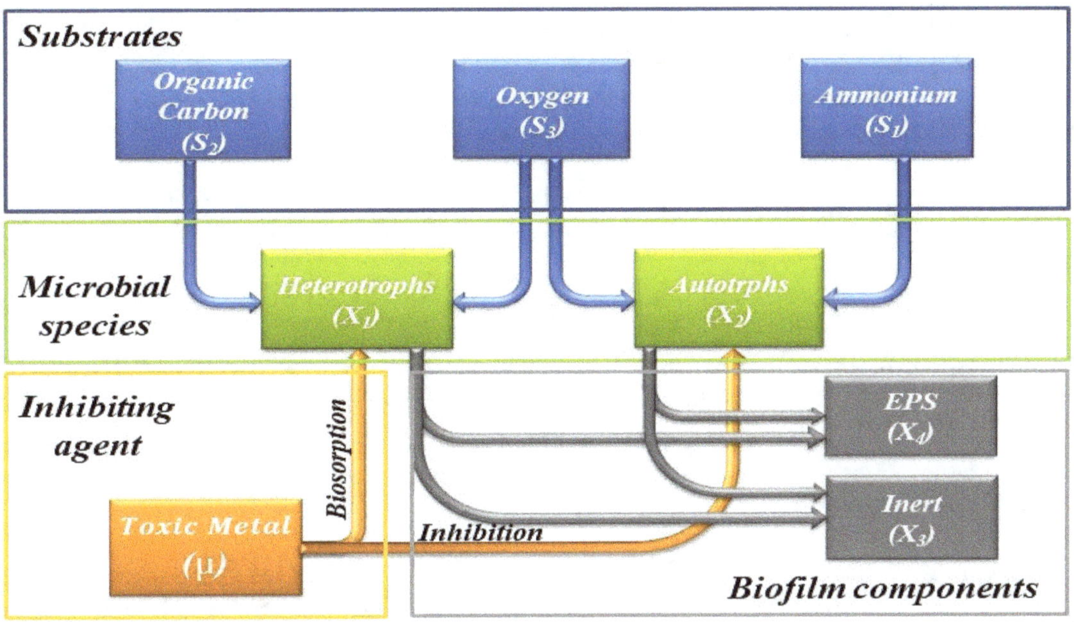

Figure 1. Schematic representation of the kinetic process.

The inhibiting agent μ (orange in Figure 1) was assumed to interact with the biofilm in two different ways: it can adsorb on a specific biofilm component, e.g., the heterotrophic biomass X_1, and act as inhibiting agent for an active microbial biomass, e.g., the autotrophic bacteria X_2. Note that a single inhibiting agent μ was considered in this work, but, in more complex cases, the effect of different heavy metals $\mu_k, k = 1, ..., l$ can be taken into account by using a similar approach. The concentration of heavy metals in biofilm reactors negatively affects the kinetics of autotrophic bacteria, which are typically more sensitive to contamination than heterotrophic species. Consequently, a specific inhibition term was exclusively introduced in the autotrophic growth rate function. The sorption phenomenon was modeled by directly taking into account the dynamics of the binding sites of the biofilm matrix.

The biofilm growth is governed by the following equations:

$$\frac{\partial X_i}{\partial t} + \frac{\partial}{\partial z}(uX_i) = \rho_i r_{M,i}(z, t, \mu, \mathbf{X}, \mathbf{S}),\ i = 1, ..., 4,\ 0 \le z \le L(t),\ t > 0,$$
$$X_i(z, 0) = X_{i0}(z),\ i = 1, ..., 4,\ 0 \le z \le L_0, \tag{1}$$

$$\frac{\partial u}{\partial z} = \sum_{i=1}^{4} r_{M,i}(z,t,\mu,\mathbf{X},\mathbf{S}),\ 0 < z \leq L(t),\ t \geq 0,\ u(0,t) = 0,\ t \geq 0, \tag{2}$$

$$\dot{L}(t) = u(L(t),t) + \sigma_a(t) - \sigma_d(L(t)),\ t > 0,\ L(0) = L_0, \tag{3}$$

where $X_i = \rho_i f_i(z,t)$ denotes the concentration of the four biofilm components considered, ρ_i is the constant density, $u(z,t)$ is the velocity of microbial mass displacement with respect to the biofilm substratum, $r_{M,i}(z,t,\mu,\mathbf{X},\mathbf{S})$ is the biomass growth rate, $L(t)$ is the biofilm thickness, $\mathbf{X} = (X_1, X_2, X_3, X_4)$, and $\mathbf{S} = (S_1, S_2, S_3)$. Equation (1) is derived from local mass balance considerations and governs the growth of the microbial species constituting the biofilm. The biomass expansion is modelled as an advective flux and depends on the metabolic reactions carried out by the microbial species. The reaction terms $r_{M,i}$ account for the microbial growth and decay, and EPS production. Equation (2) governs the biomass growth velocity; it is obtained summing over i Equation (1) and considering the constrain $\sum_{i=1}^{n} f_i = 1$. The biofilm thickness evolution is ruled by an ordinary differential equation (Equation (3)) that is derived from global mass balance considerations and depends on both the biomass growth velocity $u(L(t),t)$ and the detachment $\sigma_d(L(t))$ and attachment $\sigma_a(t)$ fluxes [32,33]. The latter represent the exchange fluxes between the biofilm and the bulk liquid compartment.

The kinetic terms $r_{Mi}(z,t,\mu,\mathbf{X},\mathbf{S})$ for the biofilm components X_1, X_2, X_3, and X_4 can be expressed as specified in the following lines. For the active biomass X_1 and X_2,

$$r_{M,1} = ((1-k_1)K_{\max,1}\frac{S_2}{K_{1,2}+S_2}\frac{S_3}{K_{1,3}+S_3} - b_{m,1}F_1\frac{S_3}{K_{1,3}+S_3} - (1-F_1)c_{m,1})X_1, \tag{4}$$

$$r_{M,2} = ((1-k_2)K_{\max,2}\frac{S_1}{K_{2,1}+S_1}\frac{S_3}{K_{2,3}+S_3}\frac{K_I}{K_I+\mu} - b_{m,2}F_2\frac{S_3}{K_{2,3}+S_3} - (1-F_2)c_{m,2})X_2, \tag{5}$$

and, for the inert component X_3,

$$r_{M,3} = (1-F_1)c_{m,1}X_1 + (1-F_2)c_{m,2}X_2, \tag{6}$$

while, for the EPS component X_4,

$$r_{M,4} = k_1 K_{\max,1}\frac{S_2}{K_{1,2}+S_2}\frac{S_3}{K_{1,3}+S_3}X_1 + k_2 K_{\max,2}\frac{S_1}{K_{2,1}+S_1}\frac{S_3}{K_{2,3}+S_3}\frac{K_I}{K_I+\mu}X_2, \tag{7}$$

where $K_{\max,i}$ denotes the maximum growth rate for biomass i, k_i is the coefficient associated with EPS formation, $K_{i,j}$ represents the affinity constant of substrate j for biomass i, $b_{m,i}$ denotes the endogenous rate for biomass i, $c_{m,i}$ is the decay–inactivation rate for biomass i, F_i represents the biodegradable fraction of biomass i, μ is the concentration of the heavy metal, which is supposed toxic for autotrophic bacteria X_2, and K_I is the inhibition constant. The kinetic growth rates for the inert material (Equation (6)) end EPS component (Equation (7)) are directly connected to the biological activities performed by the microbial species. These are modeled by Monod-like kinetics (Equations (4) and (5)) regulated by the availability of substrates within the biofilm.

The evolution of the free $\vartheta_i(z,t)$ and occupied $\bar{\vartheta}_i(z,t)$ binding sites is modeled by the equations

$$\frac{\partial \vartheta_i}{\partial t} + \frac{\partial}{\partial z}(u\vartheta_i) = r_{M,i}(z,t,\mu,\mathbf{X},\mathbf{S}) - r_{D,i}(z,t,\mu,\boldsymbol{\vartheta}),\ i = 1,...,4,\ 0 \leq z \leq L(t), t > 0,$$

$$\vartheta_i(z,0) = \vartheta_{i0}(z),\ i = 1,...,4,\ 0 \leq z \leq L_0, \tag{8}$$

$$\frac{\partial \bar{\vartheta}_i}{\partial l} + \frac{\partial}{\partial z}(u\bar{\vartheta}_i) = r_{D,i}(z,t,\mu,\boldsymbol{\vartheta}),\ i = 1,...,4,\ ,\ t > 0, 0 \leq z \leq L(t),$$

$$\bar{\vartheta}_i(z,0) = \bar{\vartheta}_{i0}(z),\ i = 1,...,4,\ 0 \leq z \leq L_0, \tag{9}$$

where $r_{D,i}$ denotes the sorption rate, and ϑ_{i0} and $\bar{\vartheta_{i0}}$ are the initial distribution of the free and occupied binding sites, respectively. The free binding site fractions can increase (Equation (8)) due to the generation of new biomass, or decrease due to the biosorption. A parabolic partial differential equation (PDE) describes the evolution of the adsorbing compound μ within the biofilm

$$\frac{\partial \mu}{\partial t} - \frac{\partial}{\partial z}\left(D_\mu \frac{\partial \mu}{\partial z}\right) = -Y_{ADS} N_\mu r_D(z,t,\mu,\boldsymbol{\vartheta}_i),\ 0 < z < L(t),\ t > 0,$$

$$\mu(z,0) = \mu_0(z),\ \frac{\partial \mu}{\partial z}(0,t) = 0,\ \mu(L(t),t) = \mu_L(t),\ 0 \le z \le L_0,\ t > 0, \tag{10}$$

where D_μ is the diffusivity coefficient for the adsorbing compound, N_μ denotes the binding sites density and Y_{ADS} is the yield of the adsorbing compound. The kinetic term r_D describes a non-reversible heavy metal sorption mechanism. This is expressed by

$$r_D = K_{ads}\mu\vartheta_1, \tag{11}$$

where K_{ads} denotes the adsorption constant. According to Equations (10) and (11), the dynamics of the adsorbing compound μ are regulated by the sorption rate r_D, which is multiplied by two parameters with physical meaning; Y_{ADS} is the amount of adsorbing compound allocated in each binding site, and N_μ is the number of binding sites related to the specific biofilm component. These parameters describe the sequestration ability of a specific biofilm component.

The diffusion−reaction of each substrate was modeled by the equations

$$\frac{\partial S_j}{\partial t} - \frac{\partial}{\partial z}\left(D_{S,j} \frac{\partial S_j}{\partial z}\right) = r_{S,j}(z,t,\mu,\mathbf{X},\mathbf{S}),\ j = 1,...,3, 0 < z < L(t),\ t > 0,$$

$$S_j(z,0) = S_{j0}(z),\ \frac{\partial S_j}{\partial z}(0,t) = 0,\ S_j(L(t),t) = S_{jL},\ j = 1,...,3,\ 0 \le z \le L_0,\ t > 0, \tag{12}$$

where $D_{S,j}$ is the diffusivity coefficient, and $r_{S,j}(z,t,\mu,\mathbf{X},\mathbf{S})$ is the conversion rate of substrate j. These terms are specifically expressed by

$$r_{S,1} = -\frac{1}{Y_2}((1-k_2)K_{\max,2}\frac{S_1}{K_{2,1}+S_1}\frac{S_3}{K_{2,3}+S_3}\frac{K_I}{K_I+\mu}X_2, \tag{13}$$

$$r_{S,2} = -\frac{1}{Y_1}((1-k_1)K_{\max,1}\frac{S_2}{K_{1,2}+S_2}\frac{S_3}{K_{1,3}+S_3}X_1, \tag{14}$$

$$\begin{aligned} r_{S,3} = &-(1-k_1)\frac{(1-Y_1)}{Y_1}((1-k_1)K_{\max,1}\frac{S_2}{K_{1,2}+S_2}\frac{S_3}{K_{1,3}+S_3}X_1, \\ &-(1-k_2)\frac{(1-Y_2)}{Y_2}((1-k_2)K_{\max,2}\frac{S_1}{K_{2,1}+S_1}\frac{S_3}{K_{2,3}+S_3}\frac{K_I}{K_I+\mu}X_2, \\ &- b_{m,1}F_1\frac{S_3}{K_{1,3}+S_3}X_1 - b_{m,2}F_2\frac{S_3}{K_{2,3}+S_3}X_2, \end{aligned} \tag{15}$$

where Y_i denotes the yield for biomass i. A schematic representation of the model structure is shown in Figure 1.

3. Numerical Simulation

The presented mathematical model was applied to simulate the effect of exposition to a toxic heavy metal in a multispecies biofilm with an initial thickness of 300 μm. The metal represents an adsorbing compound for one of the microbial species and acts as a toxic agent for the other. In particular, μ is supposed to be toxic for autotrophic bacteria but can be sorbed on the cellular membrane of heterotrophic bacteria. The values of the kinetic and stoichiometric parameters, and the mass transfer coefficients are reported in Table 1. They were adopted according to [30]. The initial conditions and biological parameters adopted in the simulations are reported in Table 2.

Numerical solutions to the free boundary problem stated in Section 2 were obtained by using the method of characteristics, e.g., [34–37]. Accuracy was checked by comparison to the geometric constraint $\sum_{i=1}^{4} f_i(z,t) = 1$. Simulations were performed using an original software developed on the Matlab platform.

Table 1. Kinetic parameters used for model simulations.

Parameter	Definition	Unit	Value
$K_{max,1}$	Maximum growth rate for X_1	d^{-1}	4.8
$K_{max,2}$	Maximum growth rate for X_2	d^{-1}	0.95
k_1	EPS formation by X_1	mg COD/mg COD	0.02
k_2	EPS formation by X_2	mg COD/mg COD	0.011
$K_{1,2}$	Organics half saturation constant for X_1	mg COD L^{-1}	5
$K_{1,3}$	Oxygen half saturation constant for X_1	mg L^{-1}	0.1
$K_{2,1}$	Ammonium half saturation constant for X_2	mg N L^{-1}	1
$K_{2,3}$	Oxygen half saturation constant for X_2	mg L^{-1}	0.1
$b_{m,1}$	Endogenous rate for X_1	d^{-1}	0.025
$b_{m,2}$	Endogenous rate for X_2	d^{-1}	0.0625
F_1	Biodegradable fraction of X_1	–	0.8
F_2	Biodegradable fraction of X_2	–	0.8
$c_{m,1}$	Decay-inactivation rate for X_1	d^{-1}	0.05
$c_{m,2}$	Decay-inactivation rate for X_2	d^{-1}	0.05
Y_1	Yield of X_1	$g_{biomass}/g_{substrate}$	0.4
Y_2	Yield of X_2	$g_{biomass}/g_{substrate}$	0.22
Y_{ads}	Yield of adsorbent	g_{metal}/n_{sites}	1
N_μ	Binding sites density for X_1	$n_{sites} L^{-1}$	1 and 50
K_I	Inhibition constant	mg L^{-1}	10^{-5}
ρ	Biofilm density	g m^{-3}	2500
λ	Biomass shear constant	mm d^{-1}	1250

Table 2. Initial conditions for biofilm growth.

Parameter	Symbol	Unit	Value
COD concentration at $L = L(t)$	S_{1L}	mgL^{-1}	20
Oxygen concentration at $L = L(t)$	S_{3L}	mgL^{-1}	8
Ammonium concentration at $L = L(t)$	S_{2L}	mgL^{-1}	2
Free metal concentration at $L = L(t)$	μ_L	mgL^{-1}	2
Time Simulation	T	d	100
Initial Biofilm thickness	L_0	mm	0.3
Initial Volume Fraction of Autotrophs (X_1)	$f_{1,0}(z)$	–	0.399
Initial Volume Fraction of Heterotrophs (X_2)	$f_{2,0}(z)$	–	0.5
Initial Volume Fraction of Inert (X_3)	$f_{3,0}(z)$	–	0.001
Initial Volume Fraction of EPS (X_4)	$f_{4,0}(z)$	–	0.1

For all the dissolved species, i.e., substrates and adsorbing contaminant, Dirichlet conditions on the free boundary were assumed. In Equation (3) governing the free boundary evolution, $\sigma_d(L(t))$ was assumed to be a known function of L and t

$$\sigma_d(L(t)) = \lambda L^2(t), \tag{16}$$

where λ is the share constant whose value is reported in Table 1. No attachment phenomena were considered for all the numerical simulation, thus $\sigma_a(t)$ was fixed to zero. The initial biofilm composition is defined in Table 2. In particular, the biofilm is set to be initially constituted by the autotrophic

(39.9%) and heterotrophic (50%) bacteria, EPS (10%) and inert (0.1%). The simulations reproduce a typical environmental condition occurring in the biological units of municipal wastewater treatment plants. The oxygen concentration in the bulk liquid was fixed to 8 mg/L, consistently with real scale continuous aerated systems. The concentrations of soluble organic carbon, i.e., chemical oxygen demand (COD), and ammonium, i.e., nitrogen content (N), in the bulk liquid were fixed to 20 mg/L and 2 mg/L, respectively.

The model outputs are reported in Figures 2 and 3. Numerical simulations demonstrate model capability of predicting the spatial distribution of biofilm components, the occupied and free binding site fractions, the substrate trends, the free contaminants profiles over biofilm depth and the biofilm thickness. The simulations show the effect of the biosorption phenomenon on the biological evolution of the overall system, and how the different features of the heterotrophic biomass, such as the binding site density N_μ, can substantially affect the final configuration of the biofilm and its properties.

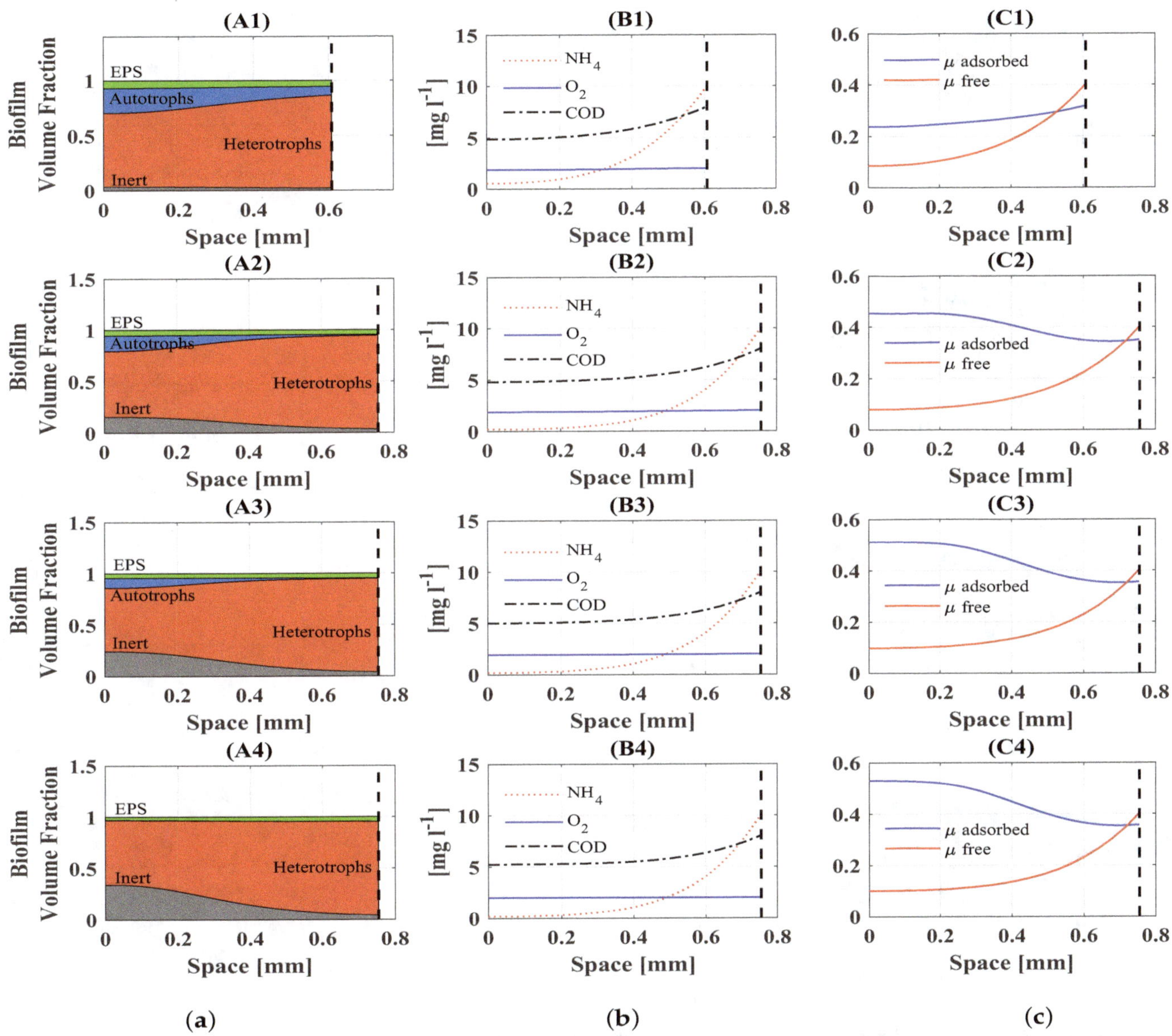

Figure 2. Effect of site density $N_\mu = 1$ on adsorption phenomenon and biological activity: (**a**) microbial species distribution (A1–A4); (**b**) substrate profiles (B1–B4); (**c**) adsorbed and free metal profile (C1–C4) after 1 (A1,B1,C1), 10 (A2,B2,C2), 20 (A3,B3,C3), and 100 (A4,B4,C4) days of simulation. The free metal concentration μ is multiplied by a factor of 10^3. The initial condition is reported in Table 2.

Two different values of the binding site density N_μ were used for numerical simulations (Figures 2 and 3). In the first simulation set, the value of site density N_μ was fixed to 1 (Figure 2), while, in the second set, N_μ was increased to 50 (Figure 3). In the first case ($N_\mu = 1$), the low site density determines a low adsorption rate resulting in a high diffusion of the free contaminant, which shows a fully penetrated profile (Figure 2, C1). The presence of μ in the inner part of the biofilm inhibits the metabolic activities of autotrophic bacteria. Despite the presence of autotrophs into the biofilm (Figure 2, A1), the ammonium is not degraded; indeed, the concentration of ammonium in the system remains constant (Figure 2, B1). The fraction of autotrophic bacteria decreases with time due to the toxic effect of μ (Figure 2, A2, A3 and A4). After 100 days of simulation, the autotrophs completely disappear from the biofilm (Figure 2, A4).

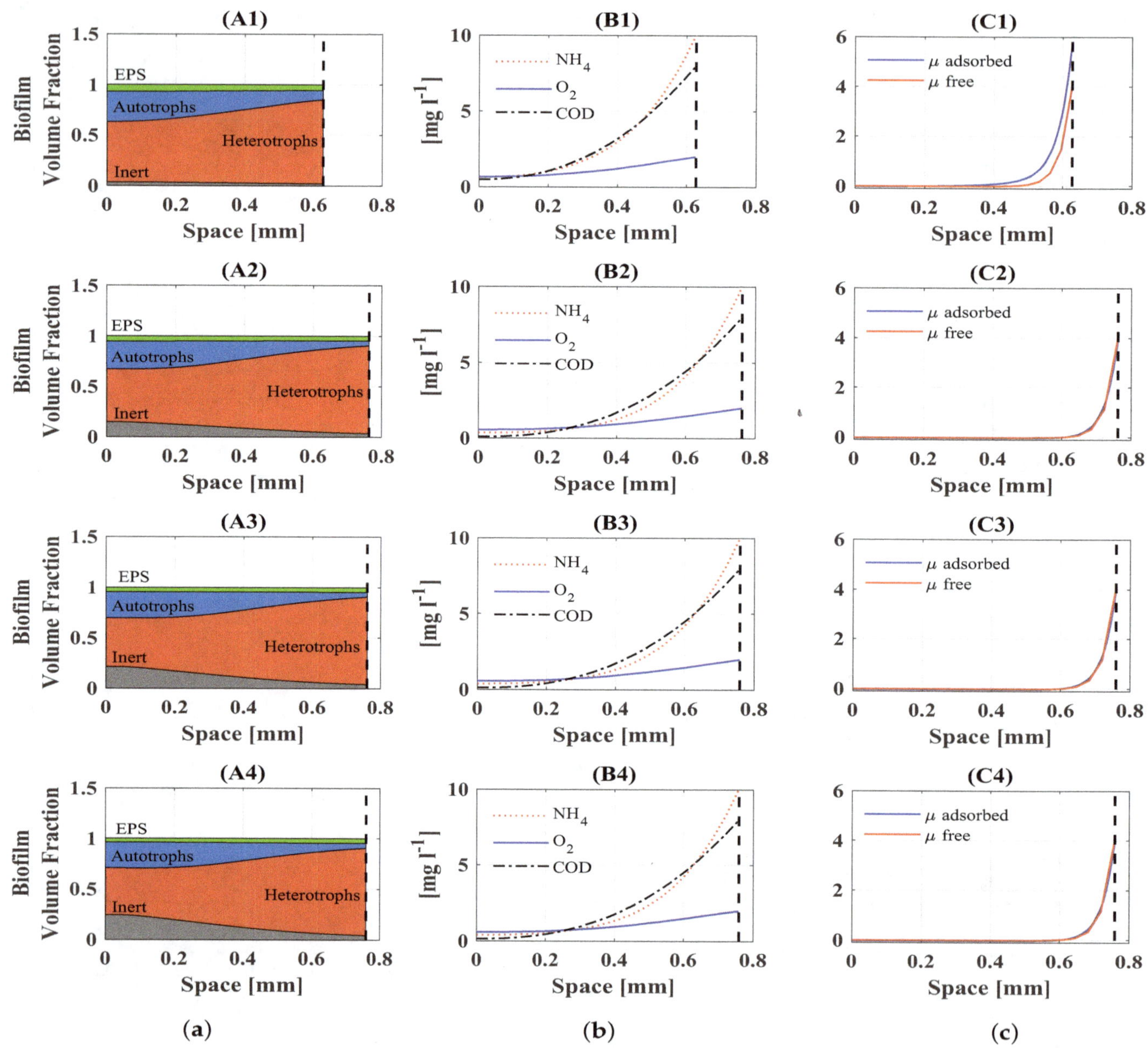

Figure 3. Effect of site density $N_\mu = 50$ on adsorption phenomenon and biological activity: (**a**) microbial species distribution (A1–A4); (**b**) substrate profiles (B1–B4); (**c**) adsorbed and free metal profile (C1–C4) after 1 (A1,B1,C1), 10 (A2,B2,C2), 20 (A3,B3,C3), and 100 (A4,B4,C4) days of simulation. The free metal concentration μ is multiplied by a factor of 10^4. The initial condition is reported in Table 2.

In the second case ($N_\mu = 50$), the higher site density determines a higher adsorption rate resulting in a lower diffusion of the heavy metal than in the first case (Figure 2, C1). It is interesting to notice that the concentration of the metal μ is essentially zero in the inner part of the biofilm, allowing the proliferation of autotrophic bacteria (Figure 3, A1). Due to the absence of μ, the metabolic activity of autotrophic bacteria is not inhibited and the ammonium is degraded (Figure 3, B1). Notably, the existence of autotrophic bacteria in the inner part of the biofilm is due to the relevant adsorption phenomenon occurring in the external part of the biofilm. Indeed, heterotrophic bacteria act as a biological shield for autotrophic bacteria, which can live and proliferate in the biofilm structure performing their biological activity. The coexistence of the two species is preserved after 100 d of simulation as it is possible to notice from the final distribution of the microbial species within the biofilm (Figure 2, A4).

Additional simulations were run by varying the inhibition constant K_I and the metal concentration within the bulk liquid μ_L. The first set of simulations was performed to test the effect of an increasing resistance of the autrotrophic component X_2 to the toxic metal. The site density was set to $N_\mu = 1$ to reproduce the case of a heterotrophic-autotrophic biofilm with a low sorption capability. The results were summarized in Figure 4.

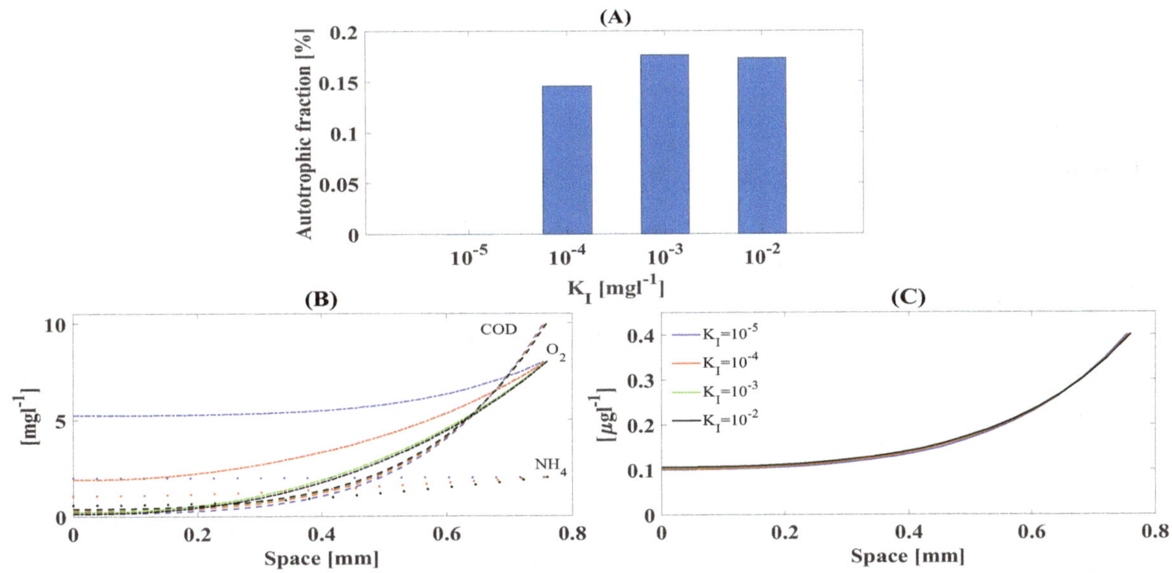

Figure 4. Effect of the inhibition constant K_I on (**A**) autotrophic fraction; (**B**) substrate profiles; (**C**) free metal trend within the biofilm with site density $N_\mu = 1$ after 100 days of simulation time.

When varying K_I from 10^{-5} to 10^{-4} (Figure 4A), the autotrophic fraction rises from 0 to 15%. By further increasing the value of K_I to 10^{-3} and 10^{-2}, the autotrophic fraction reaches 18 and 17%, respectively. This slight difference is due to the biofilm thickness, which is smaller in the case of $K_I = 10^{-3}$, and affects the diffusion of substrates within the biofilm (Figure 4B). Ammonia shows a fully penetrated profile for all values of K_I, due to the low concentration of autotrophic bacteria (<20%) within the biofilm. Oxygen is characterized by a similar trend for the lowest values of K_I. When the autotrophic fraction increases as a result of a higher K_I value, the oxygen concentration decreases all over the biofilm due to the additional consumption related to the autotrophic metabolism. No significant differences can be noted in the μ profile, which shows a fully penetrated profile for all K_I values (Figure 4C). The simulation results highlight the key role played by the inhibition constant in the definition of the biofilm composition.

The second set of simulations was performed by varying the metal concentration in the bulk liquid to test the effectiveness of the biological shield provided by heterotrophic bacteria (Figure 5). The final simulation time and the site density were set at $T = 100$ d and $N_\mu = 50$, respectively.

For $\mu_L = 4 \times 10^{-2}$, the metal showed a fully penetrated profile (Figure 5C), which determines a strong inhibition of the autotrophic species. For all the other values of μ_L, the metal concentration reaches zero within the biofilm.

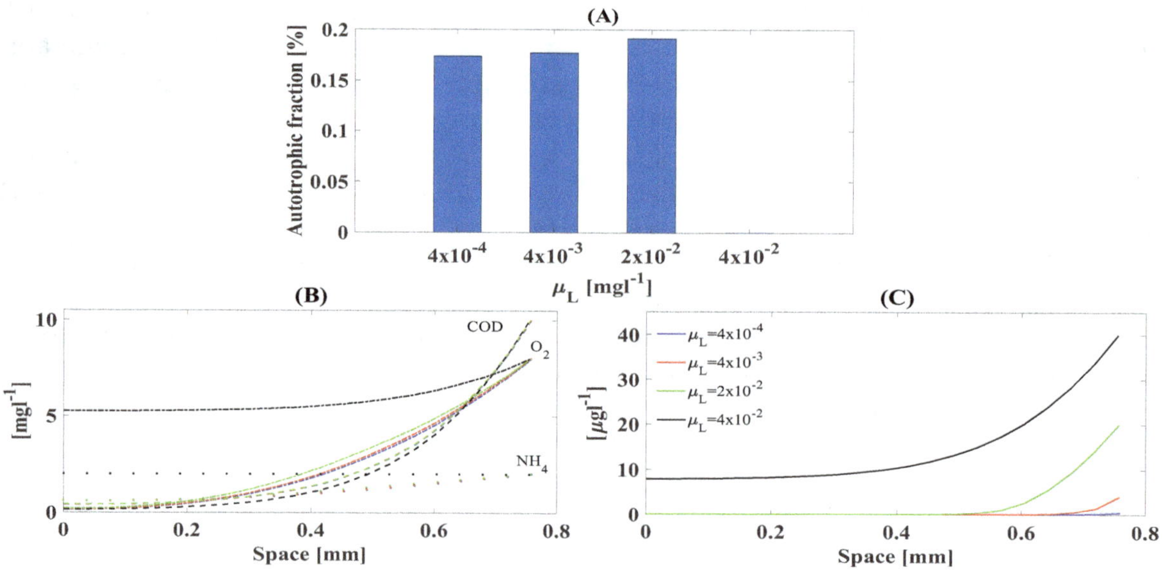

Figure 5. Effect of the free metal concentration in the bulk liquid μ_L on (**A**) autotrophic fraction; (**B**) substrate profiles; (**C**) free metal trend within the biofilm with site density $N_\mu = 50$ after 100 days of simulation time.

Increasing μ_L from 4×10^{-4} to 2×10^{-2}, the volume fraction of the autotrophic species slightly increases (Figure 5A) due to the small difference in biofilm thickness and substrates trends within the biofilm. When the autotrophic fraction is inhibited by the high metal concentration, ammonia remains constant within the biofilm and oxygen shows a fully penetrated profile. For all the simulations, the COD profile is invariant (Figure 5B). Except for $\mu_L = 4 \times 10^{-2}$, the simulation results prove the effectiveness of the heterotrophic component in protecting the autotrophic fraction from metal exposure.

Further numerical simulations were carried out to test the influence of the initial distribution of biofilm components in both the experimental cases $N_\mu = 1$ and $N_\mu = 50$ (data not shown). After 100 days of simulation time, numerical results showed a negligible variation of the biofilm components distribution and a similar biological response to the heavy metal exposition.

4. Conclusions

In this work, a free boundary problem related to biofim growth and evolution during heavy metal exposition in wastewater treatment plants has been discussed. The model highlights the dynamic interactions occurring between different biofilm components when an inhibiting compound diffuses from the bulk liquid within the biofilm structure. The biosorption phenomenon has been considered by assigning a specific binding site density to the heterotrophic biomass, which is able to act as a biological shield for autotrophic bacteria. Numerical results showed the crucial role of heterotrophic bacteria on biosorption processes occurring in wastewater treatment plants. The combined effect of heavy metals inhibition and biosorption phenomena within the biofilm structure has been newly analyzed in this study. The general form and the structure of the mathematical model allow for its application to different biological cases of high engineering interest. Simulation results demonstrated that biofilm systems can be effectively used in the context of bioremediation. The development of 1D mathematical models able to predict biofilm evolution and features under different environmental condition is highly relevant for real scale applications, such as heavy metal recovery in biofilm reactors. Further experimental studies are still required to elucidate the different interactions occurring between heavy metals and specific biofilm components. For instance, the role of EPS on heavy metals biosorption can affect the biological response of a specific multispecies biofilm. This role could be further taken into account with the presented mathematical model by assigning a specific site density for the EPS component as a function of the biosorption affinity with the diffusing heavy metal.

Author Contributions: The authors equally contributed to this work.

Acknowledgments: This paper has been performed under the auspices of the G.N.F.M. of I.N.d.A.M. It was partially supported by CARIPLO Foundation (Grant No.: 2017-0977).

References

1. Flemming, H.C.; Wingender, J.; Szewzyk, U.; Steinberg, P.; Rice, S.A.; Kjelleberg, S. Biofilms: An emergent form of bacterial life. *Nat. Rev. Microbiol.* **2016**, *14*, 563–575. [CrossRef] [PubMed]
2. D'Acunto, B.; Esposito, G.; Frunzo, L.; Mattei, M.; Pirozzi, F. Mathematical modeling of heavy metal biosorption in multispecies biofilms. *J. Environ. Eng.* **2015**, *142*, 1–14 [CrossRef]
3. Flemming, H.C.; Wingender, J. The biofilm matrix. *Nat. Rev. Microbiol.* **2010**, *8*, 623–633. [CrossRef] [PubMed]
4. Costerton, J.W.; Lewandowski, Z.; Caldwell, D.E.; Korber, D.R.; Lappin-Scott, H.M. Microbial biofilms. *Annu. Rev. Microbiol.* **1995**, *49*, 711–745. [CrossRef] [PubMed]
5. Davey, M.E.; O'toole, G.A. Microbial biofilms: From ecology to molecular genetics. *Microbiol. Mol. Biol. Rev.* **2000**, *64*, 847–867. [CrossRef]
6. Stoodley, P.; Sauer, K.; Davies, D.G.; Costerton, J.W. Biofilms as complex differentiated communities. *Annu. Rev. Microbiol.* **2000**, *56*, 187–209. [CrossRef] [PubMed]
7. Mattei, M.R.; Frunzo, L.; D'Acunto, B.; Pechaud, Y.; Pirozzi, F.; Esposito, G. Continuum and discrete approach in modeling biofilm development and structure: A review. *J. Math. Biol.* **2018**, *76*, 945–1003. [CrossRef]
8. Gaebler, H.J.; Eberl, H.J. A simple model of biofilm growth in a porous medium that accounts for detachment and attachment of suspended biomass and their contribution to substrate degradation. *Eur. J. Appl. Math.* **2018**, *29*, 1110–1140. [CrossRef]
9. D'Acunto, B.; Frunzo, L.; Mattei, M.R. Continuum approach to mathematical modelling of multispecies biofilms. *Ric. Mat.* **2017**, *66*, 153–169. [CrossRef]
10. Infante, C.D.; Castillo, F.; Pérez, V.; Riquelme, C. Inhibition of Nitzschia ovalis biofilm settlement by a bacterial bioactive compound through alteration of EPS and epiphytic bacteria. *Electron. J. Biotechnol.* **2018**, *33*, 1–10. [CrossRef]
11. Elias, S.; Banin, E. Multi-species biofilms: Living with friendly neighbors. *FEMS Microbiol. Rev.* **2012**, *36*, 990–1004. [CrossRef] [PubMed]
12. Costerton, J.W.; Lewandowski, Z.; DeBeer, D.; Caldwell, D.; Korber, D.; James, G. Biofilms, the customized microniche. *J. Bacteriol.* **1994**, *176*, 2137–2142. [CrossRef] [PubMed]
13. Tiwari, S.K.; Bowers, K.L. Modeling biofilm growth for porous media applications. *Math. Comput. Model.* **2001**, *33*, 299–319. [CrossRef]
14. Tan, L.C.; Papirio, S.; Luongo, V.; Nancharaiah, Y.V.; Cennamo, P.; Esposito, G.; van Hullebusch, E.D.; Lens, P.N. Comparative performance of anaerobic attached biofilm and granular sludge reactors for the treatment of model mine drainage wastewater containing selenate, sulfate and nickel. *Chem. Eng. J.* **2018**, *345*, 545–555. [CrossRef]
15. Beveridge, T.J.; Murray, R.G. Uptake and retention of metals by cell walls of Bacillus subtilis. *J. Bacteriol.* **1976**, *127*, 1502–1518. [PubMed]
16. Maharaj, B.C.; Mattei, M.R.; Frunzo, L.; van Hullebusch, E.D.; Esposito, G. ADM1 based mathematical model of trace element precipitation/dissolution in anaerobic digestion processes. *Bioresour. Technol.* **2018**, *267*, 666–676. [CrossRef] [PubMed]
17. Kostrytsia, A.; Papirio, S.; Frunzo, L.; Mattei, M.R.; Porca, E.; Collins, G.; Lens, P.N.; Esposito, G. Elemental sulfur-based autotrophic denitrification and denitritation: Microbially catalyzed sulfur hydrolysis and nitrogen conversions. *J. Environ. Manag.* **2018**, *211*, 313–322. [CrossRef] [PubMed]

18. Colombaro, I.; Giusti, A.; Mainardi, F. On transient waves in linear viscoelasticity. *Wave Motion* **2017**, *74*, 191–212. [CrossRef]
19. Frunzo, L.; Garra, R.; Giusti, A.; Luongo, V. Modeling biological systems with an improved fractional Gompertz law. *Commun. Nonlinear Sci.* **2019**, *74*, 260–267. [CrossRef]
20. Comte, S.; Guibaud, G.; Baudu, M. Biosorption properties of extracellular polymeric substances (EPS) towards Cd, Cu and Pb for different pH values. *J. Hazard. Mater.* **2008**, *151*, 185–193. [CrossRef]
21. Gadd, G.M. Biosorption: Critical review of scientific rationale, environmental importance and significance for pollution treatment. *J. Chem. Technol. Biotechnol.* **2009**, *84*, 13–28. [CrossRef]
22. Van Hullebusch, E.D.; Zandvoort, M.H.; Lens, P.N. Metal immobilisation by biofilms: Mechanisms and analytical tools. *Rev. Environ. Sci. Biotechnol.* **2003**, *2*, 9–33. [CrossRef]
23. Eberl, H.J.; Parker, D.F.; van Loosdrecht, M. A new deterministic spatio-temporal continuum model for biofilm development. *Comput. Math. Method Med.* **2001**, *3*, 161–175. [CrossRef]
24. Peszynska, M.; Trykozko, A.; Iltis, G.; Schlueter, S.; Wildenschild, D. Biofilm growth in porous media: Experiments, computational modeling at the porescale, and upscaling. *Adv. Water Resour.* **2015**, *95*, 288–301. [CrossRef]
25. Gokieli, M.; Kenmochi, N.; Niezgódka, M. Mathematical modeling of biofilm development. *Nonlinear Anal. Real World Appl.* **2018**, *42*, 422–447. [CrossRef]
26. Alpkvista, E.; Klapper, I. A multidimensional multispecies continuum model for heterogeneous biofilm development. *Bull. Math. Biol.* **2007**, *69*, 765–789. [CrossRef] [PubMed]
27. Wanner, O.; Eberl, H.; Morgenroth, E.; Noguera, D.; Picioreanu, C.; Rittmann, B.; van Loosdrecht, M. *Mathematical Modeling of Biofilms*; IWA Publishing: London, UK, 2006.
28. Wanner, O.; Gujer, W. A multispecies biofilm model. *Biotechnol. Bioeng.* **1986**, *28*, 314–328. [CrossRef] [PubMed]
29. D'Acunto, B.; Frunzo, L.; Mattei, M.R. On a free boundary problem for biosorption in biofilms. *Nonlinear Anal. Real World Appl.* **2018**, *39*, 120–141. [CrossRef]
30. Frunzo, L. Modeling sorption of emerging contaminants in biofilms. *arXiv* **2017**, arXiv:1706.04541.
31. Laspidou, C.S.; Rittmann, B.E. Non-steady state modeling of extracellular polymeric substances, soluble microbial products, and active and inert biomass. *Water Res.* **2002**, *36*, 1983–1992. [CrossRef]
32. Abbas, F.; Sudarsan, R.; Eberl, H.J. Longtime behavior of one-dimensional biofilm models with shear dependent detachment rates. *Math. Biosci. Eng.* **2012**, *9*, 215–239. [CrossRef] [PubMed]
33. D'Acunto, B.; Frunzo, L.; Luongo, V.; Mattei, M.R. Free boundary approach for the attachment in the initial phase of multispecies biofilm growth. *Z. Angew. Math. Phys.* **2019**, *70*, 1–16. [CrossRef]
34. D'Acunto, B.; Frunzo, L. Qualitative analysis and simulations of a free boundary problem for mulispecies biofilm models. *Math. Comput. Model.* **2011**, *53*, 1596–1606. [CrossRef]
35. D'Acunto, B.; Esposito, G.; Frunzo, L.; Pirozzi, F. Dynamic modeling of sulfate reducing biofilms. *Comput. Math. Appl.* **2011**, *62*, 2601–2608. [CrossRef]
36. D'Acunto, B.; Frunzo, L. Free boundary problem for an initial cell layer in multispecies biofilm formation. *Appl. Math. Lett.* **2012**, *25*, 20–26. [CrossRef]
37. Frunzo, L.; Mattei, M.R. Qualitative analysis of the invasion free boundary problem in biofilms. *Ric. Mat.* **2017**, *66*, 171–188. [CrossRef]

13

Evaluation of Elite Athletes Training Management Efficiency based on Multiple Criteria Measure of Conditioning using Fewer Data

Aleksandras Krylovas [1], Natalja Kosareva [1], Rūta Dadelienė [2] and Stanislav Dadelo [3,*]

1 Department of Mathematical Modelling, Vilnius Gediminas Technical University, Saulėtekio al. 11, 10221 Vilnius, Lithuania; aleksandras.krylovas@vgtu.lt (A.K.); natalja.kosareva@vgtu.lt (N.K.)

2 Department of Rehabilitation, Physical and Sports Medicine, Institute of Health Science, Vilnius University, Saulėtekio al. 11, 10221 Vilnius, Lithuania; ruta.dadeliene@mf.vu.lt

3 Department of Entertainment Industries, Vilnius Gediminas Technical University, Saulėtekio al. 11, 10221 Vilnius, Lithuania

* Correspondence: stanislav.dadelo@vgtu.lt

Abstract: Innovative solutions and techniques in the sports industry are commonly used and tested in real conditions. Elite athletes have to achieve their peak performance before the main competition of the year, which is the World Championship, and every fourth year before the Olympic Games, when the main competition of athletes takes place. The present study aims to analyze and evaluate the ability of elite kayakers to achieve the best form at the right times, with the Olympic Games taking the greatest importance. Target values for multiple measures of conditioning are compared to target values set by experts. A weighted least squares metric with weights varied by time period is developed as a measure of fulfillment of the athletes' conditioning plans. The novelty of the paper is the idea of using linear combination of polynomials and trigonometric functions for approximating the target functions and application of the proposed methodology for the optimization and evaluation of athletic training.

Keywords: athletes' condition; approximation; parameter estimation; least squares method; visualization

---

1. Introduction

Sports are a phenomenon of global importance. The investment in professional athletes is a particularly important process. Sporting events promote the development of powerful worldwide industries. A key factor in sports is the ability to develop training schedules to optimize the athletes' training conditions. The optimization and evaluation of athletic training is the most important problem for sports scientists, coaches, and athletes. Elite athletes are willing to perform more voluminous and high-quality training routines, and this process requires effective management. Elite athletes attain a world-class status in endurance sports after four to seven years of specialized training [1]. Over this period, they usually have 3000–7000 h of effective training and cover distances of 3000–4000 km/yr [2]. Winning medals in international competitions requires not only outstanding abilities and the long-term training of athletes, it is also crucial that they achieve peak performance at the right time.

The Olympic Games is the main competition for the Olympic sports athletes. There are four years of the Olympic training cycle, with its organizational structure and training methodology directed at making the successful start of athletes in the Olympic Games [3]. The elite athletes' training must be carefully planned for several years in the future and based on individual indicators (training load specification and the athlete's body adaptation to a particular training load). The process of the elite athletes' training is complex and difficult to predict because of a number of unknown factors (i.e., accidents, acclimatization, illnesses, psychological changes, the pace of recovery, etc.) [4].

The Olympic training cycle consists of four one-year macro cycles, with the main competition, the World Championship, taking place in each of these cycles. The athletes make strenuous efforts to be in their best form before the World Championship. In the last year of the Olympic cycle, some of the athletes get a chance to participate in the Olympic Games during the World Championship. It means that athletes have to achieve their peak performance twice a year. In this case, the management of training is very complicated, because the athletes need to achieve their peak performance in a short time [5,6].

Monitoring the training load and athletes' physiological adaptation is essential for optimizing training and minimizing the risk of overtraining, injuries, illnesses, etc. Usually, sports managers and scientists are restricted in publishing the data on the athletes' condition before competitions. In elite sports, monitoring is extensive, but most of the data remain confidential [7].

Scientists are searching for novel approaches to visualize and analyze the data obtained in the training sessions. One of the most relevant research topics is visualization for motion analysis, which is important for training optimization, technical and tactical improvements, as well as the prevention of injuries [8]. However, not only motion optimization is required for achieving good results. The growing need causes the necessity for making a training process much more technologically advanced, and this is an important field of interest for athletes, coaches, and scientists. Massive amounts of data are generated and used in sports medicine, including preventative care and rehabilitation. Most major professional teams today make data-driven decisions and employ the analytical staff to help prepare training plans, predict athlete risks, and prescribe personalized recovery strategies. Most of the athletes use trackers to measure, accompany, and control the data (e.g., power, velocity, duration, altitude, and heart rate) obtained in training and later perform an individual analysis and online planning [9]. The present study focuses on the visualization of the training process based on the physiological condition of athletes.

The optimization of the athletic training process is an appropriate application of operational research (OR). Prediction management and decision making in athletic training should be based on objective data analysis and parameter estimation [10].

Reviewing the literature related to the application of mathematical models and methods used for solving similar problems has shown that the authors use various approaches. For example, Armstrong, Weidner, and Walker [11] have elaborated on the analysis of variance (ANOVA) and t-test of independent samples for clinical proficiency evaluation in athletic training. A qualitative analysis of the respondents' comments has also been performed.

Li, Zhu, Chen, and Xue [12] have elaborated on the balanced approach of data envelopment analysis (DEA) to cross-efficiency evaluation. An iterative algorithm for obtaining the final optimal and balanced cross-efficiency score has been developed by the authors.

Four imbalance indicators for measuring the difference between standard distribution and the ideally balanced distribution have been proposed by Karsu and Morton [13]. The interesting phenomenon is the total componentwise proportional deviation because it is an individual-oriented measure, which is the weighted sum of fractional misallocations for each party.

The Delphi method has been used by Reefke and Sundaram [14] to identify a set of key sustainable supply chain management (SSCM) problems and the associated research opportunities. The data components have been synthesized, and the parameters have been estimated according to their relative importance based on the experts' judgments. New insights into the potential dependencies between the factors and their influence on the success of SSCM have been provided in this work. The development of the schedule for the sporting activities of the Ecuadorian football federation under a set of constraints with the use of integer programming has been proposed by Recalde, Torres, and Vaca [15]. A heuristic three-phase approach has also been adjusted to solve the considered problem. The developed methodology provides more benefits than the empirical method.

Algorithms based on multiple criteria decision making are often used in solving various problems in sports management. Dadelo, Turskis, Zavadskas, and Dadelienė [16] have described a novel framework for practical assessment and the ranking of basketball players based on the adjusted well-known Technique for Order of Preference by Similarity to Ideal Solution (TOPSIS) method. The graphical visualization of similarities provides further insights [17,18]. Visualization solutions of processes are better understood, allowing one to intuitively select a more appropriate optimization method. The visualization of processes is particularly useful for the cases with a variety of large and small factors, which should be controlled [19]. This is particularly important for inexperienced (non-professional) users.

The present study aims to analyze the ability of elite kayakers to achieve peak performance at the right times, with the Olympic Games taking the greatest importance. Real data is collected from two Olympic kayakers, and regression analysis is used (a combination of polynomial and trigonometric functions) to model how well their conditioning fits the ideal. A weighted least squares metric with weights varied by time period is developed as a measure of fulfillment of the athletes' conditioning plans.

In our knowledge, the proposed methodology for the first time was applied for sports management process optimization. However, the scope of the methodology is not limited to the field of athletic training analysis and optimization. Another challenge in this study is the limited amount of data available. The data of two athletes' Olympic training cycles are available. In addition, the data are not independent, as both athletes compete in a pair. This complicates the application of other methods, such as statistical methods, to address this problem. Therefore, in this application, the visualization approach should be defined for monitoring the condition of top kayakers during the three-year macro cycles before the Olympic Games.

Except for the Introduction, the article is organized as follows. In Section 2, the description of the experiment and the measured and expected target values of the indicators set by experts are provided. Section 3 is focused on justifying the selection of the approximation function. In Section 4, the research methodology, based on the approximation of functions by using a linear combination of polynomials and trigonometric functions, is described, and the calculated unknown coefficients are given. Section 5 describes visualization of the approximation of the target and the measured functions, and presents a measure of the difference between the target and the measurement functions, which defines the athlete's condition. Section 6 provides the discussion of the investigated problem and conclusions.

2. The Investigation Object and the Initial Data

2.1. The athletes' Participation in the Study

The research was performed in the second, third, and the fourth year of the Olympic four-year cycle, at the time of athletic training for Rio de Janeiro Olympic Games in 2016. Two (1, 2) elite flat-water kayak paddlers (a racing team), performing at the international competitive level at the distance of 1000 m (whose ages were 27 and 26 years, and body mass were 88.5 and 84.5 kg, respectively), volunteered to take part in the investigation. These two athletes gained fifth place in a 1000 m event (K-2) in the 2016 Olympic Games in Rio de Janeiro.

2.2. The Description of the Experiment

A council, consisting of 22 Lithuanian kayaking elite experts (trainers and sports researchers), with no less than 10 years of experience in execution and organization in kayaking, have rated the competences and created the dynamics of the ideal curves of indicators based on correlations between indicators and sports outcomes [20].

The training volume and intensity were carefully controlled and quantified by using the Garmin Connect Forerunner 910 XT during each training session throughout the cycle. The data obtained from the devices were sent to three delegates from the sports science experts' council, who are the creators of the athletes' training programs. During the study, the athletes were encouraged to undertake their standard training sessions but not to train on the day before each test. The athletes were acquainted with the experimental procedures prior to testing and gave a written consent to participate in the study. All experimental procedures were approved by the Lithuanian Ethics Committee.

The testing lasted for three seasons (macro cycles): in each season, the training period lasted for 8 months, the competition period lasted for 3 months, and the transition period lasted for 1 month. Physiological testing of the athletes' condition was performed at the Lithuanian University of Education Science. For each athlete, all tests were conducted at the same time of the day, between 09:00 and 11:00, and 24 h after the last training session. During the testing session, the tests were performed in the same order. Prior to beginning the study, the sports doctor examined the kayakers to exclude any medical disorders that could limit their participation in the investigation.

During the macro cycles (I, II, III), the athletes were tested in six periods of time: in (T1), in the first week of the introductory training period, in (T2), at the beginning of the general training period, in (T3), at the beginning of the specific training period, in (T4), at the beginning of the competitive training period, in (T5), at the beginning of the main competition period, and in (T6), at the time after the main competition period.

Standard methodologies were applied to determine the athletes' physical parameters, i.e., their height (cm) and body mass (kg) [21].

The resting heart rate (b/min) was determined for each athlete in the supine position and in the period after the application of the standard physical load (30 squats within 45 s) by using the Garmin Connect Forerunner 910 XT.

Hemoglobin concentration (g/L) was determined by using a Hemocue analyzer. A trace amount of blood samples was taken from the fingertips of the athletes in the resting position.

The ergometer test was performed, using Oxycon Mobile 781023-052, version 5.2 (Cardinal Health Germany 234 GmbH, Höchberg, Germany). Gas analyzers were calibrated before and verified after each test. Each athlete performed the incremental submaximal ergometer test, using a calibrated kayak ergometer (Dansprint PRO, KE001 ergo, Hvidovre, Denmark) for determining the maximum oxygen uptake (VO_2max) [22]. Five minutes before performing the submaximal ergometer test, the athletes completed a 15-min warm up. The incremental test began with the application of the initial workload of 100 W, and increments of 20 W were applied at the intervals of 30 s to bring the athlete to the limit of tolerance in 8–12 min. During the last minute, the athlete was encouraged to do as much work as he could. Then, the values were averaged over the intervals of 30 s. Pulmonary ventilation (PV) (1/min), oxygen uptake (VO_2) (1/min, mL/min/kg), work capacity (W), and speed (km/h) were recorded at the point of the critical intensity limit (CIL).

For reaching the aim of the present study, all tests were arranged in hierarchical order by the experts' board (from the most informative to the least informative indicator), which influenced the outcome of a sporting event, measured in percent (Table 1). In addition, all the testing sessions were also arranged in the hierarchical order. Their importance is shown in Table 2.

Table 1. Test indicators presented in the hierarchical order set by the experts from the most to the least important indicator.

Hierarchic Value	Test Indicator	Importance in Percent
1	Speed (km/h) at the point of the CIL	30
2	Work capacity (W), at the point of the CIL	20
3	Oxygen uptake (1/min, mL/min/kg) at the point of the CIL	15
4	Pulmonary ventilation (1/min) at the point of the CIL	10
5	Resting heart rate (b/min)	10
6	Heart rate (b/min) after the standard physical load	10
7	Haemoglobin concentration (g/L)	5

Table 2. Importance of the testing sessions presented in the chronological order from the least (1) to the most (10) important session and their hierarchical order set by the experts.

No	Date	Testing Sessions	Hierarchical Order	Athletic Training Phase
1	2013 09 13	IT1	1	Transitional period after the season
2	2013 12 18	IT2	3	The first half of the training period (without water)
3	2014 03 28	IT3	4	The second half of the training period (with water)
4	2014 05 15	IT4	5	The period of non-core competitions
5	2014 07 15	IT5	6	European Championship
6	2014 08 12	IT6	7	World Championship
7	2014 09 12	IIT1	2	Transitional period after the season
8	2014 12 18	IIT2	4	The first half of the training period (without water)
9	2015 03 28	IIT3	5	The second half of the training period (with water)
10	2015 05 13	IIT4	6	The period of non-core competitions
11	2015 07 15	IIT5	7	European Championship
12	2015 08 12	IIT6	8	World Championship
13	2015 09 29	IIIT1	4	Transitional period after the season
14	2015 12 22	IIIT2	6	The first half of the training period (without water)
15	2016 03 04	IIIT3	7	The second half of the training period (with water)
16	2016 04 04	IIIT4	8	The period of non-core competitions
17	2016 07 07	IIIT5	9	European Championship/ Olympic selection
18	2016 08 01	IIIT6	10	Olympic Games

In Table 3, the results of measurements of seven indicators for both athletes $R_i^{(l,j)}$ and the expected target values $S_i^{(l,j)}$ determined by the experts are presented. Here, l denotes the athletes ($l = 1,2$), j is the number of the indicator ($j = 1,2,\ldots,7$), and i is the number of the measurement ($i = 1,2,\ldots,18$). For example, $R_4^{(2,6)} = 105$, $R_2^{(1,3)} = 60.7$, $S_6^{(1,7)} = S_6^{(2,7)} = 168$. The expected target values $S_i^{(l,j)}$ differ for athletes 1 and 2 only in the case of $j = 4$ because of the different lung volumes of the athletes.

Table 3. Measurements of seven indicators for two athletes $R_i^{(l,j)}$ (the first row) and the expected target values $S_i^{(l,j)}$ (the second row).

i	$j=1$		$j=2$		$j=3$		$j=4$		$j=5$		$j=6$		$j=7$	
1	260	260	180	185	60.9	48	186	146	48	56	106	126	158	180
	310		180		53		170	140	56		131		140	
2	280	280	195	185	60.7	48	189	149	48	48	109	132	164	179
	315		188		55		175	147	53.5		128		145	
3	290	290	200	200	61.8	44	198	167	44	52	106	130	167	163
	323		197		58		180	155	51		124		152	
4	300	300	205	200	64	56	187	167	56	56	111	105	146	170
	328		203		60		183	160	49		122		157	
5	320	320	213	205	62	48	189	171	48	52	109	129	165	173
	336		211		63		187	166	47		118		164	
6	320	320	215	215	57.2	48	186	170	48	48	119	118	134	162
	340		215		65		190	170	46		115		168	

Table 3. *Cont.*

i	$j=1$	$j=2$	$j=3$	$j=4$	$j=5$	$j=6$	$j=7$
7	300 300 320	190 212 190	62.1 48 58	159 159 180 155	48 60 54	111 114 128	156 169 154
8	310 310 326	195 195 197	59 48 60.5	195 156 183 160	48 52 51.5	114 120 123	159 177 158
9	320 320 334	210 211 205	55 56 63.5	188 163 188 165	56 56 48.5	119 128 119	161 175 163
10	340 340 338	216 216 210	63.7 44 65	193 160 190 168	44 52 47	113 116 116	166 174 167
11	320 320 346	220 220 217	62.6 48 67	193 155 193 173	48 44 45	113 125 112	156 172 171
12	340 340 350	220 221 220	63.9 48 68	191 162 195 175	48 52 44	110 121 109	163 182 174
13	300 300 330	200 195 195	60.2 52 60	193 161 185 160	52 48 52	122 117 123	174 178 160
14	320 320 338	210 205 203	54.1 44 62.5	192 167 188.5 165	44 56 49.5	103 120 118	166 176 165
15	340 340 345	217 216 210	50 48 65	185 164 192 170	48 52 47	108 108 113	159 164 169
16	350 350 348	220 221 213	59.4 48 66	190 172 194 172	48 48 46	111 112 110	167 162 171
17	340 340 357	224 224 223	58.2 52 69	192 157 198.5 178	52 44 43	113 106 103	164 180 178
18	340 340 360	217 217 225	57.8 52 70	181 154 200 180	52 48 42	121 102 100	160 167 180

3. The Selection of the Approximating Functions

The shape of the approximating functions was determined only for the target curves, which were created according to experts' judgments. The available data were compared with the ideal (target) curves. The following factors were taken into account for the selection of the suitable functions for the curves:

1. The approximation curve should not only describe the upward or downward trend, but also show the characteristic phase of the fall and the subsequent growth phase of the considered processes.
2. The shape of the functions should be the same for all the curves and be simple enough. Its parameter values were determined by the least squares method.

Let us show how the ideal curve was constructed for the case of RN2. The values of the normalized data (t_j, s_j), $j = 1, 2, \ldots, 18$ were as follows:

(0.0,1.0), (0.091,1.044), (0.186,1.094), (0.232,1.172), (0.290,1.150), (0.316,1.194), (0.346,1.056), (0.438,1.094), (0.533,1.139), (0.576,1.167), (0.636,1.206), (0.663,1.222), (0.708,1.083), (0.788,1.128), (0.857,1.167), (0.887,1.183), (0.976,1.239), (1.0,1.250).

The shape of the selected curve with the coefficients determined by the least squares method was as follows:

$$w(t) = 1.019 + 0.245t - 0.010t^2 + 0.044\sin(2\pi t) - 0.013\cos(2\pi t)$$

The curve $w(t)$ and the linear interpolation I of the points (t_j, s_j) are depicted in Figure 1. Note that the graphical accuracy is sufficient for the purpose of the study, because the curve $w(t)$ satisfies requirements 1 and 2. On the other hand, describing all the fluctuations of the linear interpolation I does not make sense, because the similarity of these real data points can be hardly realized in practice. It should also be noted that the authors did not intend to construct the best curves, and therefore constructed the curves of sufficient approximation. However, other combinations of algebraic and trigonometric polynomials, which were acceptable, were also tested. The final decision about the choice of the formula was taken according to requirement 2 (the formula is appropriate for all the cases considered).

Figure 2 shows the functions

$$y(t) = 1.133 + 0.020t,\ z(t) = 1.046 + 0.24183t - 0.081t^2$$

with the coefficients determined by the least squares method based on the same data. We can see that the curves $y(t)$ and $z(t)$ do not satisfy requirement 1. The values of the error function were calculated as follows:

$$S_f = \sqrt{\sum_{i=1}^{18}\left(s_j - f(t_j)\right)^2},$$

where f is equal to $y(t), z(t), w(t)$. The following error function values were obtained: $S_y = 0.0642$, $S_z = 0.0487, S_w = 0.0430$. One can see that the best result was achieved by incorporating trigonometric components into the approximation function.

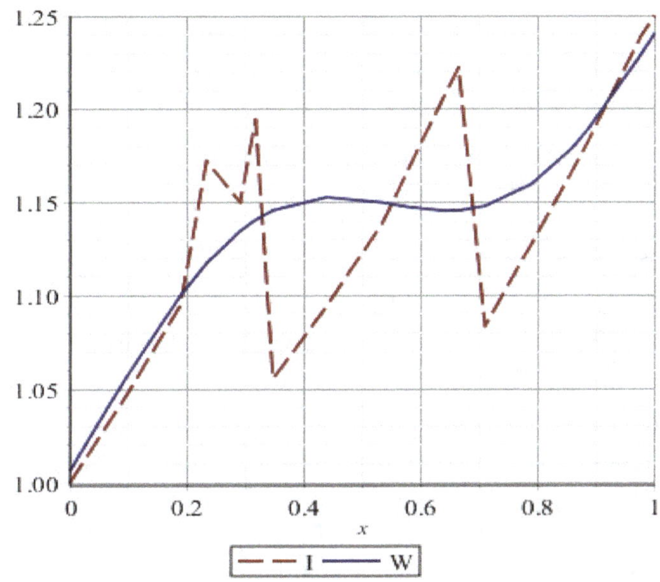

Figure 1. Approximation $w(t)$ of the target function I.

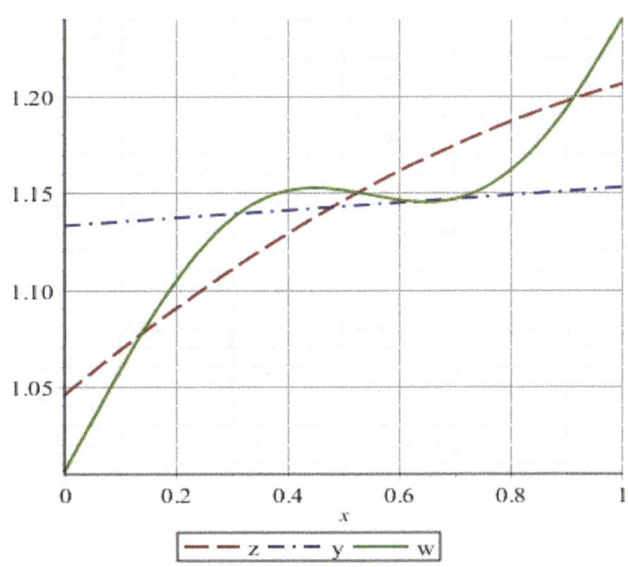

Figure 2. Approximations $y(t)$, $z(t)$, $w(t)$ of the target function I.

4. Research Methodology

In the first step, all the initial data points (Table 3) were normalized to eliminate the influence of the measurement units on the final results. The values $R_i^{(l,j)}$ and $S_i^{(l,j)}$ were normalized by the equations:

$$\widetilde{r}_i^{(l,j)} = \frac{R_i^{(l,j)}}{S_1^{(l,j)}}, \ \widetilde{s}_i^{(l,j)} = \frac{S_i^{(l,j)}}{S_1^{(l,j)}}$$

This method of data normalization, when each measurement $R_i^{(l,j)}$ and target value $S_i^{(l,j)}$ were divided by the first target value of the respective measurement, was chosen for the convenience of data representation. In this case, the graphs of all the target values of the indicators have the same initial point of 1.

The measurement time was normalized as follows:

$$t_i = \frac{T_i}{T_{18}}$$

where T_i is the number of days passed since the first measurement, $T_1 = 0$. The respective values are shown in Table 4.

Table 4. Normalized time values t_i.

T_i	0	96	196	244	305	333	364	461	561
$t_i = T_i/T_{18}$	0	0.0912	0.1861	0.2317	0.2896	0.3162	0.3457	0.4378	0.5328
T_i	607	670	698	746	830	903	934	1028	1053
$t_i = T_i/T_{18}$	0.5764	0.6363	0.6629	0.7085	0.7882	0.8576	0.8870	0.9763	1.000

The goal was to find the functions most suitable for approximating the athlete's indicators. The target values for a multiple measure of conditioning were compared to the targets set by the experts. The regression analysis was used to model how well their conditioning could fit the target. The weighted least square metric is a measure of the fulfillment of the athletes' conditioning plans. The expression with five undefined coefficients, a, b, c, A, B, was chosen for approximating the functions $\widetilde{r}^{(l,j)}(t)$, $\widetilde{s}^{(l,j)}(t)$ as follows:

$$f(t) = a + bt + ct^2 + A\sin 2\pi t + B\cos 2\pi t \quad (1)$$

where $t \in [0;1]$ is the normalized time. The unknown coefficients were determined by a standard method of the least squares [23]:

$$\sum_{i=1}^{18} (f(t_i) - f_i)^2 \to min. \quad (2)$$

The system of linear equations was obtained from Equations (1) and (2) as follows:

$$\begin{pmatrix} a_{11} & a_{12} & a_{13} & a_{14} & a_{15} \\ a_{21} & a_{22} & a_{23} & a_{24} & a_{25} \\ a_{31} & a_{32} & a_{33} & a_{34} & a_{35} \\ a_{41} & a_{42} & a_{43} & a_{44} & a_{45} \\ a_{51} & a_{52} & a_{53} & a_{54} & a_{55} \end{pmatrix} \cdot \begin{pmatrix} a \\ b \\ c \\ A \\ B \end{pmatrix} = \begin{pmatrix} b_1 \\ b_2 \\ b_3 \\ b_4 \\ b_5 \end{pmatrix}$$

where

$a_{11} = \sum_{i=1}^{18} 1 = 18,$ $a_{12} = a_{21} = \sum_{i=1}^{18} t_i,$ $a_{13} = a_{31} = \sum_{i=1}^{18} t_i^2,$ $a_{14} = a_{41} = \sum_{i=1}^{18} \sin 2\pi t_i,$ $a_{15} = a_{51} = \sum_{i=1}^{18} \cos 2\pi t_i,$

$a_{22} = \sum_{i=1}^{18} t_i^2,$ $a_{23} = a_{32} = \sum_{i=1}^{18} t_i^3,$ $a_{24} = a_{42} = \sum_{i=1}^{18} t_i \sin 2\pi t_i,$ $a_{25} = a_{52} = \sum_{i=1}^{18} t_i \cos 2\pi t_i,$ $a_{33} = \sum_{i=1}^{18} t_i^4,$

$a_{34} = a_{43} = \sum_{i=1}^{18} t_i^2 \sin 2\pi t_i,$ $a_{35} = a_{53} = \sum_{i=1}^{18} t_i^2 \cos 2\pi t_i,$ $a_{44} = \sum_{i=1}^{18} \sin^2 2\pi t_i,$ $a_{45} = a_{54} = \sum_{i=1}^{18} \sin 2\pi t_i \times \cos 2\pi t_i,$ $a_{55} = \sum_{i=1}^{18} \cos^2 2\pi t_i.$

The coefficients $b_1, b_2, \ldots, b_5$ were calculated by the equations:

$b_1 = \sum_{i=1}^{18} f_i,$ $b_2 = \sum_{i=1}^{18} f_i t_i,$ $b_3 = \sum_{i=1}^{18} f_i t_i^2,$ $b_4 = \sum_{i=1}^{18} f_i \sin 2\pi t_i,$ $b_5 = \sum_{i=1}^{18} f_i \cos 2\pi t_i,$

where the functions f_i obtained the values $\widetilde{r}^{(l,j)}(t)$ and $\widetilde{s}^{(l,j)}(t)$, i.e., the coefficients a, b, c, A, B were calculated separately for each athlete $l = 1, 2$ and each indicator $j = 1, 2, \ldots, 7$ for the functions $\widetilde{r}$ and $\widetilde{s}$.

The matrix $A = \left(a_{ij}\right)_{5\times5}$, depending only on the values t_i, was calculated only once:

$$A = \begin{pmatrix} 18 & 9.5242 & 6.6505 & -0.2485 & 0.5155 \\ 9.5242 & 6.6505 & 5.1966 & -2.7916 & 0.7607 \\ 6.6505 & 5.1966 & 4.3262 & -2.7751 & 1.6716 \\ -0.2485 & -2.7916 & -2.7751 & 9.2166 & -0.0602 \\ 0.5155 & 0.7607 & 1.6716 & -0.0602 & 8.7834 \end{pmatrix}$$

For calculating the coefficients a, b, c, A, B, the inverse matrix method is convenient to use (see, for example, [24]):

$$\begin{pmatrix} a \\ b \\ c \\ A \\ B \end{pmatrix} = A^{-1} \cdot \begin{pmatrix} b_1 \\ b_2 \\ b_3 \\ b_4 \\ b_5 \end{pmatrix} \quad (3)$$

$$A^{-1} = \begin{pmatrix} 2.7001 & -14.5727 & 13.8234 & -0.1888 & -1.5284 \\ -14.5727 & 85.2846 & -83.4043 & 0.3864 & 9.3445 \\ 13.8234 & -83.4043 & 82.7523 & -0.0333 & -9.3368 \\ -0.1888 & 0.3864 & -0.0333 & 0.2103 & -0.0146 \\ -1.5284 & 9.3445 & -9.3368 & -0.0146 & 1.7111 \end{pmatrix}$$

In Table 5, the values of the coefficients a, b, c, A, B, which were calculated by Equation (3), are presented for $j = 1, 2, \ldots, 7$, $l = 1, 2$ for the functions $\widetilde{r}$ and $\widetilde{s}$.

Table 5. Calculated values of the coefficients a, b, c, A, B.

			a	**b**	**c**	**A**	**B**
$j = 1$	$l = 1$	$\widetilde{r}$	0.835	0.529	−0.257	0.030	0.002
		$\widetilde{s}$	1.011	0.094	0.062	0.018	−0.010
	$l = 2$	$\widetilde{r}$	0.915	0.169	0.019	0.028	−0.007
		$\widetilde{s}$	1.011	0.094	0.062	0.018	−0.010
$j = 2$	$l = 1$	$\widetilde{r}$	0.976	0.651	−0.446	0.022	0.036
		$\widetilde{s}$	1.019	0.245	−0.010	0.044	−0.013
	$l = 2$	$\widetilde{r}$	1.038	0.237	−0.018	0.032	−0.023
		$\widetilde{s}$	1.019	0.245	−0.010	0.044	−0.013
$j = 3$	$l = 1$	$\widetilde{r}$	1.207	−0.323	0.240	0.007	−0.049
		$\widetilde{s}$	1.062	0.045	0.267	0.041	−0.058
	$l = 2$	$\widetilde{r}$	1.024	0.795	−0.836	0.033	0.052
		$\widetilde{s}$	1.062	0.045	0.267	0.041	−0.058

Table 5. *Cont.*

			a	b	c	A	B
$j = 4$	$l = 1$	$\widetilde{r}$	1.079	0.244	−0.279	−0.030	0.033
		$\widetilde{s}$	1.017	0.140	0.031	0.021	−0.014
	$l = 2$	$\widetilde{r}$	0.876	1.467	−1.368	0.033	0.133
		$\widetilde{s}$	1.023	0.312	−0.035	0.041	−0.021
$j = 5$	$l = 1$	$\widetilde{r}$	0.930	−0.501	0.569	0.029	−0.067
		$\widetilde{s}$	0.969	−0.145	−0.080	−0.037	0.018
	$l = 2$	$\widetilde{r}$	0.952	0.029	−0.136	−0.005	−0.005
		$\widetilde{s}$	0.969	−0.145	−0.080	−0.037	0.018
$j = 6$	$l = 1$	$\widetilde{r}$	0.907	−0.484	0.561	0.021	−0.085
		$\widetilde{s}$	0.962	0.089	−0.310	−0.030	0.031
	$l = 2$	$\widetilde{r}$	0.990	−0.058	−0.154	−0.047	0.001
		$\widetilde{s}$	0.962	0.089	−0.310	−0.030	0.031
$j = 7$	$l = 1$	$\widetilde{r}$	1.125	0.151	−0.167	−0.047	0.032
		$\widetilde{s}$	1.029	0.214	0.066	0.036	−0.029
	$l = 2$	$\widetilde{r}$	1.389	−0.779	0.692	−0.042	−0.081
		$\widetilde{s}$	1.029	0.214	0.066	0.036	−0.029

5. Visualization of the Results and Measurement of the Athlete's Condition

The graphs of approximations of the functions $\widetilde{r}(t)$ and $\widetilde{s}(t)$, with the coefficient values a, b, c, A, B, which are given in Table 5, are presented in Figures 3–9. The graphs of the approximations of the function $\widetilde{r}(t)$ are depicted in black, while the approximations of the function $\widetilde{s}(t)$ are shown in green.

Visualization of the results provides information about the quality of the athletes' training management and gives an understanding of the time periods when the maximum discrepancy between the ideal and the real curves can be observed. This allows the coaches to quickly change the athletes' workouts. To assess the numerical value of this mismatch, the measurement of the difference between the respective functions is required.

It is worth noting that five indicators have a direct relationship with the target, while two indicators, i.e., resting heart rate (5) and heart rate after the standard physical load (6), have the inverse relation with it. By analyzing speed at the point of the CIL (Figure 3), it could be observed that the athlete RN could consistently approach the target (ideal) indicators, although the AO progress was much slower. The tested athletes could not maximally synchronize their efforts (with respect to the above indicator). This can be accounted for by the specific character of their adaptation.

The evaluation of the athletes' work capacity at the point of the CIL (Figure 4) allows the authors to state that the dynamics of this indicator for the tested athletes was close to the target values, and they achieved their planned peak performance before the beginning of the Olympic Games.

Evaluation of the dynamics of the criteria describing oxygen uptake at the point of the CIL (Figure 5) showed the decreasing trend of VO_2max during the testing period and its considerable difference from the target value. This can be accounted for by the specific character of the athletes' training aimed at increasing the load in the muscles' area where glycolytic energy is generated. This phenomenon requires further investigation.

When evaluating the variation in the criteria values of pulmonary ventilation at the point of the CIL (Figure 6), a similar trend could be observed as that characteristic of the criteria describing oxygen uptake at the point of the CIL. It can be assumed that the athletes' aerobic capacity had a tendency to decrease. This can be associated with the striving of athletes to increase their capacity of energy generation in the training period.

Evaluation of the variation of criteria describing the resting heart rate (Figure 7) in a resting position and the heart rate after the application of a standard physical load (Figure 8) showed that the adaptation of the athletes' circulatory system to training was different. Thus, it was unstable for RN, while AO consistently approached the ideal values of indicators.

The analysis of the variation in hemoglobin concentration of the athletes (Figure 9) has shown that this indicator was difficult to control. In the training process, hemoglobin concentration varied considerably. As a result, the target values could not be achieved before the main competitions.

In the ideal case, when the athlete's training plan is fully realized, the curves $\widetilde{r}(t)$ and $\widetilde{s}(t)$ coincide. In practice, this objective can only be achieved partially; therefore, it is necessary to quantify the difference between the real and the ideal curves.

This difference is a measure of the athletes' condition. The fact that the differences $|\widetilde{r}(t) - \widetilde{s}(t)|$ observed in various time intervals are of various importance is taken into account. The relative importance of the time intervals was determined in the following way. The relative importance of the period prior to measuring was equal to 1, while the importance of other periods was higher, ranging from 2 to 10. The definitions of the time intervals, their relative importance, and the dates and lengths of the intervals (in percent) are given in Table 6.

Table 6. Relative importance of the time intervals.

Date	Definition of the Time Interval	Relative Importance of the Interval	Interval	The Interval's Length, in Percent
2013 09 13	The transitional period after the season	1	Till T_1	-
2013 12 18	The first half of the training period (without water)	3	$[T_1, T_2]$	9.1
2014 03 28	The second half of the training period (with water)	4	$[T_2, T_3]$	9.5
2014 05 15	The period of non-core competitions	5	$[T_3, T_4]$	4.6
2014 07 15	European Championship	6	$[T_4, T_5]$	5.8
2014 08 12	World Championship	7	$[T_5, T_6]$	2.7
2014 09 12	The transitional period after the season	2	$[T_6, T_7]$	2.9
2014 12 18	The first half of the training period (without water)	4	$[T_7, T_8]$	9.2
2015 03 28	The second half of the training period (with water)	5	$[T_8, T_9]$	9.5
2015 05 13	The period of non-core competitions	6	$[T_9, T_{10}]$	4.4
2015 07 15	European Championship	7	$[T_{10}, T_{11}]$	6.0
2015 08 12	World Championship	8	$[T_{11}, T_{12}]$	2.7
2015 09 29	The transitional period after the season	4	$[T_{12}, T_{13}]$	4.6
2015 12 22	The first half of the training period (without water)	6	$[T_{13}, T_{14}]$	7.9
2016 03 04	The second half of the training period (with water)	7	$[T_{14}, T_{15}]$	6.9
2016 04 04	The period of non-core competitions	8	$[T_{15}, T_{16}]$	2.9
2016 07 07	European Championship/Olympic selection	9	$[T_{16}, T_{17}]$	8.9
2016 08 01	Olympic Games	10	$[T_{17}, T_{18}]$	2.4

Depending on the relative importance of the time intervals and their relative length, the weights w_k were assigned to the normalized time intervals $[t_{k-1}, t_k]$. In this case, $t_0 = 0$, $t_k = \frac{k}{100}$ and $k = 1, 2, \ldots, 100$. The weights of the time intervals are given in Table 7.

Table 7. The weights of the time intervals.

w_1	w_2	$w_3 = \ldots = w_9$	$w_{10} = \ldots = w_{18}$	$w_{19} = \ldots = w_{23}$	$w_{24} = \ldots = w_{29}$	$w_{30} = \ldots = w_{32}$	w_{33}
0.01	0.02	0.03	0.04	0.05	0.06	0.07	0.06
w_{34}	w_{35}	w_{36}	w_{37}	w_{38}	$w_{39} = \ldots = w_{44}$	$w_{45} = \ldots = w_{53}$	$w_{54} = \ldots = w_{57}$
0.05	0.04	0.03	0.02	0.03	0.04	0.05	0.06
$w_{58} = \ldots = w_{63}$	$w_{64} = \ldots = w_{66}$	w_{67}	w_{68}	w_{69}	$w_{70} = \ldots = w_{71}$	w_{72}	$w_{73} = \ldots = w_{79}$
0.07	0.08	0.07	0.06	0.05	0.04	0.05	0.06
$w_{80} = \ldots = w_{86}$	$w_{87} = \ldots = w_{89}$	$w_{90} = \ldots = w_{98}$	$w_{99} = \ldots = w_{100}$				
0.07	0.08	0.09	0.10				

It should be emphasized that the fluctuations of the relative importance values in Table 6 were smoothed out. For example, $w_{66} = 0.08$ and the intermediate values $w_{67} = 0.07$, $w_{68} = 0.06$, $w_{69} = 0.05$ appeared before $w_{70} = 0.04$, since it was difficult to detect the exact moment of their sudden decrease.

Thus, the difference between the functions $\widetilde{r}(t)$ and $\widetilde{s}(t)$ was measured as the weighted sum as follows:

$$\|\widetilde{r}(t) - \widetilde{s}(t)\| = \frac{\sum_{k=1}^{100} w_k \left|\widetilde{r}(t_k) - \widetilde{s}(t_k)\right|}{\sum_{k=1}^{100} w_k \widetilde{s}(t_k)}.$$

The values $\left\|\widetilde{r}(t) - \widetilde{s}(t)\right\|$ calculated for each indicator and both athletes are given in Table 8.

Table 8. The values $\left\|\widetilde{r}(t) - \widetilde{s}(t)\right\|$ for seven indicators and two athletes.

j	$l = 1$	$l = 2$
1	0.066	0.053
2	0.011	0.015
3	0.085	0.084
4	0.046	0.024
5	0.067	0.049
6	0.019	0.062
7	0.058	0.039

The results provided in Table 8 are summarized as the weighted sums $S^{(l)}$, $l = 1,2$, with the respective criteria $j = 1,2,\ldots,7$ weights: $\widetilde{w}_1 = 0.3$, $\widetilde{w}_2 = 0.2$, $\widetilde{w}_3 = 0.15$, $\widetilde{w}_4 = \widetilde{w}_5 = \widetilde{w}_6 = 0.1$, $\widetilde{w}_7 = 0.05$ (see the data in Table 1). The weighted sums were as follows:

$$S^{(l)} = \sum_{j=1}^{7} \widetilde{w}_j \|\widetilde{r}^{(l,j)}(t) - \widetilde{s}^{(l,j)}(t)\|. \tag{4}$$

Therefore, $S^{(1)} = 0.051$, $S^{(2)} = 0.047$. The values $S^{(l)}$ have an inverse relationship with the athletic training. The higher these values, the lower the achievement of the goals by the athlete. Then, the values of $K^{(l)} = 1 - S^{(l)}$, describing the results of the athletic training, were calculated as follows:

$$K^{(1)} = 0.949, \quad K^{(2)} = 0.953.$$

The higher the value of the indicator K, the higher the level of the fulfillment of the athlete's training plan. It can be concluded that the results of the second athlete are better than the results obtained by the first athlete. The indicator K is important for the implementation and improvement (optimization) of the athlete's training plan development.

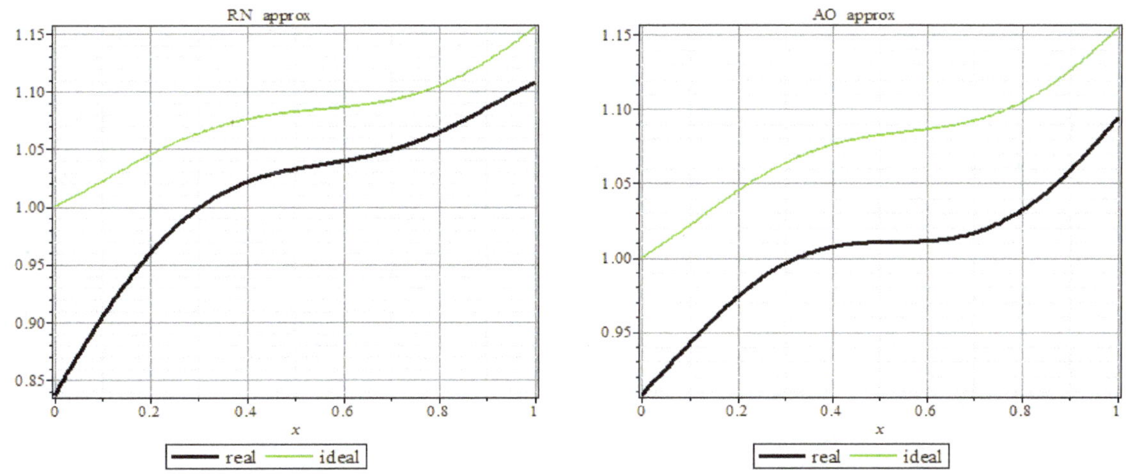

$j = 1, l = 1$ $j = 1, l = 2$

Figure 3. Approximation of functions $\widetilde{r}(t)$ and $\widetilde{s}(t)$ for two athletes according to speed (km/h) at the point of the CIL.

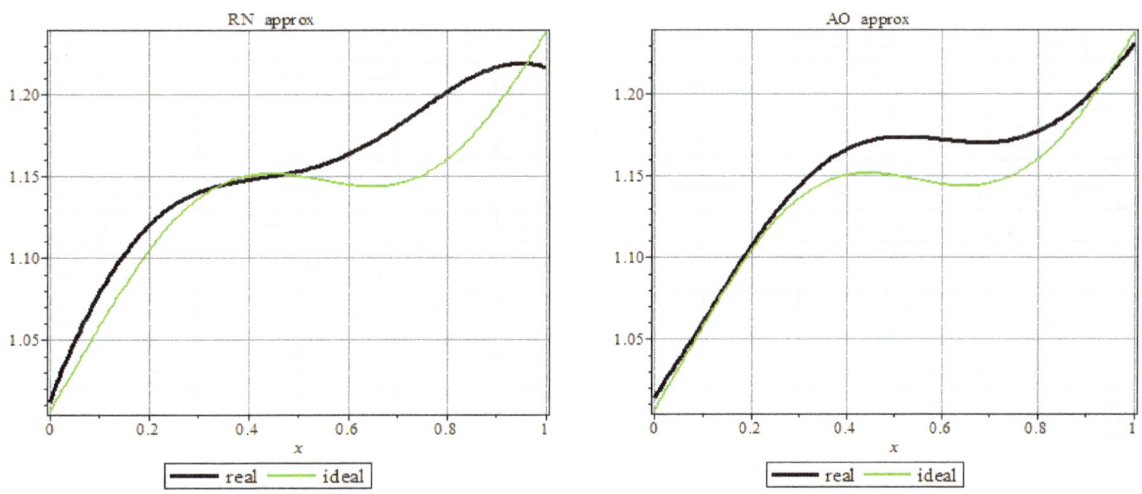

$j = 2, l = 1$ $j = 2, l = 2$

Figure 4. Approximation of functions $\widetilde{r}(t)$ and $\widetilde{s}(t)$ for two athletes based on their work capacity (W) at the point of the CIL.

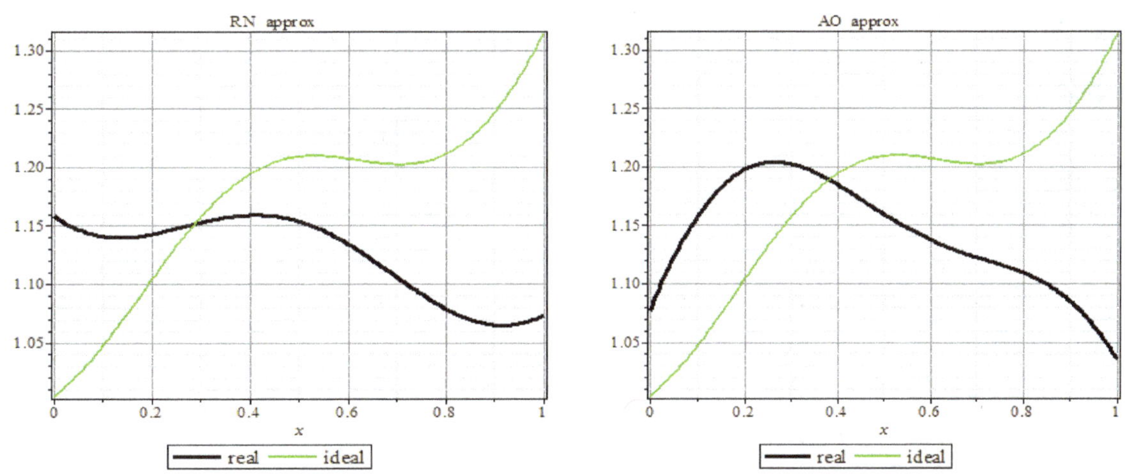

$j = 3, l = 1$ $j = 3, l = 2$

Figure 5. Approximation of functions $\widetilde{r}(t)$ and $\widetilde{s}(t)$ for two athletes based on oxygen uptake (1/min, mL/min/kg) at the point of the CIL.

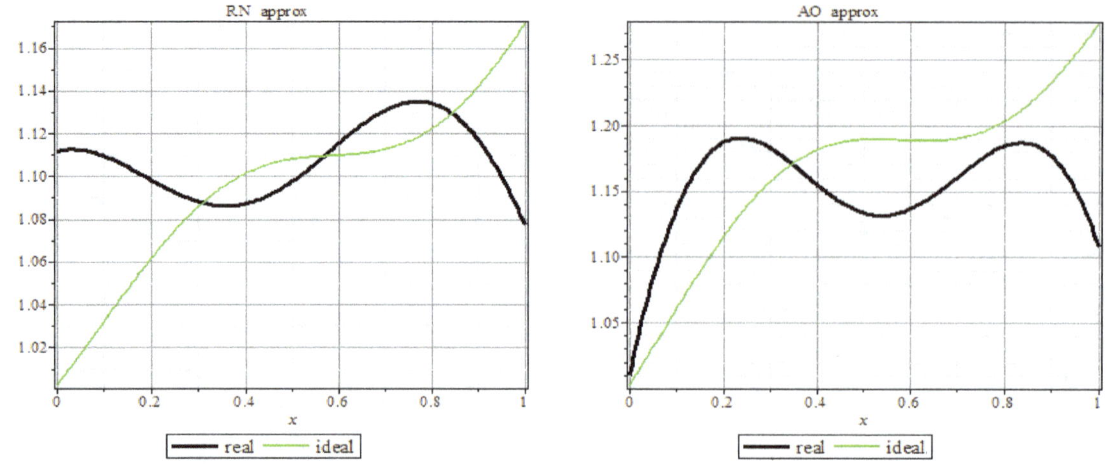

$j = 4, l = 1$ $j = 4, l = 2$

Figure 6. Approximation of functions $\widetilde{r}(t)$ and $\widetilde{s}(t)$ for two athletes based on pulmonary ventilation (1/min) at the point of the CIL.

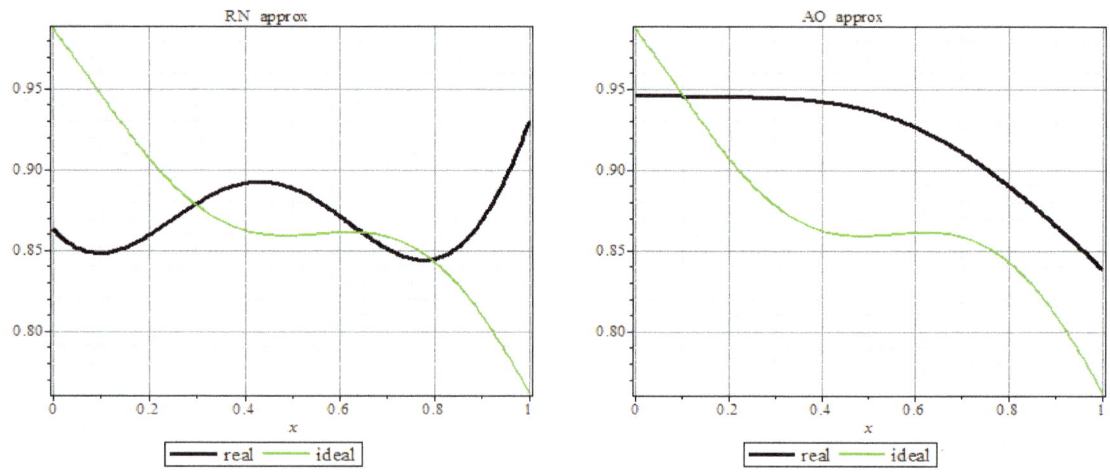

$j = 5, l = 1$ $j = 5, l = 2$

Figure 7. Approximation of the functions $\widetilde{r}(t)$ and $\widetilde{s}(t)$ for two athletes based on the resting heart rate (b/min).

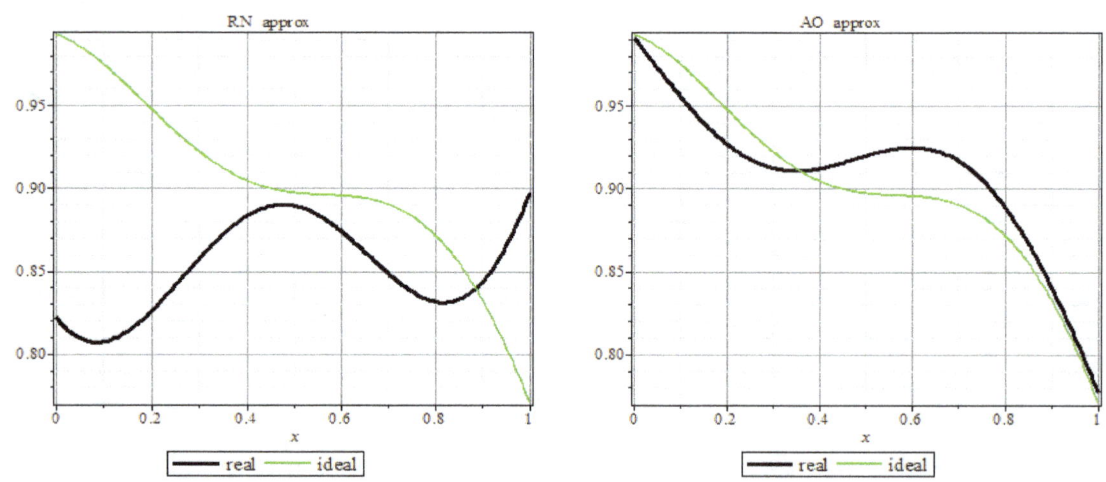

$j = 6, l = 1$ $j = 6, l = 2$

Figure 8. Approximation of the functions $\widetilde{r}(t)$ and $\widetilde{s}(t)$ for two athletes based on their heart rate (b/min) after the standard physical load.

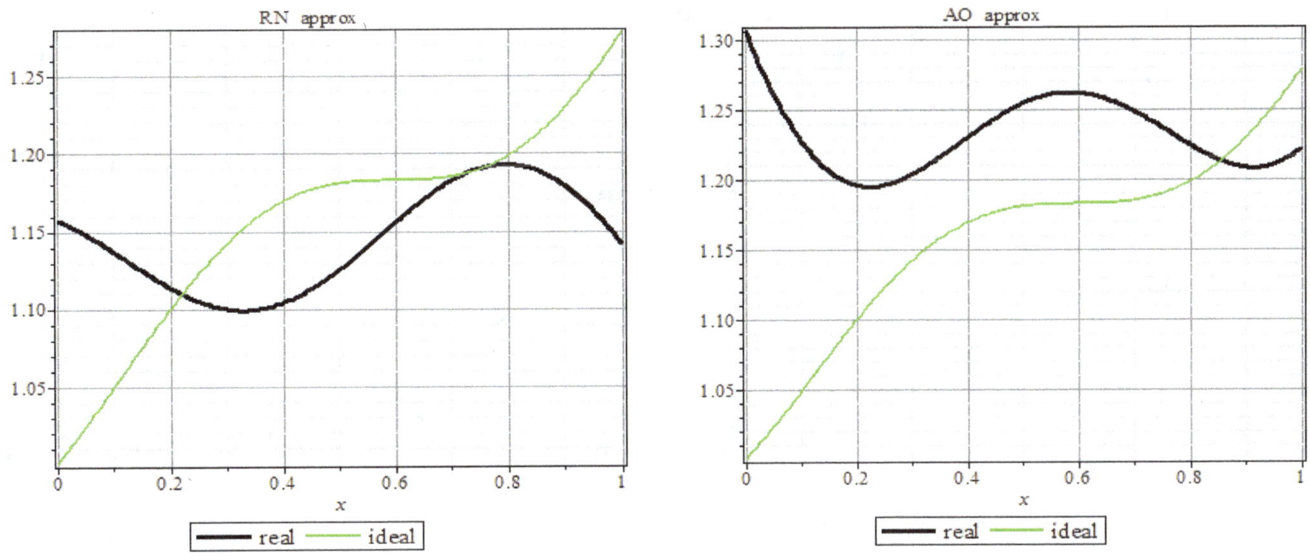

$j = 7, l = 1$ $j = 7, l = 2$

Figure 9. Approximation of functions $\widetilde{r}(t)$ and $\widetilde{s}(t)$ for two athletes based on haemoglobin concentration (g/L).

6. Discussion and Conclusions

The optimization and evaluation of athletic training has become an increasingly important problem for sports scientists, coaches, and athletes [25]. The character of the identification factors of the elite athlete's condition and their integration into the processes of strategic planning, management, and parameter estimation have not been thoroughly investigated, but they are obviously important. To govern the process of elite athletes' training, it is required to consider this process both as a whole and in detail. Not all the aspects of the executed cycle can be in compliance. Therefore, the research should not be performed in terms of the standardized compliance management. The OR methods allow us to define the criteria and the limits of their permissible pursuit [14].

The papers considered do not offer a unique view of past research, and the integration they provide is likely to be only fractional. Therefore, real research data are needed. Usually, the research data seem to be more significant than dimensions. They present (1) the constructs and their relational properties and (2) consistency in the measurement of phenomena. The combinations of these dimensions can yield the following results: (1) theoretical plurality and empirical convergence, (2) the dominant paradigm, (3) fragmentation, and (4) the convergent theory and empirical plurality. The integration will require fostering both taxonomic and methodological commensurability among the different sub-fields and contributing disciplines [26]. The method of parameter estimation of the dynamics of the factors presents a viable research approach in this context [27]. It is particularly useful for the investigation of the process cycle, e.g., its long-term effects and results [28]. Interactive data visualization is an evolving approach, providing wide opportunities for managing and decision making in the case of multidimensional decision problems and planning processes.

The proposed methodology of the athletic training visualization makes it easier to identify the advantages and disadvantages of training and shows when (at what stages of training) physical loads (or training methods and techniques) do not match the adaptive capacities of the tested athletes. It also provides new opportunities for streamlining and optimizing the key factors influencing the process by modifying its components. The authors believe that data visualization allows for achieving a better way of informing athletes and coaches about the mistakes and the right choices in the considered training period. According to Xu and Ding [29], visualization allows for identifying the deviations of the indicator values. The automated alarm system, warning about critical deviations of the process components, could help to optimize the process management and prevent hazardous incidents [30].

To the author's best knowledge, no cases of applying the methodology based on the goal function's approximation by a linear combination of polynomials and trigonometric functions can be found in the scientific literature on sports management. Meanwhile, this methodology, in conjunction with the least squares method, yields fairly accurate approximation results.

The described methodology (algorithm) has a number of advantages over other similar approaches. The visualization of measurement and target approximation functions enables sports scientists and coaches to perform the following functions:

- to supervise the athletes' training process,
- to identify undesirable tendencies,
- to correct them immediately (if necessary).

The proposed Formula (4) for measuring the differences between the indicators and the current and target functions enables the researchers to identify the athlete's condition, as well as showing his/her potential and providing more accurate predictions of sports results.

New opportunities are provided by the proposed method, which:

- Facilitate the understanding and achievement of the goal;
- Make processes clearer to the general public;
- Allow for effective management of the processes.

The provided high-quality, effective alternative solutions to the problem of achieving the final goals of the considered management process give a deeper insight into the effectiveness of a solution, which can support a decision-making process. The principles of the model of the parameter estimation provide visualization capabilities, allowing for evaluating the possibilities to reach the goal, which is the effective management process. This method also allows for predicting the future achievements of athletes.

The proposed model for evaluating the planning of athletic training is used in other sports and can be applied not only to sports practice, but also to the implementation of many various strategies. Moreover, based on the proposed visualization method, an application could be created that would make this method a more widely used technique.

Author Contributions: Conceptualization, S.D.; Data curation, A.K., S.D. and N.K.; Formal analysis, R.D.; Methodology, A.K.; Supervision, S.D.; Visualization, A.K.; Writing—original draft, N.K., R.D.; Writing—review & editing, N.K. and S.D. All authors have read and agreed to the published version of the manuscript.

References

1. Issurin, V.B. Evidence-based prerequisites and precursors of athletic talent: A review. *Sports Med.* **2017**, *47*, 1993–2010. [CrossRef] [PubMed]
2. Jurimae, J. Methods for monitoring training status and their effects on performance in rowing. *Int. SportMed J.* **2008**, *9*, 11–21.
3. Latyshev, M.; Latyshev, S.; Kvasnytsya, O.; Knyazev, A. Performance analysis of freestyle wrestling competitions of the last Olympic cycle. *J. Phys. Educ. Sport* **2017**, *17*, 590–594.
4. Guellich, A.; Seiler, S.; Emrich, E. Training methods and intensity distribution of young world-class rowers. *Int. J. Sports Physiol.* **2009**, *4*, 448–460.
5. Bompa, T.O.; Haff, G. *Periodization: Theory and Methodology of Training*, 6th ed.; Human Kinetics: Champaign, IL, USA, 2018.
6. Zatsiorsky, V.M.; Kraemer, W.J. *Science and Practice of Strength Training*; Human Kinetics: Champaign, IL, USA, 2006.
7. Halson, S.L. Monitoring training load to understand fatigue in athletes. *Sports Med.* **2014**, *44*, 139–147. [CrossRef]

8. Polak, E.; Kulasa, J.; Vencesbrito, A.; Castro, M.; Fernandes, O. Motion analysis systems as optimization training tools in combat sports and martial arts. *Rev. Artes Marciales Asiáticas* **2015**, *10*, 105–123. [CrossRef]
9. Fister, I.; Rauter, S.; Yang, X.-S.; Ljubic, K. Planning the sports training sessions with the bat algorithm. *Neurocomputing* **2015**, *149*, 993–1002. [CrossRef]
10. Dadelo, S.; Turskis, Z.; Zavadskas, E.K.; Kačerauskas, T.; Dadelienė, R. Is the Evaluation of the Students' Values Possible? An Integrated Approach to Determining the Weights of Students' Personal Goals Using Multiple-Criteria Methods. *Eurasia J. Math. Sci. Technol.* **2016**, *12*, 2771–2781. [CrossRef]
11. Armstrong, K.J.; Weidner, T.G.; Walker, S.E. Athletic training approved clinical instructors' reports of real-time opportunities for evaluating clinical proficiencies. *J. Athl. Train.* **2009**, *44*, 630–638. [CrossRef]
12. Li, F.; Zhu, Q.; Chen, Z.; Xue, H.A. A balanced data envelopment analysis cross-efficiency evaluation approach. *Expert Syst. Appl.* **2018**, *106*, 154–168. [CrossRef]
13. Karsu, Ö.; Morton, A. Incorporating balance concerns in resource allocation decisions: A bi-criteria modelling approach. *Omega* **2014**, *44*, 70–82. [CrossRef]
14. Reefke, H.; Sundaram, D. Key themes and research opportunities in sustainable supply chain management—Identification and evaluation. *Omega* **2017**, *66*, 195–211. [CrossRef]
15. Recalde, D.; Torres, R.; Vaca, P. Scheduling the professional Ecuadorian football league by integer programming. *Comput. Oper. Res.* **2013**, *40*, 2478–2484. [CrossRef]
16. Dadelo, S.; Turskis, Z.; Zavadskas, E.K.; Dadelienė, R. Multi-criteria assessment and ranking system of sport team formation based on objective-measured values of criteria set. *Expert. Syst. Appl.* **2014**, *41*, 6106–6113. [CrossRef]
17. Jones, C. Visualization and Optimization. *J. Oper. Res. Soc.* **1997**, *48*, 964. [CrossRef]
18. Laengle, S.; Merigó, M.M.; Miranda, J.; Słowiński, R.; Bomze, I.; Borgonovo, E.; Dyson, R.G.; Oliveira, J.F.; Teunter, R. Forty years of the European Journal of Operational Research: A bibliometric overview. *Eur. J. Oper. Res.* **2017**, *262*, 803–816. [CrossRef]
19. Moritz, D.; Fisher, D.; Ding, B.; Wang, C. Trust, but verify optimistic visualizations of approximate queries for exploring big data. In Proceedings of the 2017 CHI Conference on Human Factors in Computing Systems, CHI'17, Denver, CO, USA, 6–11 May 2017; pp. 2904–2915.
20. Balčiūnas, E. *Didelio Meistriškumo Baidarininkų Rengimas 200 ir 500 m Nuotoliams: Monografija/Training of Elite Kayakers for Distances of 200 and 500 m: The Monograph*; Lietuvos edukologijos universiteto leidykla: Lithuania, Vilnius, 2016. (In Lithuania)
21. Norton, K.; Olds, T. *Antropometrica*; University of New South Wales Press: Sydney, Australia, 1996.
22. Thodens, J.S. *Testing Aerobic Power. Physiological Testing of the High-Performance Athlete*; Human Kinetics: Champaign, IL, USA, 1991.
23. Legendre, A.-M. *Nouvelles Méthodes pour la Détermination des Orbites des Comètes [New Methods for the Determination of the Orbits of Comets]*; F. Didot: Paris, France, 1805. (In French)
24. Strang, G. *Linear Algebra and Its Applications*, 4th ed.; Cengage Learning: Boston, MA, USA, 2006.
25. Spotts, J.D. Global politics and the Olympic Games: Separating the two oldest games in history. *Dickinson J. Int. Law* **1994**, *13*, 103–122.
26. Durand, R.; Grant, R.M.; Madsen, T.L. The expanding domain of strategic management research and the quest for integration. *Strateg. Manag. J.* **2017**, *38*, 4–16. [CrossRef]
27. Krylovas, A.; Dadeliene, R.; Kosareva, N.; Dadelo, S. Comparative evaluation and ranking of the European countries based on the interdependence between human development and internal security indicators. *Mathematics* **2019**, *7*, 293. [CrossRef]
28. Sivarajah, U.; Kamal, M.M.; Irani, Z.; Weerakkody, V. Critical analysis of Big Data challenges and analytical methods. *J. Bus. Res.* **2017**, *70*, 263–286. [CrossRef]
29. Xu, L.; Ding, F. Recursive least squares and multi-innovation stochastic gradient parameter estimation methods for signal modelling. *Circuits Syst. Signal Process.* **2017**, *36*, 1735–1753. [CrossRef]
30. Krylovas, A.; Dadelo, S.; Kosareva, N.; Zavadskas, E.K. Entropy-KEMIRA approach for MCDM problem solution in human resources selection task. *Int. J. Inf. Technol. Decis. Mak.* **2017**, *16*, 1183–1209. [CrossRef]

14

The Role of the Central Limit Theorem in the Heterogeneous Ensemble of Brownian Particles Approach

Silvia Vitali *,†, Iva Budimir †, Claudio Runfola and Gastone Castellani

Department of Physics and Astronomy; University of Bologna, Viale Berti Pichat 6/2, 40127 Bologna, Italy; iva.budimir.os@gmail.com (I.B.); claudio.runfola@gmail.com (C.R.); gastone.castellani@unibo.it (G.C.)

* Correspondence: silvia.vitali4@unibo.it

† These authors contributed equally to this work.

Abstract: The central limit theorem (CLT) and its generalization to stable distributions have been widely described in literature. However, many variations of the theorem have been defined and often their applicability in practical situations is not straightforward. In particular, the applicability of the CLT is essential for a derivation of heterogeneous ensemble of Brownian particles (HEBP). Here, we analyze the role of the CLT within the HEBP approach in more detail and derive the conditions under which the existing theorems are valid.

Keywords: central limit theorem; anomalous diffusion; stable distribution; fractional calculus; power law

---

1. Introduction

The heterogeneous ensemble of Brownian particle (HEBP) [1,2] describes a large class of anomalous diffusion phenomena, observed in many physical and biological systems [3–6]. The HEPB approach is based on the Langevin stochastic equation of diffusion of a free particle, i.e., the mesoscopic description of Brownian motion (Bm). The relaxation time (τ) and the diffusivity (ν) of the particle constitute two important scales of Bm process. In the classic Langevin approach, τ and ν are constant parameters. Instead, in the HEPB approach, it is assumed that τ and ν are time-independent random variables. In HEPB the single-particle trajectory (SPT) follows a classic Langevin dynamics and it is characterized by a stochastic realization of the parameters (τ_i and ν_i for particle i). The random nature of the scales τ and ν of the SPTs mimics the heterogeneity of the environment and/or the heterogeneity of an ensemble of particles diffusing in the environment. In fact, the anomalous diffusion behavior described by HEPB is generated by the heterogeneity of τ and ν values in different SPT realizations.

Long time and space correlation, characteristics of many anomalous diffusion processes [7–9], are often introduced by modifying the laws of the dynamics, by including memory kernels and/or integral operators [10,11] in the equations, for example, the fractional derivatives [12]. These changes in the dynamics introduce often non-Markovianity and/or non-locality in the processes.

The HEPB approach maintains the Markovianity of the process because the fundamental process remains the classical Brownian motion (Bm), but the heterogeneity of the scales involved in the system permits to describe a process with stationary features different from the Bm [13]. We will refer to this heterogeneity as to a *population of scales*, because the values of these scales follow a given probability distribution. Furthermore, the model structure permits to keep the standard dynamical laws, with integer time derivative of physical variables like the velocity (V) and the spatial coordinate (X) of the particle, and to avoid the introduction of fractional time derivatives.

One of the random scales contributing to the anomalous behavior the Langevin description of HEPB [1] is the time scale τ. When the distribution of τ is properly chosen and ν is kept constant, the HEPB describes the same one-time one-point probability density function (PDF) of the fractional Brownian motion (fBm), i.e., a normal distribution with variance (the mean squared displacement of the process, MSD) scaling as a power law of time in the long time limit:

$$\sigma_x^2(t) = \langle (x(t+t_0) - x(t_0))^2 \rangle = D_\alpha t^\alpha \,, \tag{1}$$

where $0 < \alpha \leq 2$ and D_α is the constant playing the role of diffusion coefficient. Depending on the value of the exponent α, it is possible to distinguish what is called super-diffusion and sub-diffusion, associated respectively to super-linear and sub-linear values of the parameter.

The convergence of the PDF to a normal distribution depends on the applicability of the classical central limit theorem (CLT). We will demonstrate later that by choosing properly the population of the time scales according to certain PDFs, both the Gaussian shape of the PDF and the anomalous scaling of the variance can be guaranteed.

The CLT represents a cornerstone in probability theory. It states that when a large amount of one -or multi-dimensional, real-valued and independent (or weakly dependent [14]) random variables are summed, the probability distribution of their sum will tend to the Gaussian distribution $\mathcal{G}$, defined by its characteristic function:

$$\hat{g}_{\mathcal{G}}(k) = exp(-i\mu k - \frac{k^2\sigma}{2}) \,. \tag{2}$$

This result has been generalized by the generalized CLT to a larger class of stable distributions described by the following characteristic function [15]:

$$\hat{g}_\alpha(k) = exp(-i\mu k - C|k|^\alpha[1 + i\beta(sign(k))\omega(k,\alpha)]), \tag{3}$$

where $\alpha, \beta, \mu, C \in \mathbb{R}$, $\omega(k,\alpha) = tan(\alpha\pi/2)$ if $\alpha \neq 1$, else $\omega(k,\alpha) = 2/\pi ln(|k|)$. The Gaussian distribution can be found to be a special yet fundamental case when $\alpha = 2$.

The generalized CLT [15] describes the convergence of the sum of stable variables with also infinite variance, for example, the symmetric Levy stable distribution. The stability property of the symmetric Levy stable distribution is fundamental to obtain a random walk with infinite large displacements as the well known Lévy–Feller diffusion process [8,16,17]. The PDF of this process converges in fact to the non-Gaussian but symmetric Levy stable distributions.

In the following sections, we briefly review the CLT formulation, then we introduce the problem of the convergence in the distribution of a mixture of Gaussian random components with random variances when the variance distribution is particularly extreme. Thereafter we recall the HEBP model formulation and define the sufficient conditions over τ to obtain a fBm-like process.

2. The Classical CLT Formulation

For completeness, we summarize here the most famous versions of the CLT and introduce some useful notation and definitions.

For parameters $\mu \in \mathbb{R}$ and $\sigma \in \mathbb{R}_+$, a normal (or Gaussian) distribution $\mathcal{N}(\mu, \sigma^2)$ is a continuous probability distribution defined by its density function

$$f(x \mid \mu, \sigma^2) = \frac{1}{\sqrt{2\pi\sigma^2}} e^{-\frac{(x-\mu)^2}{2\sigma^2}} \,, \tag{4}$$

where μ and σ^2 are the expectation and variance of the distribution, respectively. For $\mu = 0$ and $\sigma = 1$, we obtain what is usually called the standard normal distribution $\mathcal{N}(0,1)$.

For the sequence of random variables $(X_n)_{n\geq 1}$, we define random variables $(S_n)_{n\geq 1}$ as partial sums $S_n = X_1 + X_2 + \cdots + X_n$. The theory of central limit theorem derives conditions for which there exist sequences of constants $(a_n)_{n\geq 1}, a_n > 0$, and $(b_n)_{n\geq 1}$ such that the sequence $\left(\frac{S_n - b_n}{a_n}\right)_{n\geq 1}$ converges in distribution to a non-degenerate random variable. In particular, CLT describes the convergence to standard normal distribution with constants defined as $a_n^2 = \sum_{k=1}^n Var[X_k]$ and $b_n = \sum_{k=1}^n \mathbb{E}[X_k]$.

Different constraints on variables $X_1, X_2, \ldots$ lead to different versions of the CLT. We will briefly review the most prominent results of the theory of central limit theorems. For a more pedagogical and/or historical perspective, see [18–24].

We start with the case when variables $X_1, X_2, \ldots$ are independent and identically distributed. With additional requirements of finite mean μ and positive and finite variance σ^2, we obtain:

$$\frac{S_n - n\mu}{\sigma\sqrt{n}} \xrightarrow{d} \mathcal{N}(0,1) \quad \text{as} \quad n \to \infty. \tag{5}$$

Dealing with independent, but not necessary identically distributed, random variables $X_1, X_2, \ldots$ with finite variance, we define $\mu_k = \mathbb{E}X_k$, $\sigma_k^2 = VarX_k$ and $s_n^2 = \sum_{k=1}^n \sigma_k^2$ for every $k \geq 1$. To obtain the main result, we need two Lindeberg's conditions:

$$L_1(n) = \max_{1\leq k\leq n} \frac{\sigma_k^2}{s_n^2} \to 0 \quad \text{as} \quad n \to \infty, \tag{6}$$

and

$$L_2(n) = \frac{1}{s_n^2} \sum_{k=1}^{n} \mathbb{E}\left|X_k - m_k\right|^2 I\left\{|X_k - m_k| > \epsilon s_n\right\} \to 0 \quad \text{as} \quad n \to \infty \quad (\text{for every } \epsilon > 0). \tag{7}$$

The Lindeberg–Lévy–Feller theorem provides sufficient and necessary conditions for the following result:

$$\frac{S_n - \mathbb{E}S_n}{s_n} \xrightarrow{d} \mathcal{N}(0,1) \quad \text{as} \quad n \to \infty. \tag{8}$$

Lindeberg and Lévy proved (using different techniques) that if (7) holds, so do (6) and (8). Feller proved that if both (6) and (8) are satisfied, then so is (7).

Since Lindeberg's condition (7) can be hard to verify, we can instead use the Lyapounov's condition which assumes that for some $\delta > 0$, $\mathbb{E}\left|X_k\right|^{2+\delta} < \infty$ (for all $k \geq 1$) and

$$\frac{1}{s_n^{2+\delta}} \sum_{k=1}^{n} \mathbb{E}\left|X_k - \mu_k\right|^{2+\delta} \to 0 \quad \text{as} \quad n \to \infty. \tag{9}$$

If for independent random variables $X_1, X_2, \ldots$ the Lyapounov's condition is satisfied, then the central limit theorem (8) holds. Since the Lyapounov's condition implies the Linderberg's second condition this result follows directly from the Lindeberg–Lévy theorem.

In all versions of the CLT mentioned so far, the assumption of finite variance was crucial. To extend our observations to the case when variance does not exist (or is infinite), we introduce the notion of domains of attraction. We are observing a sequence $X, X_1, X_2, \ldots$ of independent, identically distributed random variables. We say that X, or, equivalently, its distribution function F_X, belongs to the domain of attraction of the (non-degenerate) distribution G if there exist normalizing sequences $(a_n)_{n\geq 1}, a_n > 0$, and $(b_n)_{n\geq 1}$, such that

$$\frac{S_n - b_n}{a_n} \xrightarrow{d} G \quad \text{as} \quad n \to \infty. \tag{10}$$

Another important concept is the one of stable distribution. Retaining the same notion, the distribution X is stable if there exist constants $(c_n)_{n\geq 1}, c_n > 0$, and $(d_n)_{n\geq 1}$ such that $S_n \stackrel{d}{=} c_n X + d_n$ (for all $n \geq 1$).

It can be shown that only stable distributions possess a domain of attraction [18]. The most notable stable distribution is Gaussian and by the classical CLT we know that all distributions X with finite variance belong to the domain of attraction of the Gauss Law. However, there are also some distributions with infinite variance that belong to it. More precisely, it can be shown [25] that random variable X with the distribution function F_X belongs to the domain of the attraction of the Gauss law if and only if

$$\lim_{x \to +\infty} \frac{x^2 \left[1 - F_X(x) + F_X(-x)\right]}{\int_{-x}^{x} t^2 dF_X(t)} = 0. \tag{11}$$

3. CLT for a Population of Gaussian Random Variables

We reviewed the fundamental theorems related to the classical CLT, having the Gaussian distribution as limit distribution of the sum of random variables S_n. The recurrent and sufficient (but not necessary) condition leading to the classical CLT description is that the variance of the i.i.d. random variables that are summed should be finite. However, there exist distributions with infinite variance that fall in the Gaussian domain of attraction [15,25]. In this paragraph, we provide a preparatory example to introduce the role of the CLT in the HEBP. The sum of a population of Gaussian variables with random variances (which may tend to infinity), is here rewritten as the sum of i.i.d. random variables defined as the mixture of random Gaussian components with random variances. If such a mixture has finite variance, the standard CLT conditions are satisfied. In fact, as it will be explained in more details hereafter, the convergence in distribution of the sum to a Gaussian is not always guaranteed when some extreme distribution of the random variance is considered.

Let us consider partial sums of independent Gaussian random variables

$$S_n = \sum_{k=1}^{n} X_k \,, \tag{12}$$

where, denoting with $f_k(x_k)$ the PDF of X_k, we have:

$$f_k(x) \sim N(0, \sigma_k^2)\,. \tag{13}$$

The distribution of the sum of n random variables can be exploited in term of a convolution integral. Thus, we can derive explicitly the limit distribution of Equation (12) by inverting the characteristic function $\phi(\omega)$ of S_n, which corresponds to the product of the characteristics $\phi_k(\omega)$ of X_k:

$$\phi(\omega) = \Pi_{k=1}^{n} \phi_k(\omega), \tag{14}$$

which gives

$$\phi(\omega) = \Pi_{k=1}^{n} \left(e^{-\frac{\omega^2}{2}\sigma_k^2}\right) \tag{15}$$

$$= e^{-\frac{\omega^2}{2}\sum_{k=1}^{n}\sigma_k^2}\,. \tag{16}$$

Let us assume $\sigma_k \sim \sqrt{\Lambda}$, with Λ distributed according to a generic PDF $f(\lambda)$. If the first moment of Λ exists in the limit of large n, by applying the law of large numbers, we can well approximate the Equation (16) in terms of $\mathbb{E}\Lambda$:

$$\phi(\omega) = e^{-\frac{\omega^2}{2}\cdot n \cdot \mathbb{E}\Lambda}\,, \tag{17}$$

which is indeed the characteristic function of a Gaussian distribution with variance $n \cdot \mathbb{E}\Lambda$ for finite expectation of $f(\lambda)$ even if the supremum of Λ does not exists.

The convergence of S_n can be proven using the CLT for the sequence of independent, identically distributed random variables $X, X_1, X_2, \ldots$ with $X \sim \mathcal{N}(0, \Lambda)$. These variables, in general, will not have a Gaussian shape and can equivalently be defined as the product of the independent random variables:

$$X = \sqrt{\Lambda} \cdot Z, \tag{18}$$

where $Z \sim f_1(z) = N(0,1)$, $\Lambda \sim f_2(\lambda)$, $\Lambda \in \mathbf{R}_+$. The PDF $f(x)$ of X can be represented by the integral form [26]

$$f(x) = \int_0^\infty f_1(x/\sqrt{\lambda}) f_2(\lambda) \frac{d\lambda}{\sqrt{\lambda}}. \tag{19}$$

Since Z is a Gaussian distribution, it follows that $\frac{1}{\sqrt{\lambda}} f_1(x/\sqrt{\lambda}) = N(0,\lambda)$. Using Fubini's theorem, now it is easy to compute the second moment of X:

$$VarX = \int_{-\infty}^{\infty} x^2 \int_0^\infty f_1(x/\sqrt{\lambda}) f_2(\lambda) \frac{d\lambda}{\sqrt{\lambda}} dx \tag{20}$$

$$= \int_0^\infty \lambda f_2(\lambda) d\lambda = \mathbb{E}\Lambda. \tag{21}$$

If $\mathbb{E}\Lambda$ is finite the partial sums $S_n = X_1 + \cdots + X_n$ of i.i.d. random variables X_k converge in distribution to a Gaussian

$$\frac{S_n}{\sqrt{n}} \xrightarrow{d} \mathcal{N}(0, \mathbb{E}\Lambda). \tag{22}$$

If $\mathbb{E}\Lambda$ is not finite, the distribution $f(x)$ may fall out of the Gaussian domain of attraction. For example, by choosing Λ (the random variance) to be the extremal Lévy density distribution, it follows that $f(x)$ (the mixture defined by Equation (19)) corresponds to the symmetric Lévy stable distribution [27]. In fact in the case $f(x)$ is itself a stable distribution, like the Levy stable distribution is, its sum belongs to its own domain of attraction.

However, infinite variance is not a synonym of stability. In fact, despite the presence of infinite $\mathbb{E}\Lambda$, under certain constraints on the tail of the distribution $f(x)$, $f(x)$ still satisfies (11) and falls in Gaussian domain of attraction, for example if its PDF for large x is proportional to $x^{-3}, x^{-3}log(x), x^{-3}/log(x)$ [15].

4. Application of the CLT in the HEBP

In the HEBP Langevin model [1] the anomalous time scaling of the ensemble-averaged MSD is generated by the superposition of a population of Bm processes in a similar way to equation (12), where each single process is characterized by its own independent timescale, and with frequency of appearance of such timescale described by the same PDF.

CLT applicability guarantees that after averaging over a properly chosen timescale distribution the shape of the PDF will remain Gaussian despite the time scaling will change from being linear in time to be a power low of time in the long time limit, following Equation (1). In order to show this applicability let first introduce the HEPB construction.

Let us start with the classic Langevin equation describing the dynamics of a free particle moving in a viscous medium (or Bm):

$$dV = -\frac{1}{\tau} V dt + \sqrt{2\nu} dW, \tag{23}$$

where V is the random process representing the particle velocity, τ in classical approach corresponds to the characteristic time scale of the process, i.e., the scale of decorrelation of the system. In the classic Langevin description the timescale is defined by the ratio $\frac{m}{\gamma}$, with m being the mass of the diffusing particle and γ the Stoke's drag force coefficient of the velocity. The multiplicative constant of the Wiener noise increment dW in the square root, ν, represents the velocity diffusivity and is related to the drag term by the fluctuation-dissipation theorem (FDT) [28]. This relation does not depend on the mass of the particle but on the average energy of the environment (the fluid) and the cross-sectional interaction between the medium and the particle moving. The Wiener increment dW is the increment per infinitesimal time induced by the presence of a Gaussian white noise with unit variance and is hence fully characterized by its first two moments:

$$\langle dW(t) \rangle = 0 , \qquad \langle dW(t)^2 \rangle = t . \tag{24}$$

The presence of Gaussian increments in the stochastic equation leads to the stationary state $V \sim N(0, kT/m)$ and, being $V = dX/dt$, to the stationary increments process $X(t) \sim N(x_0, \sigma_x^2(t))$, with $\sigma_x^2(t) = \nu\tau^2 t$.

Let now the parameters ν and τ be time independent random variables: $\nu \sim p_\nu(\nu)$ and $\tau \sim p_\tau(\tau)$. The way it will affect $V(t)$ and $X(t)$ is clear in the case of ν, but more complicated to specify in the case of τ.

Let us consider the velocity defined as a product of random variables $V = \sqrt{\nu}V'$. It is easy to show that $\sqrt{\nu}$ factorizes out from the stochastic differential equation, resulting in the following description of the evolution of V':

$$dV' = -\frac{1}{\tau}V' dt + \sqrt{2} dW . \tag{25}$$

Therefore, the PDF associated to the processes $V(t)$ and $X(t)$ can be derived by applying the same integral formula of Equation (19), eventually producing non-Gaussian PDF and weak ergodicity breaking stochastic processes as result [29–31].

Dealing with random timescales is much more tricky because the variable τ is embedded in the correlation functions and is not possible to factorize it out without simultaneously transforming the time variable. Furthermore, because of the time variable transformation different realizations of the process would not be comparable directly anymore without reverse transformation.

To avoid these complications, we define V' as the superposition of N_τ independent Bm processes each with its own timescale:

$$V'(t) = \frac{1}{N_\tau} \sum_\tau V''(t|\tau) , \tag{26}$$

where $V''(t)$ can still be described by the Equation (25). If the resulting process $V'(t)$ is still a Gaussian process, the previously described approach to derive $V = \sqrt{\nu}V'$ can be applied without further changes. However, all the correlation functions of V' and moments will become the sum of the correlation functions of the single processes $V''(t|\tau)$, which is equivalent to averaging with respect to $p_\tau(\tau)$. A careful choice of this distribution permits to obtain non-linear time scaling of the MSD of V'.

Let us demonstrate the applicability of the CLT explicitly making use of the Equation (17). Assuming that a global stationary state (in the sense of stationary increments) has been reached, the relation between the MSD and the VACF determined by the free particle Langevin dynamics can be expressed by:

$$\sigma_x^2(t, \tau) = 2 \int_0^t (t - s) R(s, \tau) \, ds , \tag{27}$$

where $R(t, \tau) = \nu\tau e^{-t/\tau}$, with ν being an arbitrary constant, is the stationary VACF of the process associated to the realization τ of the timescale, $V''(t|\tau)$.

By omitting time dependence for sake of conciseness, we can define $\lambda = \sigma_x^2 = f(\tau)$, which can be considered as a random variable itself distributed according to the PDF $P(\lambda) = p_\tau(f^{-1}(\lambda)) \cdot \partial_\lambda(f^{-1}(\lambda))$. The average over λ is thus equivalent to computation of the expectation $\langle f(\tau) \rangle$ with respect to τ.

In principle we may compute the expectation after the integration of Equation (27), however, it is much easier to compute it before performing the integration to avoid self-canceling terms:

$$\langle \lambda \rangle = 2 \int_0^t (t - s) \langle R(s, \tau) \rangle_\tau \, ds , \tag{28}$$

For a generic PDF $p_\tau(\tau)$ we obtain:

$$\langle R(t,\tau)\rangle_\tau = \nu \int_0^\infty \tau e^{-t/\tau} p_\tau(\tau) d\tau. \tag{29}$$

This expression is finite for any value of time only if $\langle \tau \rangle$ is finite. Moreover, this is a very important physical condition. In fact, when times goes to zero, $\langle R(t=0,\tau)\rangle_\tau$ is proportional to the average kinetic energy of the system.

The distribution $p_\tau(\tau)$ should have a power-law tail to introduce the desired anomalous time scaling of λ but a finite value of the first moment of τ to maintain CLT applicability. The importance of this assumption can be seen explicitly by solving the integral in Equation (29) for the distribution employed in [1]:

$$p_\tau(\tau) = \frac{\alpha}{\Gamma(1/\alpha)} \frac{1}{\tau} L_\alpha^\alpha(\frac{\tau}{C}), \tag{30}$$

where the constant $C = \langle \tau \rangle \frac{\Gamma(1/\alpha)}{\alpha}$ serves to control the value of the average and $L_\alpha^\alpha(\cdot)$ is the extreme Levy density distribution [10].

By considering the integral representation of the extremal Lévy density distribution and the Euler's gamma function with some more simplification (see Section 3.5, equation 3.109 in [32]), the result in (29) can be represented by the integral form:

$$R(t) = \nu \langle \tau \rangle \frac{1}{2\pi i} \int_{\gamma - i\infty}^{\gamma + i\infty} \frac{\Gamma(z/\alpha+1)\Gamma(-z)}{\Gamma(z+1)} \left(\frac{t}{C}\right)^z dz\,. \tag{31}$$

This expression can be solved through the residues theorem considering the poles $z/\alpha + 1 = -n$ or $z = n$, with $n = 0, 1, 2, \ldots, \infty$, to obtain the short or the long time scaling of the variable. An interested reader can verify the explicit derivation in [1,32]. By plugging this result in Equation (28), without any assumption about time values, we observe that the condition of finite $\langle \tau \rangle$ is necessary to guarantee $\langle \lambda \rangle$ to be finite too, ensuring the Gaussian form of the PDF.

5. Discussions

The CLT has a fundamental role in the HEBP approach and, generally, in the theory of stochastic processes. The domain of attraction of the distribution of the increments determines the shape of the PDF of the stochastic process in the long time limit. In this work, we reviewed the main conditions of the classical CLT, by including also the less known case of distributions with infinite variance which fall in the Gaussian domain (with slower convergence). We proposed and analyzed a preparatory exercise to give the mathematical foundations to understand the approach used in HEBP to generate PDFs with Gaussian shape and non-linear scaling of the variance in time for the long time limit. It is shown that the sum of such a population of Gaussian random variables is mathematically defined by the sum of a more complex, and in general non-Gaussian, i.i.d. random variables. The population of Gaussian distributions can be interpreted, within a Bayesian approach, as the likelihood modulated by the prior distribution of a parameter of the model. The formal randomization of the parameter of the distribution (Equation (19)) is equivalent to the computation of the marginal likelihood, which corresponds indeed to the PDF of the i.i.d. random variables. This approach could be easily generalized to other distributions and parameters for statistical application purposes. The role of CLT in HEPB is then clarified. After recalling the derivation of the model, the conditions obtained in the preparatory example have been explicitly proven.

Author Contributions: Conceptualization, S.V. and G.C.; methodology, S.V. and I.B.; formal analysis, S.V. and I.B.; investigation, S.V., I.B. and C.R.; writing—original draft preparation, S.V., I.B. and C.R.

funded through the Innovative Medicines Initiative (IMI) 2 Joint Undertaking and listed under grant agreement No. 116026.

Acknowledgments: All the authors acknowledge F. Mainardi, G. Pagnini and P. Paradisi for useful discussions and for their past contributions to the theory behind the model analyzed in the present paper.

Abbreviations

The following abbreviations are used in this manuscript:

MDPI	Multidisciplinary Digital Publishing Institute
CLT	Central Limit Theorem
MSD	Mean squared displacement
VACF	Velocity auto-correlation function
Bm	Brownian motion
HEPB	Heterogeneous ensemble of Brownian particles
PDF	Probability density function

References

1. Vitali, S.; Sposini, V.; Sliusarenko, O.; Paradisi, P.; Castellani, G.; Pagnini, G. Langevin equation in complex media and anomalous diffusion. *J. R. Soc. Interface* **2018**, *5*, 20180282. [CrossRef]
2. Sliusarenko, O.Y.; Vitali, S.; Sposini, V.; Paradisi, P.; Chechkin, A.; Castellani, G.; Pagnini, G. Finite–energy Lévy–type motion through heterogeneous ensemble of Brownian particles. *J. Phys. A Math. Theor.* **2019**, *52*, 9. [CrossRef]
3. Golding, I.; Cox, E.C. Physical nature of bacterial cytoplasm. *Phys. Rev. Lett.* **2006**, *96*, 098102. [CrossRef] [PubMed]
4. Hofling, F.; Franosch, T. Anomalous transport in the crowded world of biological cells. *Rep. Prog. Phys.* **2013**, *76*, 046602. [CrossRef] [PubMed]
5. Javer, A.; Kuwada, N.; Long, Z.; Benza, V.; Dorfman, K.; Wiggins, P.; Cicuta, P.; Lagomarsino, M. Persistent super-diffusive motion of escherichia coli chromosomal loci. *Nat. Commun.* **2014**, *5*, 3854. [CrossRef] [PubMed]
6. Caspi, A.; Granek, R.; Elbaum, M. Enhanced diffusion in active intracellular transport. *Phys. Rev. Lett.* **2000**, *85*, 5655–5658. [CrossRef] [PubMed]
7. Mandelbrot, B.B.; Van Ness, J.W. Fractional brownian motions, fractional noises and applications. *SIAM Rev.* **1968**, *10*, 422–437. [CrossRef]
8. Dubkov, A.A.; Spagnolo, B.; Uchaikin, V.V. Lévy flight superdiffusion: An introduction. *Int. J. Bifurc. Chaos* **2008**, *18*, 2649–2672. [CrossRef]
9. Eliazar, I.I.; Shlesinger, M.F. Fractional motions. *Phys. Rep.* **2013**, *527*, 101–129. [CrossRef]
10. Mainardi, F; Luchko, Y.; Pagnini, G. The fundamental solution of the space-time fractional diffusion equation. *Fract. Calc. Appl. Anal.* **2001**, *4*, 153–192.
11. Lutz, E. Fractional Langevin equation. *Phys. Rev. E* **2001**, *64*, 051106. [CrossRef] [PubMed]
12. Mainardi, F. *Fractional Calculus and Waves in Linear Viscoelasticity*; Imperial College Press: London, UK, 2010; p. 340, ISBN 978-1-84816-329-4.
13. D'Ovidio, M.; Vitali, S.; Sposini, V.; Sliusarenko, O.; Paradisi, P.; Castellani, G.; Pagnini, G. Centre-of-mass like superposition of Ornstein–Uhlenbeck processes: A pathway to non-autonomous stochastic differential equations and to fractional diffusion. *Fract. Calc. Appl. Anal.* **2018** *21*, 1420–1435. [CrossRef]
14. Fischer, J.W. *A History of Central Limit Theorem—from Classical to Modern Probability Theory*; Buchwald, J.Z., Berggren, J.L., Lützen, J., Eds; Springer: New York, NY, USA, 2011.
15. Bouchaud, P.; Georges, A. Anomalous diffusion in disordered media: statistical mechanics, models and physical applications. *Phys. Rep. (Rev. Sec. Phys. Lett.)* **1990**, *195*, 127–293. [CrossRef]
16. Montroll, E.W.; Shlesinger, O.F. The wonderful world of random walks. In *Non Equilibrium Phenomena. From Stochastics to Hydrodynamics*; Lebow tz, J.L., Montroll, E.W., Eds.; North-Holland Physics Publishing (Studies in Statistical Mechanics): Amsterdam, The Netherlands, 1984; Volume 11.

17. Gorenflo, R.; De Fabritiis, G.; Mainardi, F. Discrete random walk models for symmetric Lévy–Feller diffusion processes. *Physica A* **1999**, *269*, 79–89. [CrossRef]
18. Gut, A. *Probability: A Graduate Course*; Casella, G., Fienberg, S., Olkin, I., Eds.; Springer: New York, NY, USA, 2005.
19. Le Cam, L.M. The Central Limit Theorem around 1935. *Stat. Sci.* **1986**, *1*, 78–96. [CrossRef]
20. Chebyshev, P.L. Sur deux théorèmes relatifs aux probabilités. *Acta Math.* **1891**, *14*, 305–315.
21. Markov, A.A. Sur les racines de l'équation $e^{x^2}\frac{d^l e^{-x^2}}{dx^l}$. *Bull. Acad. Impériale Sci. St. Pètersbourg* **1898**, *9*, 435–446.
22. Seneta, E. The Central Limit Problem and Linear Least Squares in Pre-Revolutionary Russia: The Background. *Math. Sci.* **1983**, *9*, 37–77.
23. Lindeberg, J.W. Über das Gauss'sche Fehlergesetz. *Skand. Aktuarietidskr.* **1922**, *5*, 217–234.
24. Billingsley, P. The Central Limit Theorem. In *Probability and Measure*, 3rd ed.; John Wiley & Sons, Inc.: Toronto, ON, Canada, 1995; pp. 357–363.
25. Mainardi, F.; Rogosin, S.V. *The Legacy of A.Ya. Khintchine's Work in Probability Theory*; Cambridge Scientific Publishers: Cambridge, UK, 2010.
26. Mainardi, F.; Pagnini, G.; Gorenflo, R. Mellin transform and subordination laws in fractional diffusion processes. *Fract. Calc. Appl. Anal.* **2003**, *6*, 441–459.
27. Pagnini, G.; Paradisi, P. A stochastic solution with Gaussian stationary increments of the symmetric space-time fractional diffusion equation. *Fract. Calc. Appl. Anal.* **2016**, *19*, 408–440. [CrossRef]
28. Kubo, R. The fluctuation-dissipation theorem. *Rep. Progr. Phys.* **1966**, *29*, 255. [CrossRef]
29. Molina-García, D.; Minh Pham, T.; Paradisi, P.; Manzo, C.; Pagnini, G. Fractional kinetics emerging from ergodicity breaking in random media. *Phys. Rev. E* **2016**, *94*, 052147. [CrossRef]
30. Mura, A. Non-Markovian Stochastic Processes and their Applications: from Anomalous Diffusion to Time Series Analysis. *Alma Mater. Tesi* **2008**, *25*, 207.
31. Sposini, V.; Chechkin, A.; Metzler, R. First passage statistics for diffusing diffusivity. *J. Phys. A Math. Theor.* **2018**, *52*, 4. [CrossRef]
32. Vitali, S. Modeling of Birth-Death and Diffusion Processes in Biological Complex Environments. Ph.D. Thesis, University of Bologna, Bologna, Italy, 2018.

15

An Introduction to Space–Time Exterior Calculus

Ivano Colombaro *, Josep Font-Segura and Alfonso Martinez

Department of Information and Communication Technologies, Universitat Pompeu Fabra, 08018 Barcelona, Spain; josep.font@upf.edu (J.F.-S.); alfonso.martinez@upf.edu (A.M.)

* Correspondence: ivano.colombaro@upf.edu.

Abstract: The basic concepts of exterior calculus for space–time multivectors are presented: Interior and exterior products, interior and exterior derivatives, oriented integrals over hypersurfaces, circulation and flux of multivector fields. Two Stokes theorems relating the exterior and interior derivatives with circulation and flux, respectively, are derived. As an application, it is shown how the exterior-calculus space–time formulation of the electromagnetic Maxwell equations and Lorentz force recovers the standard vector-calculus formulations, in both differential and integral forms.

Keywords: exterior calculus; exterior algebra; electromagnetism; Maxwell equations; differential forms; tensor calculus

---

1. Introduction

Vector calculus has, since its introduction by J. W. Gibbs [1] and Heaviside, been the tool of choice to represent many physical phenomena. In mechanics, hydrodynamics and electromagnetism, quantities such as forces, velocities and currents are modeled as vector fields in space, while flux, circulation, divergence or curl describe operations on the vector fields themselves.

With relativity theory, it was observed that space and time are not independent but just coordinates in space–time [2] (pp. 111–120). Tensors like the Faraday tensor in electromagnetism were quickly adopted as a natural representation of fields in space–time [3] (pp. 135–144). In parallel, mathematicians such as Cartan generalized the fundamental theorems of vector calculus, i.e., Gauss, Green, and Stokes, by means of differential forms [4]. Later on, differential forms were used in Hamiltonian mechanics, e. g. to calculate trajectories as vector field integrals [5] (pp. 194–198).

A third extension of vector calculus is given by geometric and Clifford algebras [6], where vectors are replaced by multivectors and operations such as the cross and the dot products subsumed in the geometric product. However, the absence of an explicit formula for the geometric product hinders its widespread use. An alternative would have been the exterior algebra developed by Grassmann which nevertheless has received little attention in the literature [7]. An early work in this direction was Sommerfeld's presentation of electromagnetism in terms of six-vectors [8].

We present a generalization of vector calculus to exterior algebra and calculus. The basic notions of space–time exterior algebra, introduced in Section 2, are extended to exterior calculus in Section 3 and applied to rederive the equations of electromagnetism in Section 4. In contrast to geometric algebra, our interior and exterior products admit explicit formulations, thereby merging the simplicity and intuitiveness of standard vector calculus with the power of tensors and differential forms.

2. Exterior Algebra

Vector calculus is constructed around the vector space $\mathbf{R}^3$, where every point is represented by three spatial coordinates. In relativity theory the underlying vector space is $\mathbf{R}^{1+3}$ and time is treated as a coordinate in the same footing as the three spatial dimensions. We build our theory in space–time with k time dimensions and n space dimensions. The number of space–time dimensions is thus $k+n$

and we may refer to a (k,n)- or $(k+n)$-space–time, $\mathbf{R}^{k+n}$. We adopt the convention that the first k indices, i.e., $i = 0,\ldots,k-1$, correspond to time components and the indices $i = k,\ldots,k+n-1$ represent space components and both k and n are non-negative integers. A point or position in this space–time is denoted by $\mathbf{x}$, with components $\{x_i\}_{i=0}^{k+n-1}$ in the canonical basis $\{\mathbf{e}_i\}_{i=0}^{k+n-1}$, that is

$$\mathbf{x} = \sum_{i=0}^{k+n-1} x_i \mathbf{e}_i. \tag{1}$$

Given two arbitrary canonical basis vectors $\mathbf{e}_i$ and $\mathbf{e}_j$, then their dot product in space–time is

$$\mathbf{e}_i \cdot \mathbf{e}_j = \begin{cases} -1, & i = j,\, 0 \leqslant i \leqslant k-1, \\ +1, & i = j,\, k \leqslant i \leqslant k+n-1, \\ 0, & i \neq j. \end{cases} \tag{2}$$

For convenience, we define the symbol $\Delta_{ij} = \mathbf{e}_i \cdot \mathbf{e}_j$ as the metric diagonal tensor in Minkowski space–time [2] (pp. 118–120), such that time unit vectors $\mathbf{e}_i$ have negative norm $\Delta_{ii} = -1$, whereas space unit vectors $\mathbf{e}_i$ have positive norm $\Delta_{ii} = +1$. The **dot product** of two vectors $\mathbf{x}$ and $\mathbf{y}$ is the extension by linearity of the product in Equation (2), namely

$$\mathbf{x} \cdot \mathbf{y} = \sum_{i=0}^{k+n-1} x_i y_i \Delta_{ii} = -\sum_{i=0}^{k-1} x_i y_i + \sum_{i=k}^{k+n-1} x_i y_i. \tag{3}$$

2.1. Grade, Multivectors, and Exterior Product

In addition to the $(k+n)$-dimensional vector space $\mathbf{R}^{k+n}$ with canonical basis vectors $\mathbf{e}_i$, there exist other natural vector spaces indexed by ordered lists $I = (i_1,\ldots,i_m)$ of m non-identical space and time indices for every $m = 0,\ldots,k+n$. As there are $\binom{k+n}{m}$ such lists, the dimension of this vector space is $\binom{k+n}{m}$. We shall refer to m as **grade** and to these vectors as **multivectors** or grade-m vectors if we wish to be more specific. A general multivector can be written as

$$\mathbf{v} = \sum_I v_I \mathbf{e}_I, \tag{4}$$

where the summation extends to all possible ordered lists with m indices. If $m = 0$, the list is empty and the corresponding vector space is $\mathbf{R}$. The direct sum of these vector spaces for all m is a larger vector space of dimension $\sum_{m=0}^{k+n}\binom{k+n}{m} = 2^{k+n}$, the exterior algebra. In tensor algebra, multivectors correspond to antisymmetric tensors of rank m. In this paper, we study vector fields $\mathbf{v}(\mathbf{x})$, namely multivector-valued functions $\mathbf{v}$ varying over the space–time position $\mathbf{x}$.

The basis vectors for any grade m may be constructed from the canonical basis vectors $\mathbf{e}_i$ by means of the exterior product (also known as wedge product), an operation denoted by $\wedge$ [9] (p. 2). We identify the vector $\mathbf{e}_I$ for the ordered list $I = (i_1, i_2, \ldots, i_m)$ with the exterior product of $\mathbf{e}_{i_1}, \mathbf{e}_{i_2}, \ldots, \mathbf{e}_{i_m}$:

$$\mathbf{e}_I = \mathbf{e}_{i_1} \wedge \mathbf{e}_{i_2} \wedge \cdots \wedge \mathbf{e}_{i_m}. \tag{5}$$

In general, we may compute the exterior product as follows. Let two basis vectors $\mathbf{e}_I$ and $\mathbf{e}_J$ have grades $m = |I|$ and $m' = |J|$, where $|I|$ and $|J|$ are the lengths of the respective index lists. Let $(I,J) = \{i_1,\ldots,i_m,j_1,\ldots,j_{m'}\}$ denote the concatenation of I and J, let $\sigma(I,J)$ denote the signature of the permutation sorting the elements of this concatenated list of $m+m'$ indices, and let $\varepsilon(I,J)$ denote the resulting sorted list, which we also denote by $I+J$. Then, the exterior product $\mathbf{e}_I$ of $\mathbf{e}_J$ is defined as

$$\mathbf{e}_I \wedge \mathbf{e}_J = \sigma(I,J)\mathbf{e}_{\varepsilon(I,J)}. \tag{6}$$

The **exterior product** of vectors $\mathbf{v}$ and $\mathbf{w}$ is the bilinear extension of the product in Equation (6),

$$\mathbf{v} \wedge \mathbf{w} = \sum_{I,J} v_I w_J \, \mathbf{e}_I \wedge \mathbf{e}_J. \tag{7}$$

Since permutations with repeated indices have zero signature, the exterior product is zero if $m + m' > k + n$ or more generally if both vectors have at least one index in common. Therefore, the exterior product is either zero or a vector of grade $m + m'$. Further, the exterior product is a skew-commutative operation, as we can also write Equation (6) as $\mathbf{e}_I \wedge \mathbf{e}_J = (-1)^{|I||J|}\mathbf{e}_J \wedge \mathbf{e}_I$.

At this point, we define the dot product $\cdot$ for arbitrary grade-m basis vectors $\mathbf{e}_I$ and $\mathbf{e}_J$ as

$$\mathbf{e}_I \cdot \mathbf{e}_J = \Delta_{I,J} = \Delta_{i_1,j_1}\Delta_{i_2,j_2}\cdots\Delta_{i_m,j_m}, \tag{8}$$

where I and J are the ordered lists $I = (i_1, i_2, \ldots, i_m)$ and $J = (j_1, j_2, \ldots, j_m)$. As before, we extend this operation to arbitrary grade-m vectors by linearity.

Finally, we define the complement of a multivector. For a unit vector $\mathbf{e}_I$ with grade m, its Grassmann or Hodge **complement** [10] (pp. 361–364), denoted by $\mathbf{e}_I^{\mathcal{H}}$, is the unit $(k + n - m)$-vector

$$\mathbf{e}_I^{\mathcal{H}} = \Delta_{I,I}\sigma(I, I^c)\mathbf{e}_{I^c}, \tag{9}$$

where I^c is the complement of the list I, namely the ordered sequence of indices not included in I. As before, $\sigma(I, I^c)$ is the signature of the permutation sorting the elements of the concatenated list (I, I^c) containing all space–time indices. In other words $\mathbf{e}_{I^c}$ is the basis vector of grade $k + n - m$ whose indices are in the complement of I. In addition, we define the inverse complement transformation as

$$\mathbf{e}_I^{\mathcal{H}^{-1}} = \Delta_{I^c,I^c}\sigma(I^c, I)\mathbf{e}_{I^c}. \tag{10}$$

We extend the complement and its inverse to general vectors in the space–time algebra by linearity.

2.2. Interior Products

While the exterior product of two multivectors is an operation that outputs a multivector whose grade is the addition of the input grades, the dot product takes two multivectors of identical grade and subtracts their grades, yielding a zero-grade multivector, i.e., a scalar. We say that the exterior product raises the grade while the dot product lowers the grade. In this section, we define the left and right interior products of two multivectors as operations that lower the grade and output a multivector whose grade is the difference of the input multivector grades.

As always, we start by defining the operation for the canonical basis vectors. Let $\mathbf{e}_I$ and $\mathbf{e}_J$ be two basis vectors of respective grades $|I|$ and $|J|$. The left interior product, denoted by $\lrcorner$, is defined as

$$\mathbf{e}_I \lrcorner \mathbf{e}_J = \Delta_{I,I}\sigma\big(\varepsilon(I, J^c)^c, I\big)\mathbf{e}_{\varepsilon(I,J^c)^c}. \tag{11}$$

If I is not a subset of J, that is when there are elements in I not present in J, e. g. for $|I| > |J|$, the signature of the permutation sorting the concatenated list $\big(\varepsilon(I, J^c)^c, I\big)$ is zero as there are repeated indices in the list to be sorted, and the left interior product is zero. Otherwise, if I is a subset of J, the permutation rearranges the indices in J in such a way that the last $|I|$ positions coincide with I and $\varepsilon(I, J^c)^c$ represents the first $|J| - |I|$ elements in the rearranged sequence, that is $\varepsilon(I, J^c)^c = J \backslash I$.

The right interior product, denoted by $\llcorner$, of two basis vectors $\mathbf{e}_I$ and $\mathbf{e}_J$ is defined as

$$\mathbf{e}_I \llcorner \mathbf{e}_J = \Delta_{J,J}\sigma\big(J, \varepsilon(I^c, J)^c\big)\mathbf{e}_{\varepsilon(I^c,J)^c}. \tag{12}$$

As with the left interior product, if J is a subset of I, $\varepsilon(I^c, J)^c = I \backslash J$ then the permutation rearranges the indices in I so that the first $|J|$ positions coincide with J, otherwise the right interior product is zero.

In general, we have that $\mathbf{e}_I \rfloor \mathbf{e}_J = \mathbf{e}_J \lfloor \mathbf{e}_I (-1)^{|I|(|J|-|I|)}$, as verified in Appendix A.1. We note that these interior products are not commutative, unless either $|J| - |I|$ or $|I|$ is an even number, e. g. when $|I| = |J|$, in which case both interior products coincide with the dot product of the two vectors. The interior products may therefore be seen as generalizations of the dot product.

As with the dot and the exterior products, the value of the interior products does not depend on the choice of basis and we may thus compute the **left interior product** of two vectors **v** and **w** as

$$\mathbf{v} \rfloor \mathbf{w} = \sum_{I,J} v_I w_J \, \mathbf{e}_I \rfloor \mathbf{e}_J \,, \tag{13}$$

and a similar expression holds for the **right interior product** $\mathbf{v} \lfloor \mathbf{w}$. Both are grade-lowering operations, as the left (resp. right) interior product is either zero or a multivector of grade $m' - m$ (resp. $m - m'$).

The interior products are not independent operations from the exterior product, as they can be expressed in terms of the latter, the Hodge complement and its inverse (proved in Appendix A.2):

$$\mathbf{e}_I \rfloor \mathbf{e}_J = \left(\mathbf{e}_I \wedge \mathbf{e}_J^{\mathcal{H}}\right)^{\mathcal{H}^{-1}}, \tag{14}$$

$$\mathbf{e}_I \lfloor \mathbf{e}_J = \left(\mathbf{e}_I^{\mathcal{H}^{-1}} \wedge \mathbf{e}_J\right)^{\mathcal{H}}. \tag{15}$$

If **u** and **v** are 1-vectors and **w** is an r-vector, then we have the following expression

$$\mathbf{u} \rfloor (\mathbf{v} \wedge \mathbf{w}) = (-1)^r (\mathbf{u} \cdot \mathbf{v}) \mathbf{w} + \mathbf{v} \wedge (\mathbf{u} \rfloor \mathbf{w}), \tag{16}$$

as proved in Appendix A.3. This expression can be seen as a generalization of the vectorial expression

$$\mathbf{a} \times (\mathbf{b} \times \mathbf{c}) = (\mathbf{a} \cdot \mathbf{c})\mathbf{b} - (\mathbf{a} \cdot \mathbf{b})\mathbf{c} \tag{17}$$

in the vector space $\mathbf{R}^3$, i.e., a $k = 0$, $n = 3$ space–time. This fact is built of the realization that the cross product between two vectors **v** and **w** can be expressed in the following alternative ways

$$\mathbf{v} \times \mathbf{w} = (\mathbf{v} \wedge \mathbf{w})^{\mathcal{H}^{-1}} = \mathbf{v} \rfloor \mathbf{w}^{\mathcal{H}^{-1}} = \mathbf{v} \rfloor \mathbf{w}^{\mathcal{H}}. \tag{18}$$

Whenever it holds that $I \subseteq J$, the interior and exterior products are related by the following:

$$(\mathbf{e}_I \rfloor \mathbf{e}_J) \wedge \mathbf{e}_I = \Delta_{I,J} \mathbf{e}_J, \tag{19}$$

$$\mathbf{e}_I \wedge (\mathbf{e}_J \lfloor \mathbf{e}_I) = \Delta_{I,J} \mathbf{e}_J. \tag{20}$$

Having introduced the basic notions of space–time exterior algebra, the next section focuses on operations with elements in the exterior algebra, namely integrals and derivatives of vector fields.

3. Integrals and Derivatives of Vector Fields: Circulation and Flux

3.1. Oriented Integrals

Integrals are, together with derivatives, the fundamental mathematical objects of **calculus**. For example, operations on vectors fields lying in exterior algebra such as the flux and the circulation are expressed in terms of integrals over high-dimensional geometric objects. The integral of an m-graded vector field **v** over a hypersurface $\mathcal{V}^m$ of the same dimension, denoted as

$$\int_{\mathcal{V}^m} \mathrm{d}^m\mathbf{x} \cdot \mathbf{v}, \tag{21}$$

is the limit of the Riemann sums for the dot product $\mathrm{d}^m\mathbf{x} \cdot \mathbf{v}$ over points in the hypersurface, where $\mathrm{d}^m\mathbf{x}$ is an m-dimensional infinitesimal vector element. For any $\ell = 0, \ldots, k + n$, the infinitesimal vector element $\mathrm{d}^\ell \mathbf{x}$ is given by the sum of all possible differentials for ℓ-dimensional hypersurfaces in a (k, n) space–time, and is represented in the canonical basis as

$$\mathrm{d}^\ell \mathbf{x} = \sum_{I=(i_1,\ldots,i_\ell)} \mathrm{d}x_I \mathbf{e}_I, \tag{22}$$

where for a given list $I = (i_1, \ldots, i_\ell)$ each differential is given by $dx_I = \mathrm{d}x_{i_1} \cdots \mathrm{d}x_{i_\ell}$.

As in traditional calculus, the integral in Equation (21) exhibits coordinate invariance, while the integrand $\mathrm{d}^m\mathbf{x} \cdot \mathbf{v}$ is regarded as an oriented object. Orientation is well defined for integrals along a curve from one point to another, or integrals over a surface oriented at the direction of the normal to the surface. Switching the extreme points of the curve, or taking the opposite direction of the normal would induce a change of sign in the line and surface integrals. In our generalization of vector calculus, a positive orientation is implicit in the ordering of the canonical basis. The skew-symmetry property of the exterior product Equation (6) may introduce sign changes to compensate an eventual change of orientation after changes of coordinates such as permutations of the space–time components.

For a given hypersurface $\mathcal{V}^m$, a convenient transformation for solving the integral in Equation (21) is one such that, at a given point $\mathbf{x}$ in the hypersurface, the infinitesimal vector element $\mathrm{d}^m\mathbf{x}$ has one component that is *tangent* to the hypersurface at that point. Let $\mathbf{e}_\parallel$ be a unit m-graded vector parallel to $\mathcal{V}^m$ at point $\mathbf{x}$, and let $\mathbf{e}'_0, \ldots, \mathbf{e}'_{k+n-1}$ form an orthonormal basis of $\mathbf{R}^{k+n}$ such that $\mathbf{e}_\parallel = \mathbf{e}'_{k+n-m} \wedge \cdots \wedge \mathbf{e}'_{k+n-1}$ for the given point $\mathbf{x}$ in $\mathcal{V}^m$. This change of coordinates from the canonical basis to the new basis is described by a unitary matrix U, dependent on $\mathbf{x}$, and that satisfies

$$\mathbf{e}_0 \wedge \cdots \wedge \mathbf{e}_{k+n-1} = \det(U)\, \mathbf{e}'_0 \wedge \cdots \wedge \mathbf{e}'_{k+n-1}. \tag{23}$$

Being a unitary matrix, the determinant of U is ± 1. Assuming an orientation-preserving change of coordinates, that is $\det(U) = 1$, the infinitesimal vector element in Equation (22) for $\ell = m$ can be expressed as

$$\mathrm{d}^m\mathbf{x} = \mathrm{d}x_\parallel \mathbf{e}_\parallel + \sum_{I=(i_1,\ldots,i_m)\,:\, I\cap\perp\neq\varnothing} \mathrm{d}x_I \mathbf{e}'_I, \tag{24}$$

where $\perp = \{0, \ldots, k+n-m-1\}$ is the set of indices for the unit vectors in the new basis orthogonal to $\mathcal{V}^m$. Since all elements in the summation in Equation (24) have at least one differential element lying outside the integration hypersurface, their integrals vanish and therefore

$$\int_{\mathcal{V}^m} \mathrm{d}^m\mathbf{x} = \int_{\mathcal{V}^m} dx_\parallel \mathbf{e}_\parallel. \tag{25}$$

In analogy to $\mathbf{e}_\parallel$, a multivector of grade m, we define a unit $(k+n-m)$-grade vector $\mathbf{e}_\perp$ normal to $\mathcal{V}^m$ at point $\mathbf{x}$ such that $\mathbf{e}_\perp \wedge \mathbf{e}_\parallel = \mathbf{e}_0 \wedge \cdots \wedge \mathbf{e}_{k+n-1}$. From Equation (10), we see that one such normal multivector with the correct orientation is

$$\mathbf{e}_\perp = \frac{\mathbf{e}_\parallel^{\mathcal{H}^{-1}}}{\mathbf{e}_\parallel^{\mathcal{H}^{-1}} \cdot \mathbf{e}_\parallel^{\mathcal{H}^{-1}}}. \tag{26}$$

For the common spaces considered in vector calculus, $\mathbf{R}^2$ and $\mathbf{R}^3$, and according to Equation (23), orientation-preserving changes of coordinates must respectively satisfy $\mathbf{e}_\perp \wedge \mathbf{e}_\parallel = \mathbf{e}_0 \wedge \mathbf{e}_1$ and $\mathbf{e}_\perp \wedge \mathbf{e}_\parallel = \mathbf{e}_0 \wedge \mathbf{e}_1 \wedge \mathbf{e}_2$, where $\mathbf{e}_\perp$ is the basis element normal to $\mathcal{V}^m$. These two equalities turn out to describe the counterclockwise (resp. right-hand rule) orientation when $\mathbf{e}_\perp$ conventionally points outside an integration path for $\mathbf{R}^2$ (resp. a surface for $\mathbf{R}^3$) [5] (pp. 184–185).

Building on the concepts and operations of circulation and flux in vector calculus, the right and left interior products lead to general definitions of circulation and flux of multivector fields in exterior algebra along and across hypersurfaces of arbitrary number of dimensions.

3.2. Circulation and Flux of Multivector Fields

Definition 1. *The circulation of a vector field $\mathbf{v}(\mathbf{x})$ of grade m along an ℓ-dimensional hypersurface $\mathcal{V}^\ell$, denoted by $\mathcal{C}(\mathbf{v}, \mathcal{V}^\ell)$, is given by*

$$\mathcal{C}(\mathbf{v}, \mathcal{V}^\ell) = \int_{\mathcal{V}^\ell} \mathrm{d}^\ell \mathbf{x} \llcorner \mathbf{v}. \tag{27}$$

Expressing the vector field in the canonical basis and using the definition of $\mathrm{d}^\ell \mathbf{x}$ in Equation (22), the circulation can be specified in some cases of interest. For $\ell = m$, the circulation reads

$$\int_{\mathcal{V}^m} \mathrm{d}^m \mathbf{x} \cdot \mathbf{v} = \sum_{I=(i_1,\dots,i_m)} \Delta_{I,I} \int_{\mathcal{V}^m} dx_I v_I. \tag{28}$$

For instance, for $\ell = m = 1$ and $\mathbf{R}^n$, this formula recovers the definition the circulation of a vector field along a closed path with the appropriate orientation.

Alternatively, using Equation (25), we note that $\mathbf{v}$ is integrated along the direction of $\mathbf{e}_{||}$, tangential to the hypersurface, in an orientation-preserving change of coordinates, that is

$$\int_{\mathcal{V}^m} \mathrm{d}^m \mathbf{x} \llcorner \mathbf{v} = \int_{\mathcal{V}^m} \mathrm{d}x_{||}\, \mathbf{e}_{||} \llcorner \mathbf{v}. \tag{29}$$

Intuitively, the circulation Equation (27) measures the alignment of an m-vector field $\mathbf{v}$ with respect to $\mathcal{V}^\ell$ for any ℓ and m, with the circulation being an $(\ell - m)$-vector if $\ell \geqslant m$ and zero otherwise.

Definition 2. *The flux of a vector field $\mathbf{v}(\mathbf{x})$ of grade m across an ℓ-dimensional hypersurface $\mathcal{V}^\ell$, denoted by $\mathcal{F}(\mathbf{v}, \mathcal{V}^\ell)$, is given by*

$$\mathcal{F}(\mathbf{v}, \mathcal{V}^\ell) = \int_{\mathcal{V}^\ell} \mathrm{d}^\ell \mathbf{x}^{\mathcal{H}^{-1}} \lrcorner \mathbf{v}. \tag{30}$$

Expressing both $\mathbf{v}$ and $\mathrm{d}^\ell \mathbf{x}$ in the canonical basis, and using the inverse Hodge operation in Equation (10), the flux in the special case of $\ell = k + n - m$ can be written as

$$\int_{\mathcal{V}^\ell} \mathrm{d}^\ell \mathbf{x}^{\mathcal{H}^{-1}} \cdot \mathbf{v} = \sum_{I=(i_1,\dots,i_m)} \sigma(I, I^c) \int_{\mathcal{V}^\ell} dx_{I^c} v_I. \tag{31}$$

As an example in $\mathbf{R}^3$, the flux of a vector field $\mathbf{v}$ through a surface $\mathcal{V}^2$ reads

$$\int_{\mathcal{V}^2} \mathrm{d}^2 \mathbf{x}^{\mathcal{H}^{-1}} \cdot \mathbf{v} = \int_{\mathcal{V}^2} \sum_{I, i \notin I} \mathrm{d}x_I \sigma(i, I) \mathbf{e}_i \cdot \mathbf{v}. \tag{32}$$

The right-hand side of Equation (32) is a conventional surface integral, upon the identification of $\sum_{I, i \notin I} \mathrm{d}x_I \sigma(i, I) \mathbf{e}_i$ as an infinitesimal surface element $\mathrm{d}\mathbf{S}$.

Alternatively, using the analogous of Equation (25) for the differential vector element $\mathrm{d}^\ell \mathbf{x}^{\mathcal{H}^{-1}}$, the equivalent to Equation (29) for the flux is

$$\int_{\mathcal{V}^\ell} \mathrm{d}^\ell \mathbf{x}^{\mathcal{H}^{-1}} \lrcorner \mathbf{v} = \int_{\mathcal{V}^\ell} \mathrm{d}x_{||}\, \mathbf{e}_{||}^{\mathcal{H}^{-1}} \lrcorner \mathbf{v}. \tag{33}$$

This equation implies that $\mathbf{v}$ is integrated along a normal component to the hypersurface since $\mathbf{e}_{||}^{\mathcal{H}^{-1}}$ is a multivector of grade $k + n - \ell$ orthogonal to $\mathcal{V}^\ell$. Intuitively, the flux Equation (30) measures the magnitude of the multivector field crossing the hypersurface. In general, the flux is a vector of grade $(m + \ell - n - k)$ if $\ell \geqslant k + n - m$ and zero otherwise. For instance, if $\ell = k + n$, the flux of $\mathbf{v}$ over an

$(k+n)$-dimensional hypersurface $\mathcal{V}^{k+n}$ gives the integral of $\mathbf{v}$ over $\mathcal{V}^{k+n}$, an extension of the volume integral to $\mathbf{R}^{k+n}$,

$$\int_{\mathcal{V}^{k+n}} \mathrm{d}^{k+n}\mathbf{x}^{\mathcal{H}^{-1}} \lrcorner\, \mathbf{v} = \int_{\mathcal{V}^{k+n}} \mathrm{d}x_{i_1,\cdots,i_{k+n}} \mathbf{v}, \tag{34}$$

where we used the relation $1^{\mathcal{H}} = \mathbf{e}_{i_1,\ldots,i_{k+n}}$, implying that $\mathrm{d}^{k+n}\mathbf{x}^{\mathcal{H}^{-1}} = \mathrm{d}x_{i_1,\cdots,i_{k+n}}$, and that $1 \lrcorner\, \mathbf{v} = \mathbf{v}$.

3.3. Exterior and Interior Derivatives

In vector calculus, extensive use is made of the nabla operator ∇, a vector operator that takes partial space derivatives. For instance, operations such as gradient, divergence or curl are expressed in terms of this operator. In our case, we need the generalization to (k,n) space–time to the differential vector operator $\boldsymbol{\partial}$, defined as $(-\partial_0, -\partial_2, \ldots, -\partial_{k-1}, \partial_k, \ldots, \partial_{k+n-1})$, that is

$$\boldsymbol{\partial} = \sum_{i=0}^{k+n-1} \Delta_{ii} \mathbf{e}_i \partial_i. \tag{35}$$

For a given vector field $\mathbf{v}$ of grade m, we define the **exterior derivative** of $\mathbf{v}$ as $\boldsymbol{\partial} \wedge \mathbf{v}$, namely

$$\boldsymbol{\partial} \wedge \mathbf{v} = \sum_{i=0}^{k+n-1} \sum_{I} \Delta_{ii} \partial_i v_I \sigma(i, I)\, \mathbf{e}_{\varepsilon(i,I)}. \tag{36}$$

The grade of the exterior derivative of $\mathbf{v}$ is $m+1$, unless $m = k+n$, in which case the exterior derivative is zero, as can be deduced from the fact that all signatures are zero.

In addition, we define the **interior derivative** of $\mathbf{v}$ as $\boldsymbol{\partial} \lrcorner\, \mathbf{v}$, namely

$$\boldsymbol{\partial} \lrcorner\, \mathbf{v} = \sum_{i,I:\, i\in I} \partial_i v_I \sigma(I\backslash i, i) \mathbf{e}_{I\backslash i}. \tag{37}$$

The grade of the interior derivative of $\mathbf{v}$ is $m-1$, unless $m = 0$, in which case the interior derivative is zero, as implied by the fact that the grade of $\boldsymbol{\partial}$ is larger than the grade of $\mathbf{v}$. Using Equation (16) with $\mathbf{u} = \boldsymbol{\partial}$ and assuming that $\mathbf{v}$ and $\mathbf{w}$ are 1-vectors, we obtain a generalization of Leibniz's product rule

$$\boldsymbol{\partial} \lrcorner\, (\mathbf{v} \wedge \mathbf{w}) = \mathbf{v}(\boldsymbol{\partial} \cdot \mathbf{w}) - (\boldsymbol{\partial} \cdot \mathbf{v})\mathbf{w}. \tag{38}$$

The formulas for the exterior and interior derivatives allow us express some common expressions in vector calculus. For a scalar function ϕ, its gradient is given by its exterior derivative $\nabla\phi = \boldsymbol{\partial} \wedge \phi$, while for a vector field $\mathbf{v}$, its divergence $\nabla \cdot \mathbf{v}$ is given by its interior derivative $\nabla \cdot \mathbf{v} = \boldsymbol{\partial} \lrcorner\, \mathbf{v}$. From Equation (16) we further observe that for a scalar function ϕ we recover the relation

$$\nabla \cdot (\nabla\phi) = (\nabla \cdot \nabla)\phi. \tag{39}$$

In addition, for a vector fields $\mathbf{v}$ in $\mathbf{R}^3$, taking into account Equation (18) then the curl can be variously expressed as

$$\nabla \times \mathbf{v} = (\nabla \wedge \mathbf{v})^{\mathcal{H}^{-1}} = \nabla \lrcorner\, \mathbf{v}^{\mathcal{H}^{-1}} = \nabla \lrcorner\, \mathbf{v}^{\mathcal{H}}. \tag{40}$$

This formula allows us to write the curl of a vector field $\nabla \times \mathbf{v}$ in terms of the exterior and interior products and the Hodge complement, while generalizing both the cross product and the curl to grade-m vector fields in space–time algebras with different dimensions. Moreover, from Equation (16) we can recover for $r = 1$ the well-known formula for the curl of the curl of a vector,

$$\nabla \times (\nabla \times \mathbf{v}) = \nabla(\nabla \cdot \mathbf{v}) - \nabla^2 \mathbf{v}. \tag{41}$$

It is easy to verify that the exterior derivative of an exterior derivative is zero, as is the interior derivative of an interior derivative, that is for any vector field $\mathbf{v}$, we have that

$$\partial \wedge (\partial \wedge \mathbf{v}) = 0 \tag{42}$$

$$\partial \lrcorner (\partial \lrcorner \mathbf{v}) = 0. \tag{43}$$

In regard to the vector space $\mathbf{R}^3$, and using Equation (18), these expressions imply the well-known facts that the curl of the gradient and the divergence of the curl are zero:

$$\nabla \times (\nabla \phi) = (\nabla \wedge (\nabla \wedge \phi))^{\mathcal{H}^{-1}} = 0 \tag{44}$$

$$\nabla \cdot (\nabla \times \mathbf{v}) = \nabla \lrcorner (\nabla \lrcorner \mathbf{v}^{\mathcal{H}}) = 0. \tag{45}$$

3.4. Stokes Theorem for the Circulation

In vector calculus in $\mathbf{R}^3$, the Kelvin-Stokes theorem for the circulation of a vector field $\mathbf{v}$ of grade 1 along the boundary $\partial\mathcal{V}^2$ of a bidimensional surface $\mathcal{V}^2$ relates its value to that of the surface integral of the curl of the vector field over the surface itself. In the notation used in the previous section, the surface integral is the flux of the curl of the vector field across the surface and this theorem reads

$$\int_{\partial\mathcal{V}^2} \mathrm{d}\mathbf{x} \cdot \mathbf{v} = \int_{\mathcal{V}^2} \mathrm{d}^2\mathbf{x}^{\mathcal{H}^{-1}} \cdot (\nabla \times \mathbf{v}). \tag{46}$$

Taking into account the identity $\nabla \times \mathbf{v} = (\nabla \wedge \mathbf{v})^{\mathcal{H}^{-1}}$ in Equation (40), we rewrite the right-hand side in Equation (46) as

$$\begin{aligned} \textstyle\int_{\mathcal{V}^2} \mathrm{d}^2\mathbf{x}^{\mathcal{H}^{-1}} \cdot (\nabla \times \mathbf{v}) &= \textstyle\int_{\mathcal{V}^2} \mathrm{d}^2\mathbf{x}^{\mathcal{H}^{-1}} \cdot (\nabla \wedge \mathbf{v})^{\mathcal{H}^{-1}} \\ &= \textstyle\int_{\mathcal{V}^2} \mathrm{d}^2\mathbf{x} \cdot (\nabla \wedge \mathbf{v}), \end{aligned} \tag{47}$$

where we used that $\mathbf{u} \cdot \mathbf{w} = \mathbf{u}^{\mathcal{H}^{-1}} \cdot \mathbf{w}^{\mathcal{H}^{-1}} = \mathbf{u}^{\mathcal{H}} \cdot \mathbf{w}^{\mathcal{H}}$ for vectors $\mathbf{u}$, $\mathbf{w}$. The *flux* of the curl of the vector field *across* a surface is also the *circulation* of the exterior derivative of the vector field *along* that surface.

The generalized Stokes theorem for differential forms [4] (p. 80) allows us to extend the Kelvin-Stokes theorem to multivectors of any grade m as we do in the following theorem.

Theorem 1. *The circulation of a grade-m vector field $\mathbf{v}$ along the boundary $\partial\mathcal{V}^\ell$ of an ℓ-dimensional hypersurface $\mathcal{V}^\ell$ is equal to the circulation of the exterior derivative of $\mathbf{v}$ along $\mathcal{V}^\ell$:*

$$\mathcal{C}(\mathbf{v}, \partial\mathcal{V}^\ell) = \mathcal{C}(\partial \wedge \mathbf{v}, \mathcal{V}^\ell). \tag{48}$$

As hinted at above, the role of the vector curl in the right-hand side of Equation (46) is played by the exterior derivative in this generalized theorem.

Proof. We start by stating the generalized Stokes Theorem for differential forms [4] (p. 80)

$$\int_{\partial\mathcal{V}^\ell} \omega = \int_{\mathcal{V}^\ell} \mathrm{d}\omega, \tag{49}$$

where ω is a differential form and $\mathrm{d}\omega$ its exterior derivative, represented by the operator

$$\mathrm{d} = \sum_j \mathrm{d}x_j \partial_j. \tag{50}$$

Expressing the circulations in Equation (48) by means of the integrals in Equation (27), we obtain

$$\int_{\partial\mathcal{V}^\ell} \mathrm{d}^{\ell-1}\mathbf{x} \llcorner \mathbf{v} = \int_{\mathcal{V}^\ell} \mathrm{d}^\ell\mathbf{x} \llcorner (\partial \wedge \mathbf{v}). \tag{51}$$

In the integral in the left-hand side of Equation (51), the integrand is a differential form $\omega = \mathrm{d}^{\ell-1}\mathbf{x} \llcorner \mathbf{v}$. After expanding the interior product using the definitions of $\mathrm{d}^{\ell-1}\mathbf{x}$ and $\mathbf{v}$ we obtain

$$\omega = \Big(\sum_{J_{\ell-1}} \mathrm{d}x_J \, \mathbf{e}_J\Big) \llcorner \Big(\sum_{I_m} v_I \, \mathbf{e}_I\Big) = \sum_{J_{\ell-1}, I_m \,:\, I \subseteq J} \Delta_{I,I}\sigma(I, \varepsilon(I, J^c)^c) v_I \, \mathrm{d}x_J \, \mathbf{e}_{\varepsilon(I,J^c)^c}. \tag{52}$$

Then, computing the exterior derivative of this form with Equation (50) gives

$$\mathrm{d}\omega = \sum_{J_{\ell-1}, I_m \,:\, I \subseteq J} \sum_{j \notin J} \Delta_{I,I}\sigma(I, \varepsilon(I, J^c)^c) \partial_j v_I \sigma(j, J) \, \mathrm{d}x_{\varepsilon(j,J)} \, \mathbf{e}_{\varepsilon(I,J^c)^c}. \tag{53}$$

We next write down the integrand in the right-hand side of Equation (51), $\mathrm{d}^\ell\mathbf{x} \llcorner (\partial \wedge \mathbf{v})$, that is

$$\begin{aligned}
\mathrm{d}^\ell\mathbf{x} \llcorner (\partial \wedge \mathbf{v}) &= \Big(\textstyle\sum_{K_{\ell+1}} \mathrm{d}x_K \, \mathbf{e}_K\Big) \llcorner \Big(\textstyle\sum_{I_m} \sum_{j \notin I} \Delta_{j,j}\partial_j v_I \, \sigma(j, I) \, \mathbf{e}_{\varepsilon(j,I)}\Big) \\
&= \textstyle\sum_{K_\ell, I_m \,:\, \varepsilon(j,I) \subseteq K_\ell} \sum_{j \notin I} \Delta_{j,j}\partial_j v_I \, \mathrm{d}x_K \, \sigma(j, I) \Delta_{\varepsilon(j,I),\varepsilon(j,I)} \, \sigma(\varepsilon(j, I), \varepsilon(K^c, \varepsilon(j, I))^c) \, \mathbf{e}_{\varepsilon(K^c,\varepsilon(j,I))^c} \\
&= \textstyle\sum_{K_\ell, I_m \,:\, \varepsilon(j,I) \subseteq K_{\ell+1}} \sum_{j \notin I} \Delta_{I,I}\partial_j v_I \, \mathrm{d}x_K \, \sigma(j, I) \sigma(\varepsilon(j, I), \varepsilon(K^c, \varepsilon(j, I))^c) \, \mathbf{e}_{\varepsilon(K^c,\varepsilon(j,I))^c},
\end{aligned} \tag{54}$$

and verify that it coincides with exterior derivative in Equation (53). As the set of m indices I_m is included in the sets $J_{\ell-1}$ or K_ℓ in Equation (53) or (54), we may write $K_\ell = \varepsilon(J_{\ell-1}, j)$ for some $j \notin J_{\ell-1}$. Then, we obtain the following chain of equalities for the basis elements in Equations (53) and (54):

$$\mathbf{e}_{\varepsilon(K^c,\varepsilon(j,I))^c} = \mathbf{e}_{\varepsilon(K^c \cup \{j\} \cup I)^c} = \mathbf{e}_{\varepsilon(J^c \backslash \{j\} \cup \{j\} \cup I)^c} = \mathbf{e}_{\varepsilon(J^c \cup I)^c} = \mathbf{e}_{\varepsilon(I,J^c)^c}. \tag{55}$$

Therefore, and using that $\varepsilon(J^c \backslash \{j\}, \varepsilon(j, I))^c = J \backslash I$, we can write Equation (54) as

$$\mathrm{d}^\ell\mathbf{x} \llcorner (\partial \wedge \mathbf{v}) = \sum_{J_{\ell-1}, I_m, j \notin I \,:\, \varepsilon(j,I) \subseteq J \cup \{j\}} \Delta_{I,I}\partial_j v_I \, \mathrm{d}x_{\varepsilon(J,j)} \, \sigma(j, I) \sigma(\varepsilon(j, I), J \backslash I) \, \mathbf{e}_{\varepsilon(I,J^c)^c}. \tag{56}$$

Comparing Equation (56) with Equation (53), the expressions coincide if this identity holds:

$$\sigma(j, J)\sigma(I, J \backslash I) = \sigma(j, I)\sigma(j + I, J \backslash I). \tag{57}$$

To prove Equation (57) we exploit that the σ are permutation signatures and that the signature of the composition of permutations is the product of the respective signatures. We proceed with the help of a visual aid in Figure 1, which depicts the identity between two different ways of sorting the concatenated list $(j, I, J \backslash I)$. On the left column we first sort the list $(I, J \backslash I)$ to obtain J and then sort the list (j, J). On the right column, we first sort the list (j, I) and then the list $(j + I, J \backslash I)$. This proves Equation (57) and the theorem.

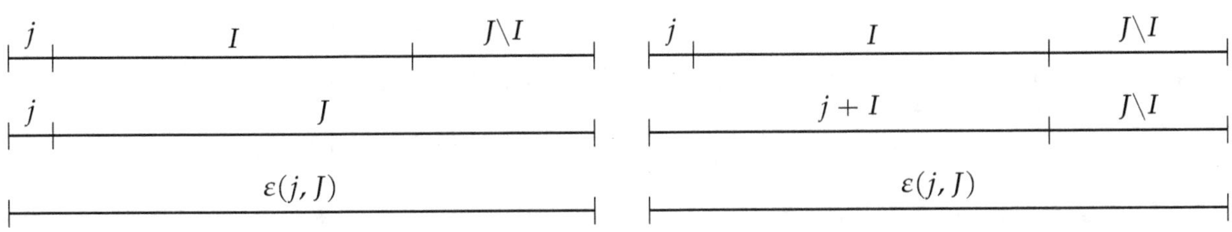

Figure 1. Visual aid for the identity among permutations in Equation (57).

Finally, we note that, had we defined the circulation with the left interior product, we would have got an incompatible relation in Equation (57), which could not be solved. □

3.5. Stokes Theorem for the Flux

In vector calculus in $\mathbf{R}^3$, the Gauss theorem relates the volume integral of the divergence of a vector field $\mathbf{v}$ over a region $\mathcal{V}^3$ to the surface integral of the vector field over the region boundary $\partial\mathcal{V}^3$. In the notation used in previous sections, and taking into account that both the surface integral and the volume integral can be expressed as fluxes for $\mathbf{R}^3$, this theorem reads

$$\int_{\partial\mathcal{V}^3} \mathrm{d}^2\mathbf{x}^{\mathcal{H}-1} \cdot \mathbf{v} = \int_{\mathcal{V}^3} \mathrm{d}^3\mathbf{x}^{\mathcal{H}-1}(\nabla \cdot \mathbf{v}). \tag{58}$$

Making use of the identity $\nabla \cdot \mathbf{v} = \nabla \lrcorner \mathbf{v}$, we can rewrite the right-hand side in Equation (58) as

$$\int_{\mathcal{V}^3} \mathrm{d}^3\mathbf{x}^{\mathcal{H}-1}(\nabla \cdot \mathbf{v}) = \int_{\mathcal{V}^3} \mathrm{d}^3\mathbf{x}^{\mathcal{H}-1}(\nabla \lrcorner \mathbf{v}). \tag{59}$$

In other words, the Gauss theorem relates the flux of the interior derivative of a vector field $\mathbf{v}$ across a region $\mathcal{V}^3$ to the flux of the vector field itself across the region boundary $\partial\mathcal{V}^3$.

The generalized Stokes theorem for differential forms allows us to extend the Gauss theorem to multivectors of any grade m as we do in the following theorem.

Theorem 2. *The flux of a grade-m vector field $\mathbf{v}$ across the boundary $\partial\mathcal{V}^\ell$ of an ℓ-dimensional hypersurface $\mathcal{V}^\ell$ is equal to the flux of the interior derivative of $\mathbf{v}$ across $\mathcal{V}^\ell$:*

$$\mathcal{F}(\mathbf{v}, \partial\mathcal{V}^\ell) = \mathcal{F}(\partial \lrcorner \mathbf{v}, \mathcal{V}^\ell). \tag{60}$$

Proof. Expressing the fluxes in Equation (60) by means of the integrals in Equation (30), we obtain

$$\int_{\partial\mathcal{V}^\ell} \mathrm{d}^{\ell-1}\mathbf{x}^{\mathcal{H}-1} \lrcorner \mathbf{v} = \int_{\mathcal{V}^\ell} \mathrm{d}^\ell\mathbf{x}^{\mathcal{H}-1} \lrcorner (\partial \lrcorner \mathbf{v}). \tag{61}$$

As in the proof of Theorem 1, we apply the Stokes theorem for differential forms in Equation (49) upon the identifications ω with $\mathrm{d}^{\ell-1}\mathbf{x}^{\mathcal{H}-1} \lrcorner \mathbf{v}$ and $\mathrm{d}\omega$ with $\mathrm{d}^\ell\mathbf{x}^{\mathcal{H}-1} \lrcorner (\partial \lrcorner \mathbf{v})$. First, for ω, we get

$$\begin{aligned}\left(\textstyle\sum_{J_{\ell-1}} \mathrm{d}x_J \Delta_{J^c,J^c}\sigma(J^c, J)\mathbf{e}_{J^c}\right) \lrcorner \left(\textstyle\sum_{I_m} v_I\, \mathbf{e}_I\right) &= \textstyle\sum_{J_{\ell-1},I_m\,:\,J^c\subseteq I} v_I\, \mathrm{d}x_J \sigma(J^c, J)\sigma(\varepsilon(J^c, I^c)^c, J^c)\,\mathbf{e}_{\varepsilon}(J^c, I^c)^c \\ &= \textstyle\sum_{J_{\ell-1},I_m\,:\,J^c\subseteq I} v_I\, \mathrm{d}x_J \sigma(J^c, J)\sigma(I\backslash J^c, J^c)\,\mathbf{e}_{I\backslash J^c}.\end{aligned} \tag{62}$$

Now, taking the exterior derivative of Equation (62), we obtain

$$\mathrm{d}\omega = \sum_{J_{\ell-1},I_m\,:\,J^c\subseteq I}\ \sum_{j\notin J} \partial_j v_I\, \mathrm{d}x_{\varepsilon(j,J)}\sigma(j, J)\sigma(J^c, J)\sigma(I\backslash J^c, J^c)\,\mathbf{e}_{I\backslash J^c}. \tag{63}$$

This quantity should be equal to $\mathrm{d}^\ell\mathbf{x}^{\mathcal{H}-1} \lrcorner (\partial \lrcorner \mathbf{v})$ in the right-hand side of Equation (61), which we expand as

$$\begin{aligned}\mathrm{d}^\ell\mathbf{x}^{\mathcal{H}-1} \lrcorner (\partial \lrcorner \mathbf{v}) &= \left(\textstyle\sum_{K_\ell} \mathrm{d}x_K \Delta_{K^c,K^c}\sigma(K^c, K)\mathbf{e}_{K^c}\right) \lrcorner \left(\textstyle\sum_{I:j\in I} \partial_j v_I \sigma(I\backslash j, j)\mathbf{e}_{I\backslash j}\right) \\ &= \textstyle\sum_{K_\ell,I_m\,:\,K^c\subseteq I\backslash j} \sum_{j\in I} \partial_j v_I\, \mathrm{d}x_K \sigma(K^c, K)\sigma(I\backslash j, j)\sigma(I\backslash j\backslash K^c, K^c)\,\mathbf{e}_{I\backslash j\backslash K^c}.\end{aligned} \tag{64}$$

We first consider the sets in the summations in the alternative expressions for $\mathrm{d}\omega$, Equations (63) and (64). Since J^c contains j and is a subset of I, but K^c does not contain j and is also a subset of I (with

$j \in I$), then we can assert that $K = J \cup \{j\}$ so that the conditions in the summations are equivalent. The basis elements coincide and so do the differentials and derivatives, and it remains to verify the identity

$$\sigma(j, J)\sigma(J^c, J)\sigma(I \backslash J^c, J^c) = \sigma(K^c, K)\sigma(I \backslash j, j)\sigma(I \backslash j \backslash K^c, K^c). \tag{65}$$

With the definition $L = I \backslash J^c$, and expressed in terms of j, J, and L, this condition gives

$$\sigma(j, J)\sigma(J^c, J)\sigma(L, J^c) = \sigma(J^c \backslash j, J + j)\sigma(J^c \backslash j + L, j)\sigma(L, J^c \backslash j). \tag{66}$$

Multiplying both sides of the equation by $\sigma(J^c, J)$, $\sigma(J^c \backslash j, J + j)$ and $\sigma(J^c \backslash j, j)$, and taking into account that the square of a signature is $+1$, we obtain

$$\sigma(J^c \backslash j, j)\sigma(L, J^c)\sigma(j, J)\sigma(J^c \backslash j, J + j) = \sigma(L, J^c \backslash j)\sigma(J^c \backslash j + L, j)\sigma(J^c \backslash j, j)\sigma(J^c, J). \tag{67}$$

We start by simplifying Equation (67) by noting that

$$\sigma(J^c \backslash j, j)\sigma(L, J^c) = \sigma(L, J^c \backslash j)\sigma(J^c \backslash j + L, j), \tag{68}$$

with help of the visual aid in Figure 2. The permutations on the left column first merge $(J^c \backslash j)$ with j and then the resulting J^c with L. Similarly, on the right column, we start with L, $(J^c \backslash j)$ and $\{j\}$, then concatenate $(L, J^c \backslash j)$ and then add j, getting the same result as the left column.

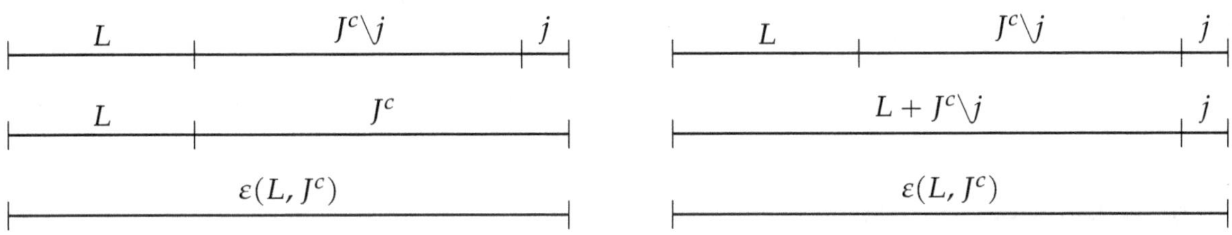

Figure 2. Visual aid for the identity $\sigma(J^c \backslash j, j)\sigma(L, J^c) = \sigma(L, J^c \backslash j)\sigma(J^c \backslash j + L, j)$.

Therefore, we have reduced Equation (67) to the simpler form

$$\sigma(j, J)\sigma(J^c \backslash j, J + j) = \sigma(J^c \backslash j, j)\sigma(J^c, J), \tag{69}$$

which we prove with the aid depicted in Figure 3. On the left column, j and J are first merged and then the concatenation $(J^c \backslash j, J + j)$ gives the sorted $\varepsilon(J^c, J)$. On the right column, after sorting $(J^c \backslash j)$ with j, merging it with J leads to the same final sequence.

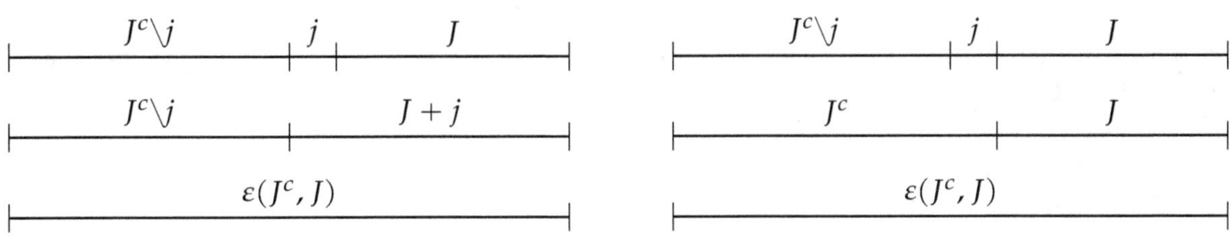

Figure 3. Visual aid for the identity $\sigma(j, J)\sigma(J^c \backslash j, J + j) = \sigma(J^c \backslash j, j)\sigma(J^c, J)$.

□

4. An Application to Electromagnetism in 1+3 Dimensions

In this section, we show how to recover the standard form of Maxwell equations and Lorentz force in $1 + 3$ dimensions from a formulation with *exterior calculus* involving an electromagnetic bivector field $\mathbf{F}$ and a 4-dimensional current density vector $\mathbf{J}$. In the appropriate units, the bivector field $\mathbf{F}$ can be decomposed as $\mathbf{F} = \mathbf{F_E} + \mathbf{F_B}$, where $\mathbf{F_E}$ contains the electric-field $\mathbf{E}$ time-space components and $\mathbf{F_B}$

contains the space-space components for the magnetic field **B**. Similarly, the current density depends on the charge density ρ and the spatial current density **j**. More specifically,

$$\mathbf{J} = \rho\mathbf{e}_0 + \mathbf{j} \tag{70}$$

$$\mathbf{F} = \mathbf{F_E} + \mathbf{F_B} = \mathbf{e}_0 \wedge \mathbf{E} + \mathbf{B}^{\mathcal{H}}. \tag{71}$$

Here the Hodge complement acts only on the space components, and $\mathbf{B}^{\mathcal{H}} = \mathbf{B}^{\mathcal{H}^{-1}}$. The bivector field **F** is closely related to the Faraday tensor, a rank-2 antisymmetric tensor.

Maxwell equations, in their differential form, constrain the divergence of the electric and the magnetic field, Equations (72) and (73), respectively, and the curl of **E** and **B**, namely Equations (74) and (75) [11] (p. 4-1).

$$\nabla \cdot \mathbf{E} = \rho \tag{72}$$

$$\nabla \cdot \mathbf{B} = 0 \tag{73}$$

$$\nabla \times \mathbf{E} = -\partial_0 \mathbf{B} \tag{74}$$

$$\nabla \times \mathbf{B} = \partial_0 \mathbf{E} + \mathbf{j}. \tag{75}$$

We refer to Equations (73) and (74) as homogeneous Maxwell equations and to Equations (72) and (75) as inhomogeneous Maxwell equations, as they include the fields and the sources given by charge and current densities. In exterior-calculus notation, both pairs of equations can be combined into simple multivector equations,

$$\boldsymbol{\partial} \wedge \mathbf{F} = 0 \tag{76}$$

$$\boldsymbol{\partial} \lrcorner \mathbf{F} = \mathbf{J}, \tag{77}$$

where $\boldsymbol{\partial}$ is the differential operator $\boldsymbol{\partial} = -\partial_0 \mathbf{e}_0 + \nabla$ for $k = 1$ and $n = 3$. As a consistency check, note that the wedge product raises the grade of **F**, and the zero in Equation (76) is the zero trivector; also, as the left interior product lowers the grade of **F**, both sides of Equation (77) relate space–time vectors.

Next to Maxwell equations, the Lorentz force density $\boldsymbol{f}$ characterizes, after integrating over the appropriate region, the force exerted by the electromagnetic field upon a system of charges described by the charge and current densities ρ and **j** [11] (pp. 13-1–13-3),

$$\boldsymbol{f} = \rho\mathbf{E} + \mathbf{j} \times \mathbf{B}. \tag{78}$$

In relativistic form, the Lorentz force density becomes a four-dimensional vector **f** [2] (pp. 153–157). The time component of this vector is $\mathbf{j} \cdot \mathbf{E}$, the power dissipated per unit of volume, or after integrating over the appropriate region, the rate of work being done on the charges by the fields. In exterior-calculus notation, the Lorentz force density vector can be computed as a left interior product, namely

$$\mathbf{f} = \mathbf{J} \lrcorner \mathbf{F}. \tag{79}$$

4.1. Equivalence of the Lorentz Force Density

In this section, we prove that Equation (79) indeed recovers the relativistic Lorentz force density by verifying that its components in both vector-calculus and exterior-calculus coincide. From the definitions of **J** and **F**, and using the distributive property of the interior product, we get

$$\begin{aligned} \mathbf{f} &= (\rho\mathbf{e}_0 + \mathbf{j}) \lrcorner (\mathbf{F_E} + \mathbf{F_B}) \\ &= \rho\mathbf{e}_0 \lrcorner \mathbf{F_E} + \rho\mathbf{e}_0 \lrcorner \mathbf{F_B} + \mathbf{j} \lrcorner \mathbf{F_E} + \mathbf{j} \lrcorner \mathbf{F_B} \\ &= \rho\mathbf{e}_0 \lrcorner (\mathbf{e}_0 \wedge \mathbf{E}) + \rho\mathbf{e}_0 \lrcorner \mathbf{B}^{\mathcal{H}} + \mathbf{j} \lrcorner (\mathbf{e}_0 \wedge \mathbf{E}) + \mathbf{j} \lrcorner \mathbf{B}^{\mathcal{H}^{-1}}. \end{aligned} \tag{80}$$

Some straightforward calculations give $\mathbf{e}_0 \,\lrcorner\, (\mathbf{e}_0 \wedge \mathbf{E}) = \mathbf{E}$, $\mathbf{e}_0 \,\lrcorner\, \mathbf{B}^{\mathcal{H}} = 0$, and $\mathbf{j} \,\lrcorner\, (\mathbf{e}_0 \wedge \mathbf{E}) = \mathbf{e}_0 \mathbf{j} \cdot \mathbf{E}$. In addition, the formula for the left interior product in Equation (18) gives $\mathbf{j} \,\lrcorner\, \mathbf{B}^{\mathcal{H}^{-1}} = (\mathbf{j} \wedge \mathbf{B})^{\mathcal{H}^{-1}} = \mathbf{j} \times \mathbf{B}$, where the cross product is only valid for three dimensions. With these calculations, we obtain

$$\mathbf{f} = \rho \mathbf{E} + \mathbf{e}_0 \mathbf{j} \cdot \mathbf{E} + \mathbf{j} \times \mathbf{B}, \tag{81}$$

namely, a time-component $\mathbf{j} \cdot \mathbf{E}$ and a spatial component equal to the Lorentz force density $\rho \mathbf{E} + \mathbf{j} \times \mathbf{B}$.

4.2. Equivalence of the Differential Form of Maxwell Equations

In this section, we prove that Equation (76) indeed recovers the homogeneous Maxwell equations and that Equation (77) recovers the inhomogeneous Maxwell equations.

First, we observe that the exterior derivative $\boldsymbol{\partial} \wedge \mathbf{F}$ gives a trivector with 4 components, while the homogeneous Maxwell equations are a scalar, Equation (73), and a vector, Equation (74). We shall verify that the scalar equation turns out to be given by the trivector component $\mathbf{e}_{123}$ of $\boldsymbol{\partial} \wedge \mathbf{F}$, while the vector equation is given by the trivector components $\mathbf{e}_{012}$, $\mathbf{e}_{013}$, and $\mathbf{e}_{023}$ of the exterior derivative.

We evaluate the exterior derivative $\boldsymbol{\partial} \wedge \mathbf{F}$ using the decomposition of $\mathbf{F}$ in Equation (71),

$$\begin{aligned} \boldsymbol{\partial} \wedge \mathbf{F} &= -\partial_0 \mathbf{e}_0 \wedge \mathbf{e}_0 \wedge \mathbf{E} - \partial_0 \mathbf{e}_0 \wedge \mathbf{B}^{\mathcal{H}} + \nabla \wedge \mathbf{e}_0 \wedge \mathbf{E} + \nabla \wedge \mathbf{B}^{\mathcal{H}} \\ &= -\partial_0 \mathbf{e}_0 \wedge \mathbf{B}^{\mathcal{H}} - \mathbf{e}_0 \wedge (\nabla \wedge \mathbf{E}) + \nabla \wedge \mathbf{B}^{\mathcal{H}} \\ &= -\mathbf{e}_0 \wedge (\partial_0 \mathbf{B}^{\mathcal{H}} + \nabla \wedge \mathbf{E}) + \nabla \wedge \mathbf{B}^{\mathcal{H}}, \end{aligned} \tag{82}$$

where we used that $\mathbf{e}_0 \wedge \mathbf{e}_0 = 0$ and that $\nabla \wedge \mathbf{e}_0 = -\mathbf{e}_0 \wedge \nabla$ in the second step of Equation (82). Taking advantage of Equation (40) we have the equality $\nabla \wedge \mathbf{E} = (\nabla \times \mathbf{E})^{\mathcal{H}}$, while $\nabla \wedge \mathbf{B}^{\mathcal{H}} = (\nabla \cdot \mathbf{B})^{\mathcal{H}}$, and

$$\boldsymbol{\partial} \wedge \mathbf{F} = -\mathbf{e}_0 \wedge (\partial_0 \mathbf{B}^{\mathcal{H}} + (\nabla \times \mathbf{E})^{\mathcal{H}}) + (\nabla \cdot \mathbf{B})^{\mathcal{H}}. \tag{83}$$

Indeed, the first summand vanishes when $\partial_0 \mathbf{B}^{\mathcal{H}} + (\nabla \times \mathbf{E})^{\mathcal{H}} = 0$ or, taking the inverse Hodge complement, when Equation (74) holds. In terms of components, the spatial Hodge complement in this equation transforms a spatial vector into a bivector with components $\mathbf{e}_{12}$, $\mathbf{e}_{13}$, and $\mathbf{e}_{23}$ only and this equation recovers the homogeneous Maxwell equation in Equation (74). After taking the exterior product with $\mathbf{e}_0$, we obtain the trivector components $\mathbf{e}_{012}$, $\mathbf{e}_{013}$, and $\mathbf{e}_{023}$. Similarly, the second term vanishes for $(\nabla \cdot \mathbf{B})^{\mathcal{H}} = 0$, recovering Equation (73). In terms of components, the spatial Hodge complement directly transforms a scalar into a trivector with a unique component $\mathbf{e}_{123}$, recovering the homogeneous Maxwell equation in Equation (73).

We move on to the inhomogeneous Maxwell equations. We compute the interior derivative $\boldsymbol{\partial} \,\lrcorner\, \mathbf{F}$,

$$\begin{aligned} \boldsymbol{\partial} \,\lrcorner\, \mathbf{F} &= (-\partial_0 \mathbf{e}_0 + \nabla) \,\lrcorner\, (\mathbf{F}_{\mathbf{E}} + \mathbf{F}_{\mathbf{B}}) \\ &= -\partial_0 \mathbf{E} - \partial_0 \mathbf{e}_0 \,\lrcorner\, \mathbf{B}^{\mathcal{H}} + \mathbf{e}_0 \nabla \cdot \mathbf{E} + \nabla \,\lrcorner\, \mathbf{B}^{\mathcal{H}} \\ &= -\partial_0 \mathbf{E} + \mathbf{e}_0 \nabla \cdot \mathbf{E} + \nabla \,\lrcorner\, \mathbf{B}^{\mathcal{H}}, \end{aligned} \tag{84}$$

since $\mathbf{e}_0 \,\lrcorner\, \mathbf{B}^{\mathcal{H}} = 0$. The interior derivative $\boldsymbol{\partial} \,\lrcorner\, \mathbf{F}$ gives a space–time vector with 4 components, while the inhomogeneous Maxwell equations are a scalar, Equation (72), and a spatial vector, Equation (75).

We can verify that the scalar equation turns out to be given by the vector component $\mathbf{e}_0$ of $\boldsymbol{\partial} \,\lrcorner\, \mathbf{F}$, while the spatial vector equation is given by the spatial vector components $\mathbf{e}_1$, $\mathbf{e}_2$, and $\mathbf{e}_3$ of $\boldsymbol{\partial} \,\lrcorner\, \mathbf{F}$. Indeed, if we match this expression with the current density vector $\mathbf{J}$, then the time component $\mathbf{e}_0$ of $\boldsymbol{\partial} \,\lrcorner\, \mathbf{F}$ gives Equation (72). Selecting the space components of $\boldsymbol{\partial} \,\lrcorner\, \mathbf{F}$, the differential equation is

$$-\partial_0 \mathbf{E} + \nabla \,\lrcorner\, \mathbf{B}^{\mathcal{H}} = \mathbf{j}, \tag{85}$$

which, using the relation $\nabla \,\lrcorner\, \mathbf{B}^{\mathcal{H}} = \nabla \times \mathbf{B}$ can be written as Equation (75).

4.3. Equivalence of the Integral Form of Maxwell Equations

After studying the exterior-calculus differential formulation of Maxwell equations, we recover the standard integral formulation. Applying the Stokes Theorem 1 to Equation (76), we find that *the circulation of the bivector field* $\mathbf{F}$ *along the boundary of any three-dimensional space–time volume* $\mathcal{V}^3$ *is zero*:

$$\int_{\partial\mathcal{V}^3} \mathrm{d}^2\mathbf{x}\cdot\mathbf{F} = \int_{\mathcal{V}^3} \mathrm{d}^3\mathbf{x}\cdot(\partial\wedge\mathbf{F}) = 0. \tag{86}$$

At this point, Equation (86) is a scalar equation and we obtain the pair of homogeneous Maxwell equations by considering two different hypersurfaces $\mathcal{V}^3$.

First, let the domain $\mathcal{V}^3 = V$ contain only spatial coordinates. There are no tangential components to V with time indices and the contribution of $\mathbf{F_E}$ to the circulation of $\mathbf{F}$ over $\partial\mathcal{V}^3$ in Equation (86) is zero, i.e.,

$$\int_{\partial V} \mathrm{d}^2\mathbf{x}\cdot\mathbf{F} = \int_{\partial V} \mathrm{d}^2\mathbf{x}\cdot\mathbf{F_B}. \tag{87}$$

Using that $\mathbf{u}\cdot\mathbf{w} = \mathbf{u}^{\mathcal{H}^{-1}}\cdot\mathbf{w}^{\mathcal{H}^{-1}}$ for any vectors $\mathbf{u}$, $\mathbf{w}$, and therefore $\mathrm{d}^2\mathbf{x}\cdot\mathbf{F} = \mathrm{d}^2\mathbf{x}^{\mathcal{H}^{-1}}\cdot\mathbf{F}_{\mathbf{B}}^{\mathcal{H}^{-1}}$ and the definition $\mathbf{F_B} = \mathbf{B}^{\mathcal{H}}$, the integral in the right-hand side of Equation (87) becomes

$$\begin{aligned}\textstyle\int_{\partial V} \mathrm{d}^2\mathbf{x}\cdot\mathbf{F_B} &= \textstyle\int_{\partial V} \mathrm{d}^2\mathbf{x}^{\mathcal{H}^{-1}}\cdot\mathbf{B} \\ &= \textstyle\int_{\partial V} \mathrm{d}\mathbf{S}\cdot\mathbf{B},\end{aligned} \tag{88}$$

where we used Equation (32) to write the last surface integral. Substituting Equation (88) back into Equation (86) gives the Gauss law for the magnetic field [11] (pp. 1-5–1-9).

Let now $\mathcal{V}^3$ be a time-space domain $(t_0, t_1)\times S$, where S is a two-dimensional spatial surface. With no real loss of generality we assume that S lies on the $\mathbf{e}_1\wedge\mathbf{e}_2$ plane. Its boundary $\partial\mathcal{V}^3$ is the union of the sets $(t_0, t_1)\times\partial S$, $t_0\times S$ and $t_1\times S$. For the first set, we choose $\mathbf{e}_\perp$ as the vector normal to ∂S pointing outwards on the plane defined by S and $\mathbf{e}_\parallel = \mathbf{e}_0\wedge\mathbf{e}_{\partial S}$, where $\mathbf{e}_{\partial S}$ is a vector tangent to ∂S with a counterclockwise orientation, so that $\mathbf{e}_\perp\wedge\mathbf{e}_\parallel = -\mathbf{e}_{012}$. Further, since $\mathbf{e}_\parallel$ is a time-space bivector, the contribution of $\mathbf{F_B}$ to the circulation of $\mathbf{F}$ over this first set in Equation (86) is zero, and

$$\int_{(t_0,t_1)\times\partial S} \mathrm{d}^2\mathbf{x}\cdot\mathbf{F} = -\int_{(t_0,t_1)\times\partial S} \mathrm{d}^2\mathbf{x}_\parallel\cdot\mathbf{F_E}. \tag{89}$$

Writing the differential vector as $\mathrm{d}^2\mathbf{x}_\parallel = \mathrm{d}t\,\mathrm{d}x\mathbf{e}_{0x}$, parameterizing the line integral over the boundary ∂S by the variable x with unit vector $\mathbf{e}_{0x}$, and using that $\mathbf{e}_{0x}\cdot\mathbf{F_E} = -\mathbf{e}_x\cdot\mathbf{E}$ and therefore $\mathrm{d}x\,\mathbf{e}_{0x}\cdot\mathbf{F_E} = -\,\mathrm{d}\mathbf{x}\cdot\mathbf{E}$, the integral of the right-hand side of Equation (89) becomes

$$\int_{t_0}^{t_1} \mathrm{d}t \int_{\partial S} \mathrm{d}\mathbf{x}\cdot\mathbf{E}. \tag{90}$$

For the second and third sets the normal vector to the integration surface pointing outwards are $\mathbf{e}_\perp = -\mathbf{e}_0$ and $\mathbf{e}_\perp = \mathbf{e}_0$ respectively. Since $\mathbf{e}_\parallel$ is a space-space bivector in both cases, then the contribution of $\mathbf{F_E}$ to the circulation is zero. We express the circulations of $\mathbf{F_B}$ as fluxes of $\mathbf{B}$ and surface integrals as done in Equation (88).Using these observations the integral for the circulation of $\mathbf{F}$ over these two sets in Equation (86) is given by

$$\int_{t_0\times S} \mathrm{d}^2\mathbf{x}\cdot\mathbf{F} + \int_{t_1\times S} \mathrm{d}^2\mathbf{x}\cdot\mathbf{F} = -\int_S \mathrm{d}\mathbf{S}\cdot\mathbf{B}(t_0) + \int_S \mathrm{d}\mathbf{S}\cdot\mathbf{B}(t_1). \tag{91}$$

Combining Equations (90) and (91) in Equation (86) we recover the integral over time of the so called Faraday law [11] (pp. 17-1–17-2). Equivalently, taking the time derivative recovers the usual Faraday law, namely

$$\int_{\partial S} d\mathbf{x} \cdot \mathbf{E} + \partial_t \int_S d\mathbf{S} \cdot \mathbf{B} = 0. \tag{92}$$

In regard to the inhomogeneous Maxwell equations, applying the Stokes Theorem 2 to Equation (77), we find that *the flux of the bivector field* $\mathbf{F}$ *across the boundary of any three-dimensional space–time volume is equal to the flux of the current density* $\mathbf{J}$ *across the three-dimensional space–time volume*:

$$\int_{\partial\mathcal{V}^3} d^2\mathbf{x}^{\mathcal{H}-1} \cdot \mathbf{F} = \int_{\mathcal{V}^3} d^3\mathbf{x}^{\mathcal{H}-1} \cdot (\boldsymbol{\partial} \lrcorner \mathbf{F}) = \int_{\mathcal{V}^3} d^3\mathbf{x}^{\mathcal{H}-1} \cdot \mathbf{J}. \tag{93}$$

As with the homogeneous Maxwell equations, the scalar Equation (93) yields the inhomogeneous Maxwell equations by considering two different hypersurfaces $\mathcal{V}^3$.

First, let the integration domain $\mathcal{V}^3$ be a spatial volume V. Since there are no normal components to V with space indices only, the contribution of $\mathbf{F_B}$ to the flux is zero so that Equation (93) becomes

$$\int_{\partial V} d^2\mathbf{x}^{\mathcal{H}-1} \cdot \mathbf{F_E} = \int_V d^3\mathbf{x}^{\mathcal{H}-1} \cdot \mathbf{J}. \tag{94}$$

From the definition of inverse Hodge complement in Equation (10), we write the differential vectors

$$d^2\mathbf{x}^{\mathcal{H}-1} = -\sum_{I, i \notin I} d^2x_I \sigma(0i, I)\mathbf{e}_{0i} \tag{95}$$

$$d^3\mathbf{x}^{\mathcal{H}-1} = -dV\mathbf{e}_0. \tag{96}$$

Plugging these expressions in Equation (94), using the definitions of $\mathbf{F_E}$ and $\mathbf{J}$, and computing the dot products on both sides of the equality, we obtain that Equation (94) simplifies as

$$-\int_{\partial V} \left(\sum_{I, i \notin I} d^2x_I \sigma(0i, I)\mathbf{e}_{0i} \right) \cdot \left(\sum_j E_j \mathbf{e}_{0j} \right) = -\int_V dV\mathbf{e}_0 \cdot (\rho\mathbf{e}_0 + \mathbf{j}) \tag{97}$$

$$\int_{\partial V} \sum_{I, i \notin I} d^2x_I \sigma(i, I) E_i = \int_V dV\rho \tag{98}$$

$$\int_{\partial V} d\mathbf{S} \cdot \mathbf{E} = \int_V dV\rho. \tag{99}$$

In Equation (98) we used that $\sigma(0i, I) = \sigma(i, I)$ and in Equation (99) we used that $\sum_{I, i \notin I} d^2x_I \sigma(i, I) E_i = d^2\mathbf{x}^{\mathcal{H}-1} \cdot \mathbf{E}$. Since the Hodge complement is over space, the result is a surface integral with positive orientation as in Equation (32). We recovered in Equation (99) the Gauss law for the electric field [11] (pp. 4-7–4-9).

For $\mathcal{V}^3 = (t_0, t_1) \times S$ where S is a two-dimensional surface lying on the $\mathbf{e}_1 \wedge \mathbf{e}_2$ plane, the boundary $\partial\mathcal{V}^3$ is the union of the sets $(t_0, t_1) \times \partial S$, $t_0 \times S$ and $t_1 \times S$. For the first set, since $d^2\mathbf{x}^{\mathcal{H}-1}$ has no time components, the contribution of $\mathbf{F_E}$ to this set is zero, that is

$$\int_{(t_0, t_1) \times \partial S} d^2\mathbf{x}^{\mathcal{H}-1} \cdot \mathbf{F} = \int_{(t_0, t_1) \times \partial S} d^2\mathbf{x}^{\mathcal{H}-1} \cdot \mathbf{F_B}. \tag{100}$$

As in the homogeneous case, we choose $\mathbf{e}_\perp$ as the vector normal to ∂S pointing outwards on the plane defined by S and $\mathbf{e}_\parallel = \mathbf{e}_0 \wedge \mathbf{e}_{\partial S}$, where $\mathbf{e}_{\partial S}$ is a vector tangent to ∂S with a counterclockwise orientation, such that $\mathbf{e}_\perp \wedge \mathbf{e}_\parallel = -\mathbf{e}_{012}$ introduces a change of sign. Expressing $d^2\mathbf{x}_\parallel^{\mathcal{H}-1}$ and $\mathbf{F_B}$ in the canonical basis, defining $I = (0, i)$ so that I^c contains only space indexes, and using that $\mathbf{e}_{I^c} \cdot \mathbf{e}_{i^c} = 1$ and $\sigma(I^c, I) = \sigma(I^c, 0, i) = \sigma(0, I^c, i) = \sigma(I^c, i)$, we obtain that Equation (100) simplifies to

$$\begin{aligned} \int_{(t_0,t_1)\times\partial S} \mathrm{d}^2\mathbf{x}^{\mathcal{H}-1} \cdot \mathbf{F} &= \int_{(t_0,t_1)\times\partial S} (\textstyle\sum_I \mathrm{d}x_I \sigma(I^c, I)\mathbf{e}_{I^c}) \cdot (\sum_i B_i \mathbf{e}_{i^c} \sigma(i^c, i)) \\ &= -\int_{t_0}^{t_1} \mathrm{d}t \int_{\partial S} \mathrm{d}x B_x \\ &= -\int_{t_0}^{t_1} \mathrm{d}t \int_{\partial S} \mathrm{d}\mathbf{x} \cdot \mathbf{B}. \end{aligned} \tag{101}$$

For the second and third sets we respectively choose $\mathbf{e}_\perp = -\mathbf{e}_0$ and $\mathbf{e}_\perp = \mathbf{e}_0$ pointing outside $\mathcal{V}^3$, implying that the contribution of $\mathbf{F}_\mathbf{B}$ is zero for this set as the inverse Hodge complement of $\mathbf{e}_\perp$ is a space vector. Expressing $\mathrm{d}^2\mathbf{x}^{\mathcal{H}-1}$ in Equation (95) and using similar steps as in Equations (97)–(99), the left-hand side of Equation (93) over these two sets is given by

$$\int_{t_0\times\partial S} \mathrm{d}^2\mathbf{x}^{\mathcal{H}-1} \cdot \mathbf{F} + \int_{t_1\times\partial S} \mathrm{d}^2\mathbf{x}^{\mathcal{H}-1} \cdot \mathbf{F} = -\int_S \mathrm{d}\mathbf{S} \cdot \mathbf{E}(t_0) + \int_S \mathrm{d}\mathbf{S} \cdot \mathbf{E}(t_1) \tag{102}$$

Finally, for the right-hand side of Equation (93), we choose $\mathbf{e}_\perp$ as the vector normal to $\mathcal{V}^3$ pointing outside. Since $\mathbf{e}_\parallel = \mathbf{e}_{012}$ implies that $\mathbf{e}_\perp \wedge \mathbf{e}_\parallel = -\mathbf{e}_{0123}$, we obtain that

$$\int_{\mathcal{V}^3} \mathrm{d}^3\mathbf{x}^{\mathcal{H}-1} \cdot \mathbf{J} = -\int_{t_0}^{t_1} \mathrm{d}t \int_S \mathrm{d}\mathbf{S} \cdot \mathbf{j}. \tag{103}$$

We have thusly recovered the integral form of the Ampere-Maxwell equation [11] (p. 18-1–18-4) integrated over the time interval (t_0, t_1) by combining Equations (101)–(103) into Equation (93), that is

$$\int_{t_0}^{t_1} \mathrm{d}t \int_{\partial S} \mathrm{d}\mathbf{x} \cdot \mathbf{B} = \int_{t_0}^{t_1} \mathrm{d}t \int_S \mathrm{d}\mathbf{S} \cdot \mathbf{j} + \int_S \mathrm{d}\mathbf{S} \cdot \mathbf{E}(t_1) - \int_S \mathrm{d}\mathbf{S} \cdot \mathbf{E}(t_0). \tag{104}$$

5. Summary

In this paper, we aimed at showing how exterior calculus provides a tool merging the simplicity and intuitiveness of standard vector calculus with the power of tensors and differential forms. Set in the context of a general space–time algebra with multiple space and time components, we provided the basic concepts of exterior algebra and calculus, such as multivectors, wedge product and interior products, with a distinction between left and right products, Hodge complement, and exterior and interior derivatives. While a space–time with multiple time coordinates leads to several issues from the physical point of view [12], we did not deal with these problems as this paper focuses on the mathematical constructions. We also defined oriented integrals, with two important examples being the flux and circulation of grade m-vector fields as integrals of the normal and tangent components of the field to a hypersurface respectively. These operations extend the standard circulation of a vector field as a line integral and the flux of a vector field as a surface integral in three dimensions to any number of dimensions and any vector grade.

Armed with the theory of differential forms, we proved two exterior-calculus Stokes theorems, one for the circulation and one for the flux, that generalize the Kelvin-Stokes, Gauss and Green theorems. We saw that the flux of the curl of a vector field in three dimensions across a surface is also the circulation of the exterior derivative of the vector field along that surface. In exterior calculus, these Stokes theorems hold for any number of dimensions and any vector grade and are simply expressed in terms of the exterior and interior derivatives for the circulation and flux respectively.

As an application of our tools, we showed how to recover the classical laws of electromagnetism, Maxwell equations and Lorentz force, from a exterior-calculus formalism in relativistic space–time with one temporal and three spatial dimensions. The electromagnetic field is described by a bivector field with six components, closely related to Faraday's antisymmetric tensor, containing both electric and magnetic fields. The differential form of Maxwell equations relates the exterior derivative of the bivector field with the zero trivector and the interior derivative of the field with the current density vector. In the integral form, these equations correspond to the statements that the circulation of the bivector field along the boundary of any three-dimensional space–time volume is zero, and that the flux of the bivector field across the boundary of any three-dimensional space–time volume is equal to the flux of the current density across the same space–time volume.

Author Contributions: Conceptualization, A.M.; Methodology, I.C., J.F.-S., A.M.; Investigation, I.C.; Writing—Original Draft Preparation, I.C.; Writing—Review & Editing, I.C., J.F.-S., A.M.

Acknowledgments: This work has been funded in part by the Spanish Ministry of Economy and Competitiveness under grants TEC2016-78434-C3-1-R and BES-2017-081360.

Appendix A. Proofs of Product Identities

In this appendix, we verify the relations about interior products introduced in Section 2.

Appendix A.1. Relation between Left and Right Interior Products

We now prove the formula

$$\mathbf{e}_I \lrcorner \mathbf{e}_J = \mathbf{e}_J \llcorner \mathbf{e}_I (-1)^{|I|(|J|-|I|)}, \tag{A1}$$

relating left and right interior products. For two lists I and J, we have

$$\mathbf{e}_I \lrcorner \mathbf{e}_J = \Delta_{I,I}\sigma(J\backslash I, I)\mathbf{e}_{J\backslash I}, \tag{A2}$$

$$\mathbf{e}_J \llcorner \mathbf{e}_I = \Delta_{I,I}\sigma(I, J\backslash I)\mathbf{e}_{J\backslash I}, \tag{A3}$$

where we assumed that $I \subseteq J$ with no loss of generality and used that $\varepsilon(I, J^c)^c = J\backslash I$ in this case. The only difference between the expressions lies in the signatures, that are related by setting $A = J\backslash I$ and $B = I$ in the following lemma.

Lemma A1. *Given two arbitrary lists A and B, of length $|A|$ and $|B|$ respectively, then the permutations sorting the concatenated lists (A, B) and (B, A) satisfy the formula*

$$\sigma(A, B) = \sigma(B, A)(-1)^{|A||B|}. \tag{A4}$$

Proof. Given a list A, let $\bar{A}$ be the reversed list, namely the list where the order of all the elements is reversed. Counting the number of position jumps needed to reverse the list, we obtain the signature of this reversing operation as

$$\sigma_{\mathrm{r}}(A) = \sigma_{\mathrm{r}}(\bar{A}) = (-1)^{|A|-1+|A|-2+\ldots+1} = (-1)^{\frac{|A|(|A|-1)}{2}}. \tag{A5}$$

The proof is based on the identity between two different ways of rearranging the concatenated list (A, B) into the ordered list $\varepsilon(A, B)$, as depicted in Figure A1.

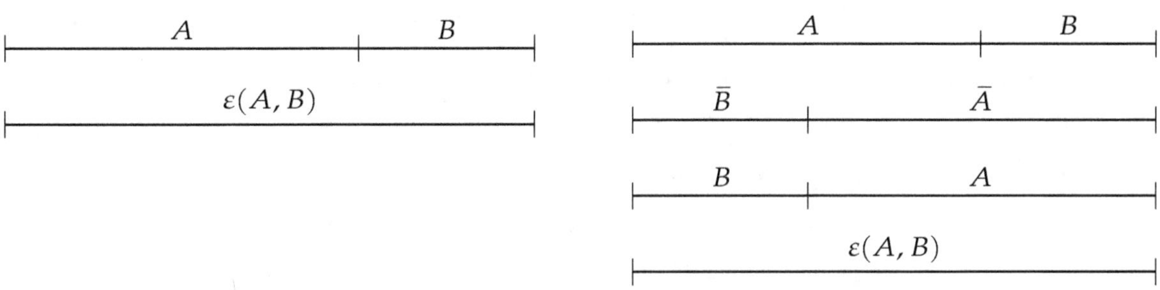

Figure A1. Visual aid for the relation between $\sigma(A, B)$ and $\sigma(B, A)$.

First, in the left column of Figure A1 we depict how a single permutation with signature $\sigma(A,B)$ orders the list (A,B). In the right column of Figure A1 we depict how a different series of permutations achieves the same result. We start by reversing the concatenated list (A,B), an operation with signature $\sigma_{\mathrm{r}}(\bar{B},\bar{A})$. Then, we separately partially reverse the lists $\bar{B}$ and $\bar{A}$, operations with respective signatures $\sigma_{\mathrm{r}}(\bar{B})$ and $\sigma_{\mathrm{r}}(\bar{A})$. A final permutation with signature $\sigma(B,A)$ orders the list (B,A) into $\varepsilon(A,B)$. Since the signature of a composition of permutations is the product of the signatures, we obtain that

$$\sigma(A,B) = \sigma_{\mathrm{r}}(\bar{B},\bar{A})\sigma_{\mathrm{r}}(\bar{A})\sigma_{\mathrm{r}}(\bar{B})\sigma(B,A). \tag{A6}$$

Using Equation (A5) in every σ_{r} in Equation (A6) and carrying out some simplifications yields Equation (A4). □

Appendix A.2. Relation between Interior and Exterior Products

We start with the expression for the left interior product Equation (14). From Equations (9) and (10), we compute

$$\begin{aligned}(\mathbf{e}_I \wedge \mathbf{e}_J^{\mathcal{H}})^{\mathcal{H}^{-1}} &= (\Delta_{J,J}\sigma(J,J^c)\mathbf{e}_I \wedge \mathbf{e}_{J^c})^{\mathcal{H}^{-1}} \\ &= \Delta_{J,J}\Delta_{\varepsilon(I,J^c)^c,\varepsilon(I,J^c)^c}\,\sigma(J,J^c)\sigma(I,J^c)\sigma(\varepsilon(I,J^c)^c,\varepsilon(I,J^c))\,\mathbf{e}_{\varepsilon(I,J^c)^c},\end{aligned} \tag{A7}$$

and since $\Delta_{\varepsilon(I,J^c)^c,\varepsilon(I,J^c)^c} = \Delta_{J\setminus I,J\setminus I}$, we can conclude that $\Delta_{J,J}\Delta_{\varepsilon(I,J^c)^c,\varepsilon(I,J^c)^c} = \Delta_{I,I}$. If we now compare the result with Equation (11), we need just to verify the identity

$$\sigma(\varepsilon(I,J^c)^c, I) = \sigma(J,J^c)\sigma(I,J^c)\sigma(\varepsilon(I,J^c)^c,\varepsilon(I,J^c)), \tag{A8}$$

or equivalently

$$\sigma(\varepsilon(I,J^c)^c, I)\sigma(J,J^c) = \sigma(\varepsilon(I,J^c)^c,\varepsilon(I,J^c))\sigma(I,J^c). \tag{A9}$$

The left-hand side of Equation (A9) corresponds to taking the sets $\varepsilon(I,J^c)^c = J\setminus I$, I and J^c, in this order, and then merging and sorting $J\setminus I$ with I and then merging and sorting the resulting set J with J^c, as shown in the left column of Figure A2. On the right-hand side, we start we the same three lists, but we first merge and sort I with J^c, and then we get the whole list by merging and sorting the result with $J\setminus I$, as represented in the right column of Figure A2. Thus, starting from the three sets and rearranging them in different ways, we get the same final ordered list, and since the signatures of the left-hand side and right-hand side are the same, and Equation (A9) is proved. As a consequence, Equation (14) is verified.

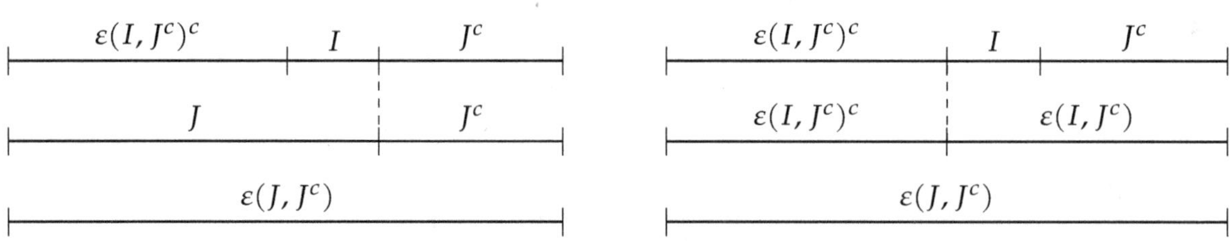

Figure A2. Visual aid for the permutations in Equation (A9).

Afterwards, we prove the formula for the right interior product Equation (15). Using Equations (9) and (10), we write

$$\begin{aligned}(\mathbf{e}_I^{\mathcal{H}^{-1}} \wedge \mathbf{e}_J)^{\mathcal{H}} &= (\Delta_{I^c,I^c}\sigma(I^c,I)\mathbf{e}_{I^c} \wedge \mathbf{e}_J)^{\mathcal{H}} \\ &= \Delta_{I^c,I^c}\sigma(I^c,I)\sigma(I^c,J)\,\mathbf{e}_{\varepsilon(I^c,J)}^{\mathcal{H}} \\ &= \Delta_{I^c,I^c}\sigma(I^c,I)\sigma(I^c,J)\Delta_{\varepsilon(I^c,J),\varepsilon(I^c,J)}\sigma(\varepsilon(I^c,J),\varepsilon(I^c,J)^c)\,\mathbf{e}_{\varepsilon(I^c,J)^c}.\end{aligned} \tag{A10}$$

Using that $\Delta_{\varepsilon(I^c,J),\varepsilon(I^c,J)}\Delta_{I^c,I^c} = \Delta_{J,J}$, in order to prove the validity of Equation (15), we need to prove the relation

$$\sigma(J,\varepsilon(I^c,J)^c) = \sigma(I^c,I)\sigma(I^c,J)\sigma(\varepsilon(I^c,J),\varepsilon(I^c,J)^c)\,. \tag{A11}$$

We can prove it applying Lemma A1 to obtain the expression Equation (A8), or by following the same procedure as before, paying attention to the difference that now list J is included in I.

Appendix A.3. Triple mixed product

Given two 1-vectors $\mathbf{u}$ and $\mathbf{v}$ and a r-vector $\mathbf{w}$, we prove the relation

$$\mathbf{u} \lrcorner (\mathbf{v}\wedge\mathbf{w}) = (-1)^r(\mathbf{u}\cdot\mathbf{v})\mathbf{w} + \mathbf{v}\wedge(\mathbf{u}\lrcorner\mathbf{w}). \tag{A12}$$

Proof. We start by evaluating $\mathbf{u} \lrcorner (\mathbf{v}\wedge\mathbf{w})$ explicitly, separating terms $i=j$ and $i\neq j$, namely

$$\mathbf{u} \lrcorner (\mathbf{v}\wedge\mathbf{w}) = \sum_{\substack{i,j,I\\ j\notin I, i\in I}} \Delta_{i,i}u_iv_jw_I\sigma(j,I)\sigma(I+j\backslash i,i)\mathbf{e}_{I+j\backslash i} + \sum_{\substack{i,I\\ i\in I}}\Delta_{i,i}u_iv_iw_I\sigma(i,I)\sigma(I,i)\mathbf{e}_I, \tag{A13}$$

then, using $\sigma(i,I)\sigma(I,i) = (-1)^r$ and adding and removing a term $(-1)^r\sum_{\substack{i,I\\ i\notin I}}\Delta_{i,i}u_iv_iw_I\mathbf{e}_I$, we get

$$\mathbf{u} \lrcorner (\mathbf{v}\wedge\mathbf{w}) = \sum_{\substack{i,j,I\\ i\in I, j\notin I\backslash i}} \Delta_{i,i}u_iv_jw_I\sigma(j,I)\sigma(I+j\backslash i,i)\mathbf{e}_{I+j\backslash i} + (-1)^r\sum_{i,I}\Delta_{i,i}u_iv_iw_I\mathbf{e}_I. \tag{A14}$$

More concretely, the left-hand side $\mathbf{u} \lrcorner (\mathbf{v}\wedge\mathbf{w})$ is given by

$$\begin{aligned}
&\sum_{\substack{i,j,I\\ j\notin I, i\in I}} \Delta_{i,i}u_iv_jw_I\sigma(j,I)\sigma(I+j\backslash i,i)\mathbf{e}_{I+j\backslash i} - (-1)^r\sum_{\substack{i,I\\ i\notin I}}\Delta_{i,i}u_iv_iw_I\mathbf{e}_I\\
&= \sum_{\substack{i,j,I\\ j\notin I, i\in I}} \Delta_{i,i}u_iv_jw_I\sigma(j,I)\sigma(I+j\backslash i,i)\mathbf{e}_{I+j\backslash i} - \sum_{\substack{i,j,I\\ j=i, j\notin I\backslash i, i\in I}} \Delta_{i,i}u_iv_jw_I\sigma(j,I)\sigma(I+j\backslash i,i)\mathbf{e}_{I+j\backslash i}\\
&= \sum_{\substack{i,j,I\\ i\in I, j\notin I\backslash i}} \Delta_{i,i}u_iv_jw_I\sigma(j,I)\sigma(I+j\backslash i,i)\mathbf{e}_{I+j\backslash i}.
\end{aligned} \tag{A15}$$

Similarly, we evaluate the right-hand side $\mathbf{v}\wedge(\mathbf{u}\lrcorner\mathbf{w})$ as

$$\mathbf{v}\wedge(\mathbf{u}\lrcorner\mathbf{w}) = \sum_{\substack{i,j,I\\ i\in I, j\notin I\backslash i}} \Delta_{i,i}u_iv_jw_I\sigma(I\backslash i,i)\sigma(j,I\backslash i)\mathbf{e}_{I+j\backslash i}. \tag{A16}$$

Comparing Equations (A15) and (A16), it remains to prove the equality, and now we prove the equality

$$\sigma(j,I)\sigma(I+j\backslash i,i) = \sigma(I\backslash i,i)\sigma(j,I\backslash i). \tag{A17}$$

We rewrite Equation (A17) multiplying both sides for $\sigma(j,I)\sigma(j,I\backslash i)$ so that we obtain

$$\sigma(j,I\backslash i)\sigma(I+j\backslash i,i) = \sigma(I\backslash i,i)\sigma(j,I) \tag{A18}$$

which we verify with the help of Figure A3. On the left column, we first merge j with $I\backslash i$ and then the resulting list with i. On the right columns, the permutations first join $I\backslash i$ and i and the resulting I is then merged with j, getting the same result in both sides of the relation.

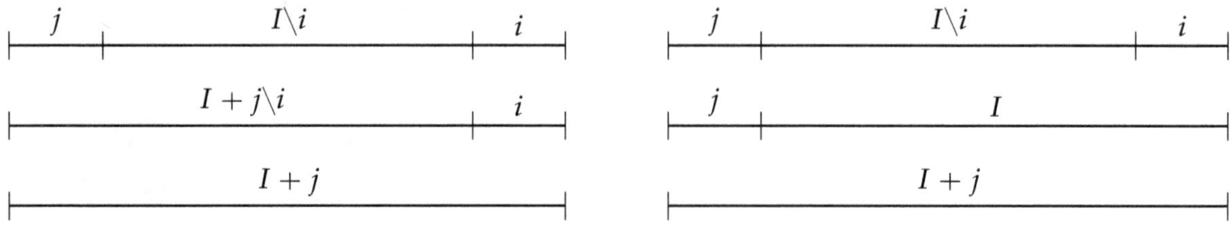

Figure A3. Visual aid for the identity $\sigma(j, I\backslash i)\sigma(I + j\backslash i, i) = \sigma(I\backslash i, i)\sigma(j, I)$.

Thus, we can write

$$\mathbf{u} \lrcorner (\mathbf{v} \wedge \mathbf{w}) = \sum_{\substack{i,j,I \\ i\in I, j\notin I\backslash i}} \Delta_{i,i} u_i v_j w_I \sigma(I\backslash i, i)\sigma(j, I\backslash i)\mathbf{e}_{I+j\backslash i} + (-1)^r \left(\sum_i \Delta_{i,i} u_i v_i\right)\left(\sum_I w_I \mathbf{e}_I\right), \tag{A19}$$

where we identify the term $(-1)^r(\mathbf{u}\cdot\mathbf{v})\mathbf{w}$, and finally conclude

$$\mathbf{u} \lrcorner (\mathbf{v} \wedge \mathbf{w}) - \mathbf{v} \wedge (\mathbf{u} \lrcorner \mathbf{w}) = (-1)^r(\mathbf{u}\cdot\mathbf{v})\mathbf{w}, \tag{A20}$$

which proves our initial formula. □

References

1. Gibbs, J.W.; Wilson, E.B. *Vector Analysis*; Yale University Press: New Haven, CT, USA, 1929.
2. Lorentz, H.A.; Einstein, A.; Minkowski, H.; Weyl, H. *The Principle of Relativity: A Collection of Original Memoirs on the Special and General Theory of Relativity*; Dover Publications: Mineola, NY, USA, 1923.
3. Ricci, M.M.G.; Levi-Civita, T. Méthodes de calcul différentiel absolu et leurs applications. *Math. Ann.* **1900**, *54*, 125–201. [CrossRef]
4. Cartan, E. *Les Systemes Differentiels Exterieurs Et Leurs Applications Geometriques*; Hermann & Cie: Paris, France, 1945.
5. Arnold, V.I.; Weinstein, A.; Vogtmann, K. *Mathematical Methods of Classical Mechanics*; Springer: Berlin, Germany, 1989.
6. Clifford, W.K. *Mathematical Papers*; Macmillan: London, UK, 1882.
7. Grassmann, H. *Extension Theory (History of Mathematics, 19)*; American Mathematical Society: Providence, RI, USA; London Mathematical Society: London, UK, 2000.
8. Sommerfeld, A. Zur Relativitätstheorie. I. Vierdimensionale Vektoralgebra. *Ann. Phys.* **1910**, *337*, 749–776. [CrossRef]
9. Winitzki, S. *Linear Algebra via Exterior Products*; Free Software Foundation: Boston, MA, USA, 2010.
10. Frankel, T. *The Geometry of Physics*, 3rd ed.; Cambridge University Press: Cambridge, UK, 2012.
11. Feynman, R.P.; Leyton, R.B.; Sands, M. *The Feynman Lectures on Physics*; Addison–Wesley: Boston, MA, USA, 1964; Volume 2.
12. Velev, M. Relativistic mechanics in multiple time dimensions. *Phys. Essays* **2012**, *25*, 403–438. [CrossRef]

Permissions

All chapters in this book were first published by MDPI; hereby published with permission under the Creative Commons Attribution License or equivalent. Every chapter published in this book has been scrutinized by our experts. Their significance has been extensively debated. The topics covered herein carry significant findings which will fuel the growth of the discipline. They may even be implemented as practical applications or may be referred to as a beginning point for another development.

The contributors of this book come from diverse backgrounds, making this book a truly international effort. This book will bring forth new frontiers with its revolutionizing research information and detailed analysis of the nascent developments around the world.

We would like to thank all the contributing authors for lending their expertise to make the book truly unique. They have played a crucial role in the development of this book. Without their invaluable contributions this book wouldn't have been possible. They have made vital efforts to compile up to date information on the varied aspects of this subject to make this book a valuable addition to the collection of many professionals and students.

This book was conceptualized with the vision of imparting up-to-date information and advanced data in this field. To ensure the same, a matchless editorial board was set up. Every individual on the board went through rigorous rounds of assessment to prove their worth. After which they invested a large part of their time researching and compiling the most relevant data for our readers.

The editorial board has been involved in producing this book since its inception. They have spent rigorous hours researching and exploring the diverse topics which have resulted in the successful publishing of this book. They have passed on their knowledge of decades through this book. To expedite this challenging task, the publisher supported the team at every step. A small team of assistant editors was also appointed to further simplify the editing procedure and attain best results for the readers.

Apart from the editorial board, the designing team has also invested a significant amount of their time in understanding the subject and creating the most relevant covers. They scrutinized every image to scout for the most suitable representation of the subject and create an appropriate cover for the book.

The publishing team has been an ardent support to the editorial, designing and production team. Their endless efforts to recruit the best for this project, has resulted in the accomplishment of this book. They are a veteran in the field of academics and their pool of knowledge is as vast as their experience in printing. Their expertise and guidance has proved useful at every step. Their uncompromising quality standards have made this book an exceptional effort. Their encouragement from time to time has been an inspiration for everyone.

The publisher and the editorial board hope that this book will prove to be a valuable piece of knowledge for researchers, students, practitioners and scholars across the globe.

List of Contributors

Marina Popolizio
Department of Electrical and Information Engineering, Polytechnic University of Bari, Via E. Orabona n.4, 70125 Bari, Italy
INdAM Research Group GNCS, Istituto Nazionale di Alta Matematica "Francesco Severi", Piazzale Aldo Moro 5, 00185 Rome, Italy
Department of Electrical and Information Engineering, Polytechnic University of Bari, Via E. Orabona 4, 70126 Bari, Italy

Vasily E. Tarasov
Skobeltsyn Institute of Nuclear Physics, Lomonosov Moscow State University, 119991 Moscow, Russia

Valentina V. Tarasova
Faculty of Economics, Lomonosov Moscow State University, 119991 Moscow, Russia
Yandex, Ulitsa Lva Tolstogo 16, 119021 Moscow, Russia

Azmin Sham Rambely and Ishak Hashim
Department of Mathematical Sciences, Faculty of Science and Technology, Universiti Kebangsaan Malaysia, 43600 UKM Bangi, Selangor, Malaysia

Awatif Jahman Alqarni
Department of Mathematical Sciences, Faculty of Science and Technology, Universiti Kebangsaan Malaysia, 43600 UKM Bangi, Selangor, Malaysia
Department of Mathematics, College of Sciences and Arts in Blqarn, University of Bisha, Bisha 61922, Saudi Arabia

Michael D. Marcozzi
Department of Mathematical Sciences, University of Nevada Las Vegas, Las Vegas, NV 89154-4020, USA

Eva Kaslik
Department of Mathematics and Computer Science, West University of Timisoara, Bd. V. Parvan 4, 300223 Timisoara, Romania

Roberto Garrappa
Department of Mathematics, University of Bari, Via E. Orabona 4, 70126 Bari, Italy
Member of the INdAM Research Group GNCS, Istituto Nazionale di Alta Matematica "Francesco Severi", Piazzale Aldo Moro 5, 00185 Rome, Italy
Dipartimento di Scienze Statistiche, "Sapienza" Università di Roma, P. le A. Moro 5, 00185 Roma, Italy

Ioannis Dassios, Andrew Keane and Paul Cuffe
School of Electrical and Electronic Engineering, University College Dublin, Dublin 4, Ireland

Natalie Baddour
Department of Mechanical Engineering, University of Ottawa, 161 Louis Pasteur, Ottawa, ON K1N 6N5, Canada

Chunlei Ruan
School of Mathematics & Statistics, Henan University of Science & Technology, Luoyang 471023, China

Luisa Beghin
Dipartimento di Scienze Statistiche, "Sapienza" Università di Roma, P. le A. Moro 5, 00185 Roma, Italy

Yuri Luchko
Department of Mathematics, Physics, and Chemistry, Beuth University of Applied Sciences Berlin, Luxemburger Str. 10, 13353 Berlin, Germany

Özlem Kaçmaz, Haci Mehmet Alakaş and Tamer Eren
Department of Industrial Engineering, Kirikkale University, Kirikkale 71450, Turkey

Berardino D'Acunto, Luigi Frunzo, Vincenzo Luongo and Maria Rosaria Mattei
Department of Mathematics and Applications, University of Naples Federico II, 80126 Naples, Italy

Aleksandras Krylovas and Natalja Kosareva
Department of Mathematical Modelling, Vilnius Gediminas Technical University, Saulėtekio al. 11, 10221 Vilnius, Lithuania

Rūta Dadelienė
Department of Rehabilitation, Physical and Sports Medicine, Institute of Health Science, Vilnius University, Saulėtekio al. 11, 10221 Vilnius, Lithuania

Stanislav Dadelo
Department of Entertainment Industries, Vilnius Gediminas Technical University, Saulėtekio al. 11, 10221 Vilnius, Lithuania

Silvia Vitali, Iva Budimir, Claudio Runfola and Gastone Castellani
Department of Physics and Astronomy; University of Bologna, Viale Berti Pichat 6/2, 40127 Bologna, Italy

Ivano Colombaro, Josep Font-Segura and Alfonso Martinez
Department of Information and Communication Technologies, Universitat Pompeu Fabra, 08018 Barcelona, Spain

Index

www.ingramcontent.com/pod-product-compliance
Lightning Source LLC
LaVergne TN
LVHW080951210926
848485LV00004B/821